Lipids and Lipid Polymers in Higher Plants

Edited by

M. Tevini and H. K. Lichtenthaler

Contributors

B. H. Davies A. H. B. Deas W. Eichenberger L. J. Goad
T. W. Goodwin E. Heinz F. W. Hemming P. J. Holloway
P. E. Kolattukudy H. K. Lichtenthaler C. Liljenberg
H. K. Mangold P. Mazliak D. Siefermann-Harms
P. Sitte F. Spener P. K. Stumpf M. Tevini

With 136 Figures

Springer-Verlag Berlin Heidelberg New York 1977

Prof. Dr. MANFRED TEVINI
Prof. Dr. HARTMUT K. LICHTENTHALER
Botanisches Institut II
Universität Karlsruhe
Kaiserstraße 12, 7500 Karlsruhe
Federal Republic of Germany

ISBN 3-540-08201-8 Springer-Verlag Berlin Heidelberg New York
ISBN 0-387-08201-8 Springer-Verlag New York Heidelberg Berlin

Library of Congress Cataloging in Publication Data. Symposium on Lipids and Lipid Polymers in Higher Plants, University of Karlsruhe, 1976. Lipids and lipid polymers in higher plants. 1. Plant lipids – Congresses. 2. Polymers and polymerization – Congresses. I. Tevini, M., 1939–. II. Lichtenthaler, H.K., 1934–. III. Davis, Brian H. IV. Title, V. Title: Lipid polymers in higher plants. QK898.L56S93.1976.581.1′9247.77-5370.

Typesetting, printing and bookbinding: Universitätsdruckerei H. Stürtz AG, Würzburg.

2131/3130-543210

Preface

This book contains a number of papers dealing with the main topics of a Symposium on "Lipids and Lipid Polymers in Higher Plants", held in July 1976 at the Botanical Institute of the University of Karlsruhe. The symposium was organized by Professors E. Heinz, H.K. Lichtenthaler, H.K. Mangold, and M. Tevini. The sponsorship by the *Deutsche Forschungsgemeinschaft* and the *Erwin-Riesch-Stiftung* is gratefully acknowledged.

The intention of the Symposium was to bring together in one place scientists working in very different fields of plant lipids, such as fatty acids, glycolipids, phospholipids, prenyllipids, sterols, and lipid polymers. The emphasis was placed on biosynthesis, distribution, function, and physiology of the various higher plant lipids and their role in biomembranes and epidermal cell walls.

By combining the major contributions in this book, we hope to give all plant scientists access to the recent developments in biochemistry and physiology of plant lipid metabolism. The editors are very grateful to the contributors, who have taken great care to present up-to-date reviews.

Karlsruhe, May 1977

M. Tevini
H.K. Lichtenthaler

Contents

Section 1. Function, Organization and Lipid Composition of Biomembranes

Chapter 1. Functional Organization of Biomembranes
P. Sitte (With 15 Figures)

Chapter 2. The Prenyllipids of the Membranes of Higher Plants
T.W. Goodwin (With 9 Figures)

Chapter 3. Glyco- and Phospholipids of Biomembranes in Higher Plants
P. MAZLIAK (With 12 Figures)

Section 2. Physiology and Biochemistry of Fatty Acids and Glycerides

Chapter 4. Lipid Biosynthesis in Developing Seeds
P.K. STUMPF (With 2 Figures)

Chapter 5. The Cyclopentenyl Fatty Acids
H.K. MANGOLD and F. SPENER (With 10 Figures)

Section 3. Physiology and Biochemistry of Plant Steroids

Chapter 8. The Biosynthesis of Plant Sterols
L.J. GOAD (With 12 Figures)

Chapter 9. Steryl Glycosides and Acylated Steryl Glycosides
W. EICHENBERGER (With 4 Figures)

Section 4. Physiology and Biochemistry of Prenyllipids

Chapter 10. The Biosynthesis and Biological Significance of Prenols and Their Phosphorylated Derivatives
F.W. HEMMING (With 10 Figures)

Chapter 11. Carotenoids in Higher Plants
B.H. Davies (With 9 Figures)

Chapter 12. The Xanthophyll Cycle in Higher Plants
D. SIEFERMANN-HARMS (With 11 Figures)

Chapter 13. Regulation of Prenylquinone Synthesis in Higher Plants
H.K. LICHTENTHALER (With 15 Figures)

Chapter 14. Chlorophyll Formation: The Phytylation Step
C. LILJENBERG (With 5 Figures)

Section 5. Lipid Polymers in Higher Plants

Chapter 15. Biosynthesis and Degradation of Lipid Polymers

P.E. KOLATTUKUDY (With 10 Figures)

Chapter 16. The Intermolecular Structure of Some Plant Cutins

List of Contributors

DAVIES, B.H., Department of Biochemistry, University College of Wales, Penglais Aberystwyth, Wales, Great Britain

DEAS, A.H.B., Long Ashton Research Station, University of Bristol, Bristol, Great Britain

EICHENBERGER, W., Institut für Biochemie der Universität Bern, Bern, Switzerland

GOAD, L.J., Department of Biochemistry, University of Liverpool, Liverpool, Great Britain

GOODWIN, T.W., Department of Biochemistry, University of Liverpool, Liverpool, Great Britain

HEINZ, E., Botanisches Institut der Universität Köln, Köln, Federal Republic of Germany

HEMMING, F.W., Department of Biochemistry, University Hospital and Medical School, Nottingham, Great Britain

HOLLOWAY, P.J., Long Ashton Research Station, University of Bristol, Bristol, Great Britain

KOLATTUKUDY, P.E., Department of Agricultural Chemistry, Washington State University, Pullman, Washington, U.S.A.

LICHTENTHALER, H.K., Botanisches Institut II der Universität Karlsruhe, Karlsruhe, Federal Republic of Germany

LILJENBERG, C., University of Göteborg, Botanical Institute, Department of Plant Physiology, Göteborg, Sweden

MANGOLD, H.K., Institut für Biochemie und Technologie, H.P. Kaufmann-Institut, Münster, Federal Republic of Germany

MAZLIAK, P., Université de Paris, Laboratoire de Physiologie cellulaire, Paris, France

SIEFERMANN-HARMS, D., Biologisches Institut der Universität Tübingen, Tübingen, Federal Republic of Germany

SITTE, P., Institut für Biologie II, Lehrstuhl für Zellbiologie, Universität Freiburg, Freiburg, Federal Republic of Germany

SPENER, F., Institut für Biochemie der Universität Münster, Münster, Federal Republic of Germany

STUMPF, P.K., Department of Biochemistry and Biophysics, University of California, Davis, U.S.A.

TEVINI, M., Botanisches Institut II der Universität Karlsruhe, Karlsruhe, Federal Republic of Germany

CHAPTER 1

Functional Organization of Biomembranes

P. SITTE

A. Introduction

Lipidology is, to a large extent, also membranology. The intracellular lipid species [34] can be classified either as storage lipids, represented chiefly by the neutral and apolar triglycerides that form spherical inclusions in plasmatic phases, or as amphiphilic structural lipids, which form extended, sheetlike structures, and which are, by and large, identical with biomembrane lipids. Biomembranes comprise up to 30% of a cell's dry weight. In energy-transducing organelles, biomembranes amount to still higher percentages.

In earlier times, when the 'elementary' membranes within the cell could neither be seen nor analyzed due to the lack of suitable methods, membrane phenomena seemed to play only a limited part in the cell's life. Since the advent of the electron microscope and of cell fractionation techniques the situation has changed drastically. Today it proves difficult to find any important phenomenon in cell biology *not* correlated with membrane action. A few, more obvious functions are: separation of metabolically different compartments; intracellular storage; specific and unidirectional translocation of certain compounds within the cell, and, consequently, a powerful regulation of metabolic activities; vectorial electron transport and ATP synthesis; impulse transmission; control of cellular movements up to the precise control of muscular contraction; translation, in the case of proteins to be secreted; at least certain aspects of DNA replication; segregation of genetic material in protocytes as well as in some more primitive eucytes; hormone action in animals and plants; phytochrome-mediated control of different plant cell activities; cell-cell recognition, communication, cooperation (and, therefore, some basic aspects of sexuality, development, and differentiation, cancer, and memory)—everywhere we find biomembranes involved directly or indirectly in vital events.

In fact, membranology has become a vast field, with hundreds of pertinent publications appearing every year (for recent reviews see, e.g., [12, 13, 16, 18, 35, 38, 39, 40, 58, 60, 82, 86, 88, 98, 111, 116, 124, 137, 156, 157, 159]). Therefore, just a few basic problems can be considered here. In particular, the following questions will be discussed:

1. What is the most general function that biomembranes serve in the living cell and how well are they fitted to serve that function?

2. How do biomembranes develop, and how do they become differentiated? Is there de novo synthesis of elementary membranes, or is there, in a certain sense, genetic continuity?

3. Which general factors govern size and shape of cells and subcellular

compartments; and to what extent is molecular self-assembly of lipid phases involved?

4. What do we really know about the molecular architecture of biomembranes?

As will be seen, biomembranes are the most important lipid-containing structures of living cells, and it is the lipid moiety of these membranes that is of particular importance.

B. Membrane Functions

I. Membrane Diversity

Every contemporary illustration of cell fine structure [21, 56, 79, 118] reminds one of the astonishing multiplicity of biomembranes in average eucytes (eukaryotic cells).

The different membranes as seen from such illustration or electron micrographs differ in thickness and spatial arrangement, as well as in their associated enzymatic activities and in their lipid composition. They also vary in their respective protein content, which in turn is strictly correlated with buoyant density, as shown in Figure 1 for some lipid carrier particles of blood [127]. Thus, the several membrane species can be separated by isopyknic centrifugation (Fig. 2) on the basis of differing protein contents. Differences in protein content are also reflected by particle density as observable on freeze-etch fracture faces of biomembranes [11, 83] (cf. Fig. 13).

Nevertheless, these membranes also have many properties in common. The generic term elementary membrane has been introduced to designate all of them—irrespective of their particular composition and function [141].

What *functions* can be served by elementary membranes?

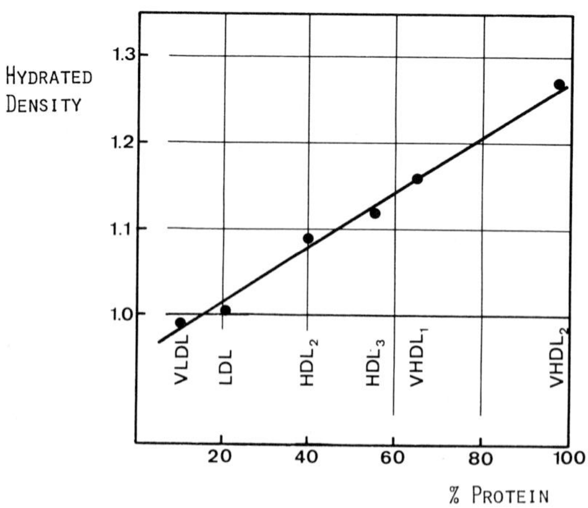

Fig. 1. Buoyant densities of lipoprotein particles from blood in correlation with their respective protein content [127]. *VLDL:* very low density lipoprotein; *LDL:* low density lipoprotein; *HDL* and *VHDL:* high density and very high density lipoprotein

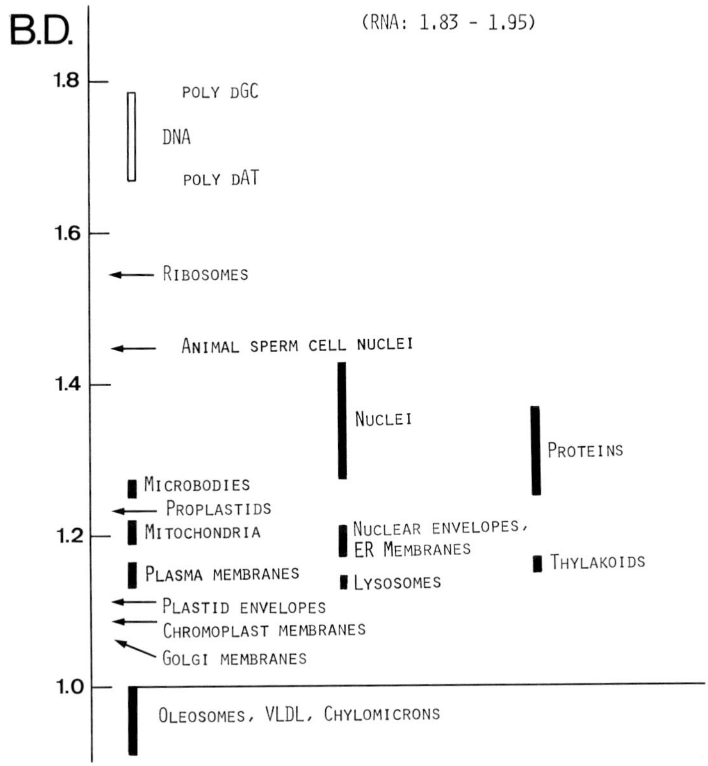

Fig. 2. Buoyant densities (*B.D.*) of some subcellular structural units, according to data in literature and own results. Author is indebted to Prof. Dr. W.W. Franke for collecting many of these data

They normally contain or carry enzymes [23, 89] and often lipophilic pigments also. Plasma membranes in particular act as chemical antennae [26, 27, 46, 50, 101], since many chemical stimuli, as mediated by hormones, transmitters [147, 152], mating substances [73, 92], or antigens, cannot penetrate the cell and are thus perceived to be at the cell surface. Yet the different *internal* elementary membranes must also possess surface specificities for membrane–membrane recognition as, for example, during membrane flow processes [28, 44, 45, 99, 109, 112, 113, 126, 132, 133, 160, 161]. Membranes also often take part in *cellular movements*. This is best demonstrated by the segregation of genetic material as mediated by the plasma membrane in bacteria [81, 110]. In eucells, membranes may serve as microtubule organizing centers or provide attraction points for contracting microfilaments (see, e.g., [4, 30, 103]).

Permeand selection in *membrane transport* often exhibits high specificity similar to enzymatic reactions [8, 9, 57, 77, 102, 141, 165]. However, as Berlin has shown [7], membrane permeases possess specificities complementary to the ones of intracellular enzymes. For example, amino acid permeases are stereospecific for the L-α-configuration, whereas the different aminoacyl-tRNA-synthetases exhibit high specificities for the different R-groups.

The fact that cellular *energy transformation* also depends, to a large part, on specialized membranes has only become clear in recent years [2, 33, 95, 96, 97, 121].

II. Membranes as Barriers

If one tries to reduce the different membrane functions just mentioned to a common denominator, one basic property common to all of them stands out, namely, the ability to impair free diffusion. Membranes principally act as barriers, both around the living cell and within it. The significance of compartmentation of the cellular interior by cytomembranes is to be seen mainly in the *separation* of certain metabolic events [93]. In higher plants, which, in contrast to higher animals, do not possess central organs for excretion, the single cell has to confine its own wastes away from metabolically active compartments [142]. In fortunate cases, such a local excretion can be seen in electron micrographs (Fig. 3).

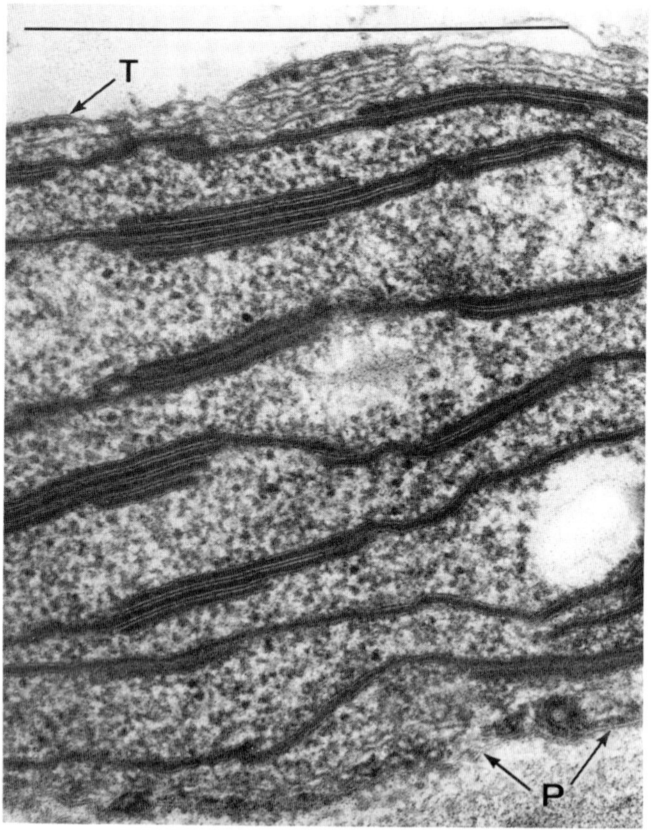

Fig. 3. Thylakoids filled with osmiophilic (possibly polyphenolic) substance. Petals of *Forsythia viridissima*, mesophyll. Plasma membrane (*P*) and tonoplast membrane (*T*) partly resolved as trilaminate unit membranes. Bar: 1 µm

Membranes can produce or maintain steep *concentration gradients* and (electro-)chemical potentials only because of their restricted permeability. Such potentials are of decisive importance in metabolism [93]. Even a few leaks, such as those produced by ionophores, channel-forming macrocyclic antibiotics, polyene antibiotics, or by complement action, inevitably kill the affected cell.

After some recent discoveries, we are just beginning to realize the role of compartmentation in the complex demands of cellular life. That steep proton concentration gradients are essential for ATP synthesis by thylakoids and mitochondrial inner membranes [95, 96, 97, 121] seems now to be generally agreed upon [52, 61, 62, 75, 115, 151, 163, 164]. Yet some totally different events are also under the control of pH gradients, as, for example, ion transport through plasma membranes [6, 155], the so-called overspill in photosynthetic light reaction as well as certain essential steps in dark reactions [3], and even lactose permease activity in *Escherichia coli* [135]. The electrochemical potential as generated by the expulsion of sodium ions from the cell is used not only in impulse transmission by nerves and muscles, but also for an active inward transport of amino acids and sugars by epithelial cells [63, 71]. Finally, there are indications of a regulation of intracellular membrane fusion during membrane flow by the local concentration of calcium ions [113, 126], which in turn might be under the control of smooth endoplasmic reticulum (ER) — as in striated muscle — or of mitochondria.

III. Lipids and Permeability

Many membrane functions require the high specificity of both intrinsic and extrinsic proteins. Yet whenever the membranes are to act as barriers against a free exchange of material, lipids must be considered: Their presence ensures restricted permeability [158]. Insulating membranes such as those of nerve myelin are particularly rich in lipids. Furthermore, the permeability properties of a given membrane depend on the kinds of lipid molecules involved. In general, it can be stated that, for an elementary membrane, the higher the content of long-chain, saturated fatty acids, the lower the permeability. Cholesterol has a similar effect, which, by means of hydrophobic interaction, causes a marked condensation effect within lipid films [29]. It is perhaps noteworthy, in this context, that plasma membranes exhibit higher percentages of both cholesterol and sphingomyelin than do the internal cytomembranes. Sphingolipids possess two saturated hydrocarbon chains whereas glycerophosphatides normally carry an *un*saturated chain in the β position.

No membrane is perfect: there is always a certain degree of leakyness, which has been termed passive permeability. This rather unspecific process is adequately described by the lipid-filter theory [158]. In Figure 4, several polar and apolar permeands are shown: among the polar molecules, urea (60 daltons) is for many biomembranes the largest permeand. Apolar particles, on the other hand, may penetrate even when they are many times larger.

The lipid-filter theory created the belief that biomembranes must contain a continuous fluid lipid sheet [108, 158]. In fact, artificial lipid layers exhibit permeability properties quite similar to those of elementary membranes [150].

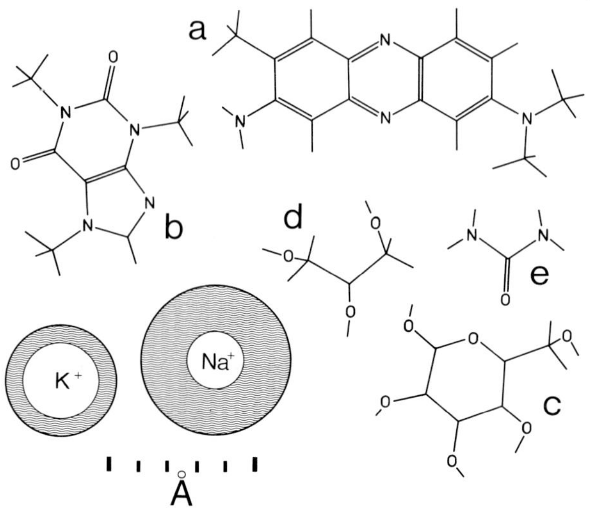

Fig. 4a–e. Sizes of different permeands: scale projections of Dreiding stereomodels (W. Büchi, CH-9230 Flawil, Switzerland). (a), neutral red (252 daltons); (b), caffeine (194 daltons); (c), glucose (180 daltons); (d), glycerol (92 daltons); (e), urea (60 daltons). (a) and (b) permeate readily due to their pronounced lipophilia. Among hydrophilic particles, only urea (e) and potassium ion are small enough to diffuse through plasma membranes. From [141]

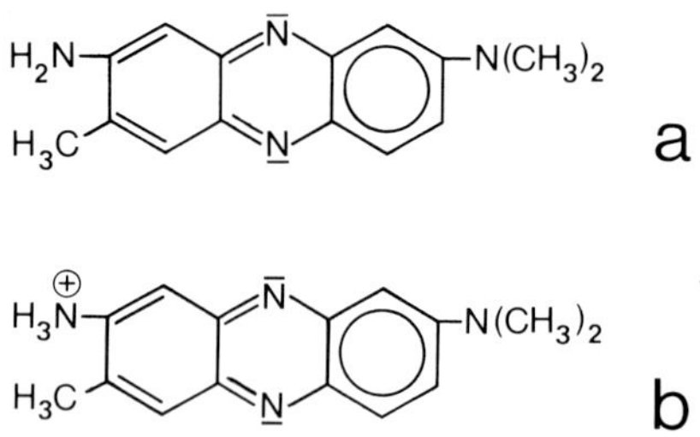

Fig. 5a and b. Neutral red molecule (a) becomes charged at low pH by attracting proton (b)

One example of true permeability that should be mentioned here is interesting in more than one respect [31, 72]. If one keeps plant cells in very dilute (1:10,000 or so), slightly alkaline solutions of neutral red (Figs. 4a, 5, and 6), one observes a heavy accumulation of this vital dye within the vacuoles (and Golgi vesicles). This is of course a remarkable concentration effect, but no active transport is involved at all. In the alkaline bathing solution, neutral red molecules prevail. As apolar particles, they pass readily into the vacuole. In the acidic cell sap, however, they attract protons and become charged. In this form, they are unable to pass back through the membrane again, and so they remain entrapped within the vacuole. At the same time, new uncharged dye molecules move in, down the still existing concentration gradient. Because no active transport is involved, it is possible to imitate this ion trap process with artificial model systems [32].

The essentials of the lipid-filter theory have best been illustrated by Höfler's two-way hypothesis [66]. Lipophilic permeands become dissolved and even concentrated in the membrane's fluid lipid phase, irrespective of particle size. For these permeands, the membrane is virtually no barrier, even when they are as large as the membrane-mobile ionophores. This is considered to be convincing evidence for the fluid character of membranes, at least as convincing

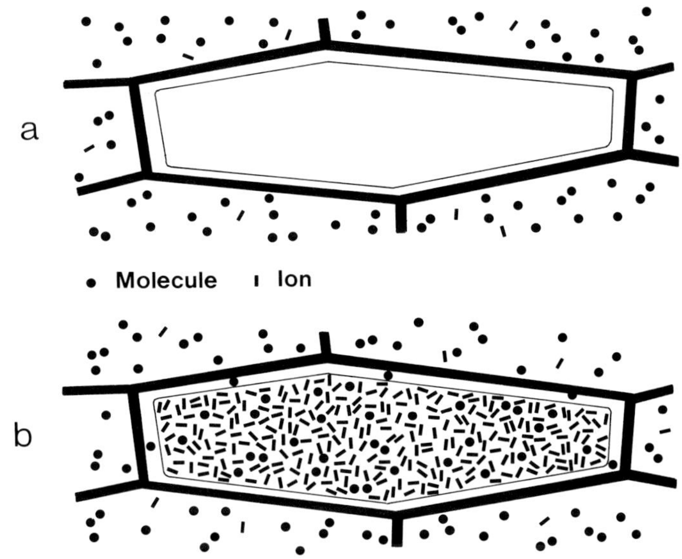

Fig. 6a and b. Ion trap mechanism of neutral red accumulation in plant cell vacuoles before (a) and after completion (b) of staining process. From [141]

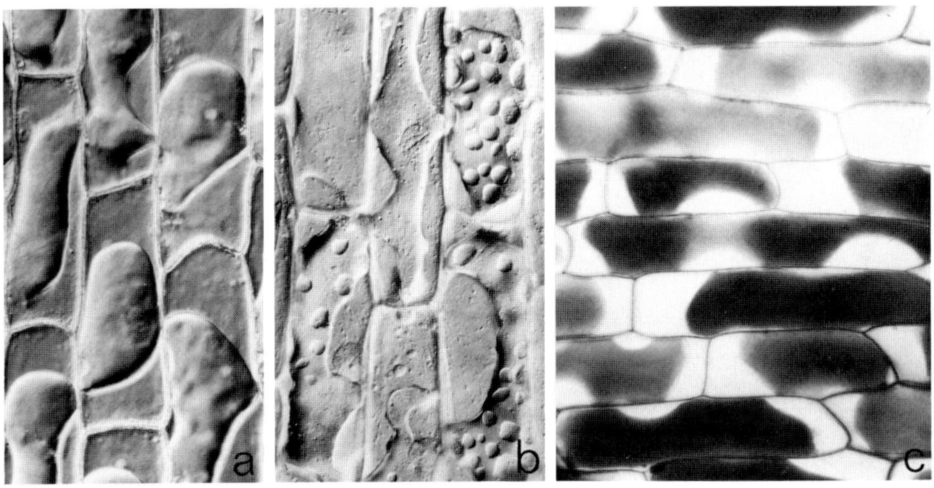

Fig. 7a–c. Plasmolysis and vital staining with neutral red (c) of glutaraldehyde-fixed onion bulb scale epidermis cells. Plasmolysis in a: 1 M KNO_3; b: 1 M $Ca(NO_3)_2$; c: 1 M sucrose. (a) and (b), Nomarski interference contrast, ×300; (c), ×200

as membrane flow or as maintenance of semipermeability after glutaraldehyde fixation [1].

[1] Glutaraldehyde (GA) does not fix lipids at all, so the membrane lipids remain in their fluid in vivo state even after prolonged fixation with GA [68, 114]. Consequently, one can plasmolyze GA-fixed (that is, entirely dead) cells; we have even been able to stain their vacuoles with neutral red according to the ion trap mechanism, which is generally regarded as proof of viability (cf. Fig. 7).

According to the two-way hypothesis, permeands with a low partition coefficient can only penetrate when they are smaller than the mean diameter of unspecific polar membrane pores. This accounts for the filtering effect of elementary membranes. The existence of such unspecific pores has been demonstrated by pronounced 'solvent drag' effects and by other experiments. The mean diameters of the pores range around 5 Å in plasma membranes; higher values have been determined for internal cytomembranes and for junctional membranes, whereas smaller pores prevail in tonoplast membranes and in mitochondrial and plastid inner membranes.

IV. Specific Transport

These unspecific pores, which are simply leaks in the membrane's fluid lipid sheet, must not be confused with specific 'channels' provided by permeases and translocases [54, 67, 165]. In specific and active transport, permeand dis-

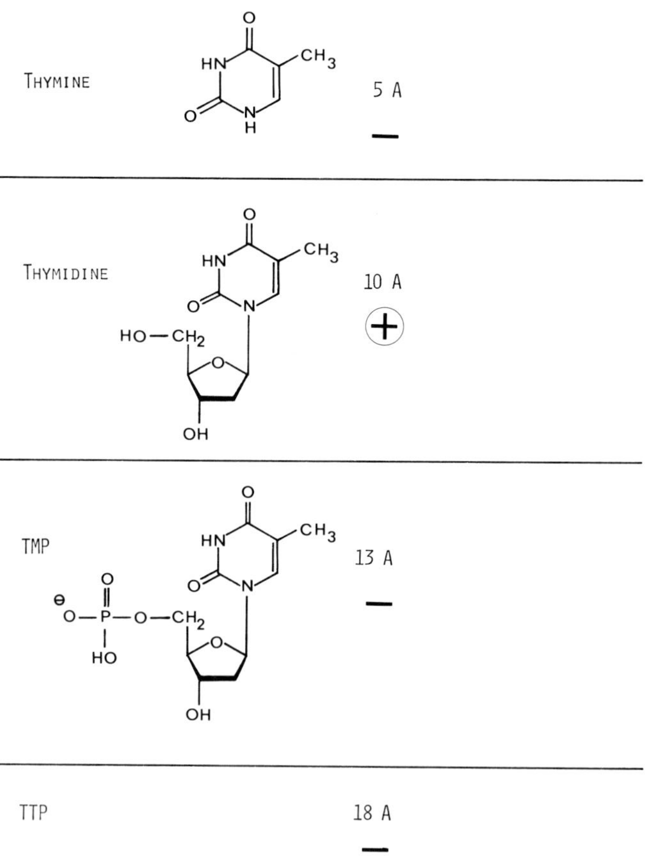

Fig. 8. Specific membrane transport: At plasma membrane of many cells, thymidine itself is transported but not the smaller thymine molecule nor the larger thymidine mono- and triphosphates (*TMP, TTP*). Molecular dimensions indicated

crimination is based primarily on particle shape, much as substrate shape is in enzymatic reactions, rather than on polarity or size. Thymidine transport is a well-known example (Fig. 8): neither the smaller thymine molecule nor the larger derivatives are transported; only the nucleoside is transported in both directions across the plasma membrane. If, however an incoming thymidine molecule becomes phosphorylated inside the cell, it cannot pass out again. This would then be a primitive form of active transport, reminiscent of the well-known phosphotransferase system in bacteria [123].

V. Membrane Flow and Membrane Families

A permease must always be larger than its permeand. Thus, the problem arises concerning the transport of macromolecules and still larger particles across membranes. Three possibilities appear to exist:

1. Formation of supramolecular transporting units such as nuclear pore complexes [41, 43, 167] and, possibly, plasmodesmata [55, 117].

2. Transitory incorporation of macromolecules into a membrane, which possibly applies to plastidal and mitochondrial proteins synthesized outside these organelles [22].

3. *Membrane flow* [5], which is not really membrane transport but formation, translocation, or fusion of *compartments* [42, 45, 109, 113, 132, 133, 160, 161]. In this case, a whole membrane becomes a 'permease', and thus an upper size limit of transported particles does not seem to exist. In fact, bulks of micromolecules (as in neurotransmitter emission, Refs. [147, 152]), macromolecules, or even whole cells may be transported. Specificity is ensured, in this case, by membrane recognition through specific sensor and receptor molecules: only certain membranes will fuse or be derived from each other, whereas others never will. There are, consequently, different membrane families (see Table 1). It is only within such a family that exchange may occur [91]. On the other hand, membrane differentiation and diversification is often mediated by spatial separation through membrane flow [42, 98, 100, 132, 160, 161].

Table 1. Membrane families[a]

Family	Characteristic constituents
1. Plasma membrane Tonoplast membrane rER, sER, Golgi membranes Cytosomes (microbodies, coated vesicles; oleosomes?) Mitochondrial and plastid outer membranes	Sterols
2. Mitochondrial inner membranes	Cardiolipin
3. Plastid inner membranes Thylakoids	Galactolipids

[a] Membrane families are defined as groups of membranes that, during membrane flow processes, may fuse or be derived from each other. Membranes not belonging to the same family never come into direct contact with each other

VI. General Principles of Cellular Compartmentation

All this can be put in a more general form. Let us start with a question: Because membranes are 'closed bags' whose interior differs from their exterior (see however, Ref. [122]), do we have in an eucyte as many sorts of cytoplasms as we have sorts of membranes? Is it really necessary to speak of a Golgi plasm, of an ER plasm, of a plast outer and inner plasm, and so on? Fortunately, it is not, as Schnepf has shown [129, 130, 131]. Nobody would speak, for example, of a vacuolar plasm, since the cell sap is normally a watery solution of metabolic wastes and storage material, and unlike the cyto- and nucleoplasm, it does not contain nucleic acids, so there is no replicating, transcribing, or translating activity within vacuoles, nor can viruses ever be multiplied within them. Consequently, the tonoplast membrane separates a plasmatic space from a nonplasmatic one. The same holds true, of course, for the plasma membrane. According to Schnepf, this can be generalized in the sense that any elementary membrane lines the border between a plasmatic and a nonplasmatic compart-

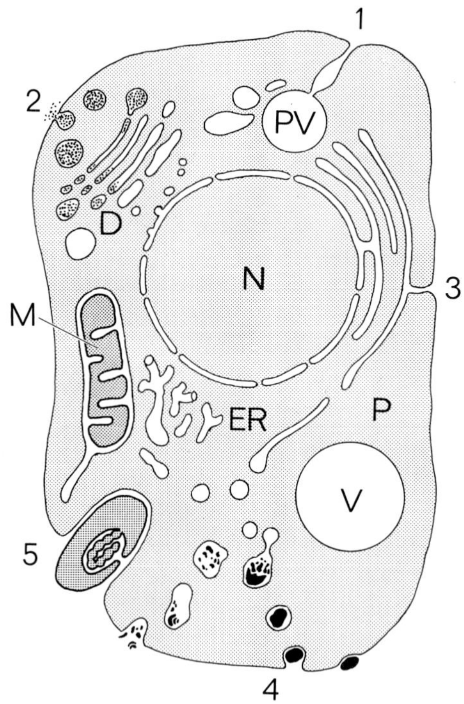

Fig. 9. Compartmentation of a typical eucyte. Plasmatic compartments stippled. *N*: nucleus; *P*: cytoplasm; *V*: vacuole; *PV*: pulsating vacuole; *ER*: endoplasmatic reticulum; *D*: dictyosome; *M*: mitochondrion; (1): discharge of the pulsating vacuole; (2): exocytosis of Golgi vesicles; (3): transient connection of ER intracisternal space with extracellular space; (4): formation of food vacuole (endosome) by endocytosis, digestion of its content by formation of heterolysosome, and extrusion of residual bodies by exocytosis; (5): formation of intracellular endosymbiosis by endocytosis of protocyte. From: Sitte, P.: Elektronenmikroskopie und Biologie — Schicksal einer Symbiose. Mikroskopie **32**, p. 145–190 (1976). With kind permission of Verlag Georg Fromme, Vienna and Munich

Table 2. Examples of the fusion of cellular compartments by membrane flow processes

Compartments	Fusion partners	Exemplified by
Plasmatic	Nucleoplasm + cytoplasm	Mitosis
	Cytoplasm + cytoplasm	Fertilization
	Plastid inner compartments	Fertilization in some unicellular algae
	Mitochondrial inner compartments	Formation of giant mitochondria in yeast, or during spermatogenesis in higher animals
Nonplasmatic	ER cisternae	Reformation of nuclear envelope after mitosis
	Vacuoles	Formation of one central vacuole by fusion of several smaller ones
	ER-derived vesicles	Formation of Golgi cisternae
	ER cisternae and mitochondrial or plastid outer compartment	See Refs. [10, 25]
	Golgi vesicles + extracellular space	Granulocrine secretion
	Endosomes + primary lysosomes	Formation of secondary lysosomes (digesting vacuoles)
	Sec. lysosome + extracellular space	Extrusion of residual bodies
Exceptional case	Cytoplasm (plasmatic) + vacuole (nonplasmatic)	Maturation of sieve tubes in phloem tissue, formation of a mictoplasm

ment. Thus, in a generalized picture of an eucyte there are only two classes of compartments, plasmatic and nonplasmatic ones (Fig. 9). *Plasmatic* is, of course, the cytoplasmic matrix and the nucleoplasm as well. Further, it includes the inner compartments of mitochondria and of plastids. On the other hand, the contents of Golgi cisternae and vacuoles, of the different cytosomes, and of the ER cisternae are always *nonplasmatic* in nature. These latter compartments may contain active enzymes[2], but they do not contain nucleic acids, or ribosomes, there is no protein synthesis within them, nor does any ATP formation or consumption occur, and they contain β-glucanes but no α-glucanes.

One consequence of Schnepf's principle is that two different plasmatic compartments cannot be separated by just one elementary membrane but only by the insertion of a nonplasmatic compartment between them, that is, by a double membrane. During membrane flow—be it endocytosis, exocytosis, or transcytosis—there is again a sharp discrimination between plasmatic and nonplasmatic compartments: only 'corresponding' compartments can merge. Some examples are listed in Table 2.

[2] These 'exoenzymes' are, however, nonregulatory and noncooperating.

Endocytosis of a protocyte without subsequent digestion within a heterolysosome (cf. [5] in Fig. 9) is of particular interest, since it leads directly to the popular hypothesis [134] of a phylogenetic origin of plasts from endosymbionts: mitochondria from primitive bacteria, and plastids from early blue-green algae. According to Schnepf's principle, the outer compartment of these plasts is nonplasmatic. On the basis of such a hypothesis one should expect distinctive differences between the outer and the inner membranes of plasts, the outer one being derived originally from the host cell's plasma membrane and the inner one from the (endo)symbiont's plasmalemma. In fact, the permeability properties of these two membranes differ dramatically [19, 49, 64]: they differ in chemical composition, enzyme complement, and in protein and lipid turnover as well [47, 98, 116, 148]. Only the outer one may fuse with ER membranes [10, 25], and it is a normal eucyte-type membrane in that it contains some sterol lipids but not cardiolipin. The inner plastid membranes, on the other hand, contain virtually *no* sterols, but—at least in the case of mitochondrial inner membranes—appreciable amounts of cardiolipin, as do many bacterial membranes [78, 91, 98, 116, 148, 153].

C. Membrane Biogenesis

This section is not intended to be a survey of the synthesis of single membrane constituents. For an excellent review, see Ref. [98].

Eucytes differ from protocytes in many respects. One impressive difference is the abundance of endo-(cyto-)membranes in eucytes in contrast to their scarcity in protocytes. A single eucyte may contain more than 10^3 separate compartments. An extreme example is provided by the rhizoid cells of *Chara*, each of which contains about 130,000 Golgi cisternae, to say nothing of all the other membrane-bound compartments of these cells.

In bacteria and blue-green algae, the situation looks quite different. The mesosomes and thylakoids of certain bacteria appear to be in continuity with the plasma membrane of these protocytes and are to be regarded therefore as local specializations of the latter [53, 106]. Membrane growth and transformation occur by insertion of new membrane constituents in between the already existing ones [106]. In the larger cells of blue-green algae, the thylakoids are also formed by invagination of the plasma membrane. However, these infoldings eventually become detached from the plasma membrane (a situation that is paralleled perfectly by the one encountered in chloroplasts). Despite all membrane growth, local or spatial membrane differentiation, and membrane separation by membrane flow, a real de novo origin of membranes has never been observed in protocytes.

This general rule seems also to apply to eucytes [47]. A few apparent exceptions have been reported, e.g., the immediate formation of a new plasma membrane around a droplet removed from a *Physarum* plasmodium. However, a reinvestigation clearly showed that this new plasma membrane is derived from the membranes of small vacuoles that fuse instantly upon rupture of the plasmodium [166]. Likewise, early reports of a supposed de novo origin of tonoplast membranes among the cytoplasm of young plant cells could not be confirmed by later investigations [14, 70, 87, 90, 94]. On the contrary, a large number of careful investigations has resulted in the ability to successfully trace membranogenesis back to preexisting membranes, most often belonging to the ER or the Golgi system [99, 160, 161]. This applies, for example, to primary lysosomes, secretion vesicles, the different sorts of microbodies, vacuoles, and also partly to coated vesicles. Extreme cases, in which membranes do not really originate from true membranes but merely from membranelike structures, have been encountered in the mitochondria of glucose-fermenting yeasts upon their return to aerobic conditions [128], and in etioplasts, when the first stroma thylakoids grow out from the tubular system of a prolamellar body upon illumination. Yet even in these borderline cases, nobody would say that cristae and thylakoids originate de novo.

The formation of a membrane seems to depend on the presence of something like a 'crystallization center', which in turn has to be a membrane. This would mean, then, that not only nucleic acids, but also membranes possess genetic continuity (although in a different sense). This is a very surprising result. Enzymes, microtubules, ribosomes, microfilaments, even lipid storage globules, etc. can all be generated in a cell that has so far been devoid of them; but *not* membranes. It may be that this enigmatic fact has a rather simple physicochemical basis, but so far nobody has considered this.

If it is true that membrane biogenesis requires membranes, then *all* the different eucyte membranes must have one common ancestral membrane, and the same must hold true for *all* protocyte and organelle membranes [146].

If this hypothesis is taken for granted, then some further considerations appear unavoidable [146]:

1. The elaborate endomembrane system of eucytes goes back phylogenetically to the one 'exomembrane' or plasma membrane, the existence of which is a prerequisite for every cell.

2. It is, therefore, understandable that all the different membranes of the eucyte-type (which include the outer organelle membranes, but not the inner ones) belong to only one membrane family [99].

3. If there is an obligate continuity of membranes, in space or in time, the same must hold true for the compartments they comprise. That means that there can be only one plasmatic compartment per cell and that all nonplasmatic compartments are derived, phylogenetically, from the extracellular space.

4. Wherever a nonplasmatic compartment encloses a plasmatic one uninterruptedly within a cell, the latter compartment must have been taken up from outside the cell by endocytosis, and it must represent *another* plasma species, which, for continuing reduplication, has to possess its own DNA and RNAs. This statement is nothing else than another formulation of the endosymbiont hypothesis.

D. Intracellular Membranes and Their Compartments

Let us now consider the size and shape of compartments, and some implications thereof concerning lipids and biomembranes.

Compartments can vary greatly. The volumes of plant cell vacuoles and of coated vesicles differ by eight orders of magnitude. Also, in addition to spherical storage compartments there are flat, extended compartments such as the thylakoids or the cisternae of the ER. Certainly, the shaping of cells and subcellular compartments is a question of functional fitness and not of membrane structure (see, e.g., Ref. [112]). It is not the membranes that dictate a compartment's shape. Instead, it is the functional demands of a compartment that direct the membrane's extension and spatial arrangement. Consequently, biomembranes can only be useful tools in the cell's life when their own intrinsic structures allow for any plasticity that may be necessary.

The whole cell is a typical input-output system and so is every compartment. Consequently, the surface-to-volume proportion might be of particular significance. By this surface-to-volume ratio (SVR), an interrelationship can be constructed between a storing or working interior and a surface membrane controlling the exchange of material between that interior and the compartment's surroundings. An optimal SVR of about 5 μm^{-1} seems to exist. This same value is approximated by protocytes [146]. One should expect that the SVR optimum is due to a corresponding proportion of specific membrane transport capacity and enzymatic capacities of the compartment's content. However, we do not yet have sufficient data to enable us to assess quantitatively such a possible interrelationship.

It is nevertheless obvious that the relatively isodiametric shape of isolation compartments such as storage vacuoles, secretion vesicles, primary lysosomes, and digesting vacuoles, is maintained in order to minimize exchange between the interior and the surroundings. That many nuclei are also spherical or lense-shaped underlines the nuclear envelope's function as a barrier between nucleoplasm and cytoplasm, i.e., between transcription and translation. If one now compares the number of nuclear pore complexes of nuclei of different sizes (Fig. 10), a positive correlation becomes evident: large nuclei possess not only more pores than smaller ones in absolute terms but also a higher pore frequency, that is, a drop in the nuclear SVR is compensated for by an increase in pore number per μm^2. From this bare fact it could have been inferred that pore complexes act as pumps in nucleocytoplasmic exchange. For geometrical reasons, pore frequency cannot amount to values above 80 pores per μm^2. Therefore, with increasing nuclear diameters, a pore frequency plateau is eventually reached. It is noteworthy in this context that extremely large nuclei like the ones of egg cells and endopolyploid cells, or the 'primary' nuclei of *Acetabularia,* consistently have their envelope enlarged by numerous evaginations, or are no longer spherical at all.

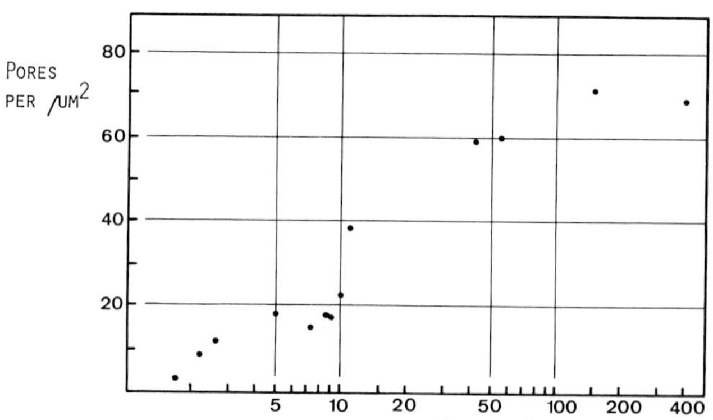

Fig. 10. Nuclear pore frequency correlated with nuclear diameter (nuclear volume). Data collected from the literature [146]

E. The Role of Lipid Self-Assembly

There are, however, known examples of cellular structures that are structurally determined by the self-assembly of lipid molecules [149].

As we have seen, storage lipids and structural lipids differ not only with regard to their role in metabolism and their chemical nature, but also with regard to the structures built by them. Whereas the former become concentrated into spherical droplets, the latter take part in the build-up of extended, very thin membranes. The SVR is at a minimum in the first case, and at a maximum in the second. This remarkable structural difference is due to corresponding differences in surface properties of the lipid bulk phases, surface tension being high in the first and low in the second case.

In fact, the molecules of storage lipids (that is, mainly triglycerides) carry almost no electric charges. Consequently, they are expelled from the aqueous cytosol and are forced, by hydrophobic interaction, to form bulk phases. The lowest possible energy state of the system is reached only when the surface area of such a bulk phase is minimized. Therefore, storage lipid bulk phases must approximate the geometric form of a sphere. This correlates well with their function as an osmotically inert energy reservoir within a plasmatic compartment.

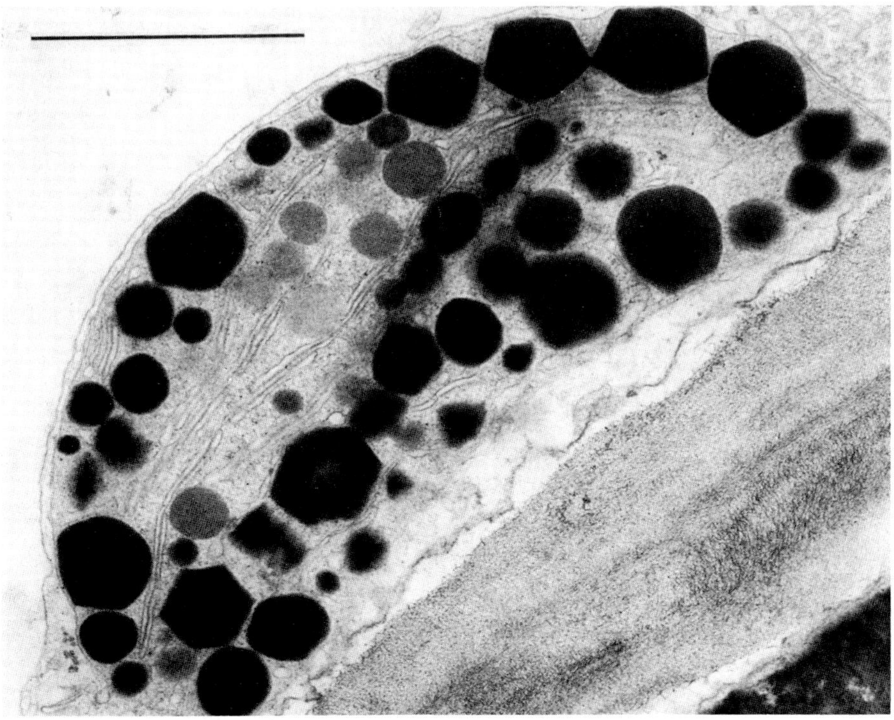

Fig. 11. Chromoplast of *Chrysosplenium alternifolium* sepal upper epidermis with partly crystallized osmiophilic plastoglobules [143], Bar: 1 μm

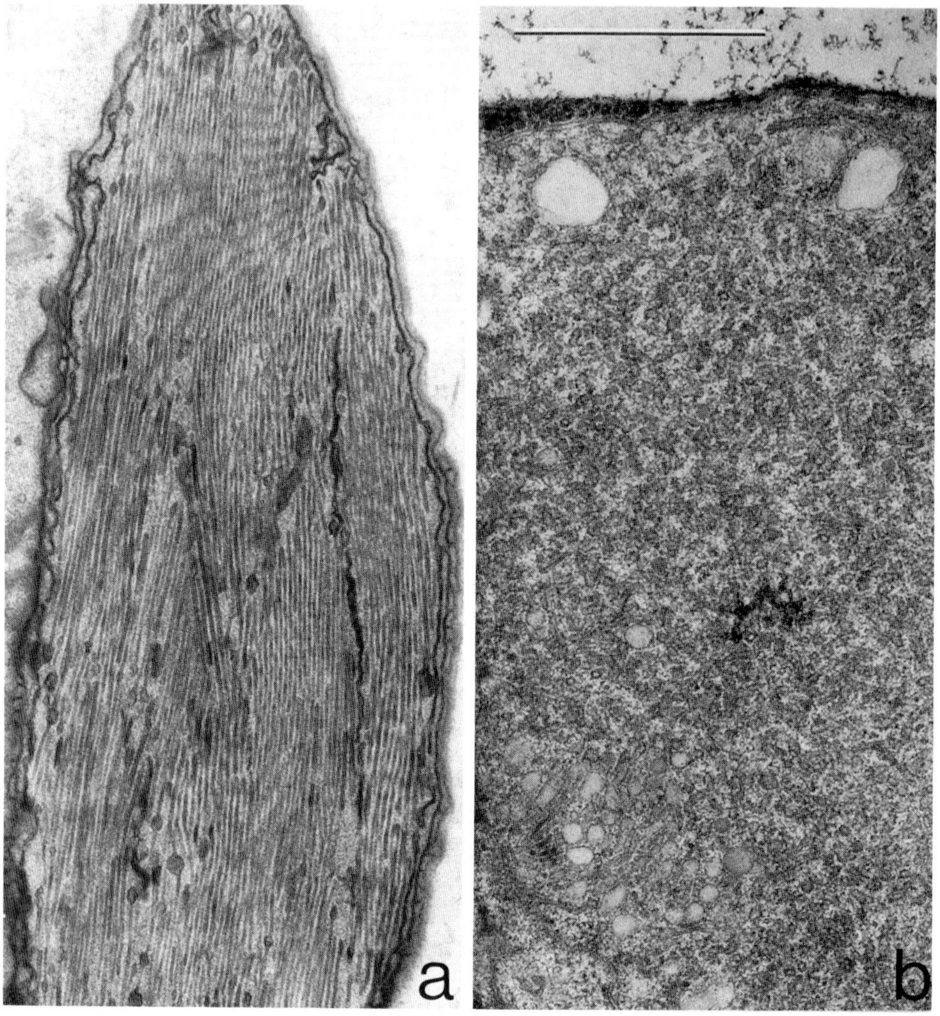

Fig. 12a and b. 'Tubulous' chromoplasts with (a) straight and (b) undulating lipoprotein structures. (a), hip of *Rosa rugosa*; (b), fruit of *Capsicum annuum*. Bar [corresponding to both (a) and (b)]: 1 μm

On the other hand, the more or less *amphiphilic structural lipids* are highly surface-active and will form, as other surfactants do, bilayers with hydrophilic surfaces in a polar medium. In this case, an extremely expanded arrangement with the lipid water interface at a *maximum* corresponds to the lowest energy level and is therefore strongly favored.

Recently, Sandermann has provided a good, though indirect, demonstration of these facts [125]. C_{55}-isoprenoid alcohol phosphokinase from *Staphylococcus aureus* is known to be an extremely hydrophobic, integral membrane protein that requires a lipid cofactor for its activity. Sandermann has now demonstrated that only those lipids that absorb 2.3 water molecules or more per molecule of lipid can act as cofactors, whereas lipids binding less than 1 H_2O molecule will not activate

Fig. 13. Freeze-etch view of 2 flower chromoplasts of the daffodil, *Narcissus pseudonarcissus*. Low particle density on chromoplast internal membranes corresponds to low buoyant density (cf. Fig. 2) and low protein content (Table 3). *ER*: endoplasmic reticulum; note higher particle density. Bar: 1 μm

the enzyme. Apparently, the kinase does not just need an apolar domain to be active but an amphipolar one, and it is therefore suitably placed in a biomembrane.

Another instructive example about how the physicochemical properties of lipid molecules can govern the shape of supramolecular entities is provided by the *carotenoid carrying structures in chromoplasts*. The carotenes as well as the xanthophylls and their esters are extremely hydrophobic. Thus, their localization in the watery stroma matrix of the plastids necessitates lipophilic carrier structures that nevertheless possess a surface that is sufficiently polar to mediate between the apolar pigment and the polar stroma matrix. Surprisingly, not just one such carrier structure has been found in nature, but four different ones [143], namely, (1) intrathylakoidal carotene crystals, (2) extrathylakoidal plastoglobules (which in some cases also eventually crystallize, cf. Fig. 11, (3) 'tubules' of filaments (Fig. 12) that are normally unbranched, straight, and parallel, yet in some cases branched and irregularly twisted, and (4) concentric membranes that have so far been found only in daffodil (Fig. 13). When one

Table 3. Chemical composition of isolated carotenoid carrier structures in different chromoplasts[a]

	Membranes[b]	Tubules[c]	Globules[d]
Apolar compounds			
triglycerides	(+)	(+)	70–90
carotenoids, carotenoid esters	4.6	21	7–12
plastid quinones	2	+	1
Semipolar compounds			
phospholipids	16	17	0.5
galactolipids	51	27	1.5
sulfolipids	2.7		
proteins	24	33	1.3
Semipolar/apolar	14.2	3.5	0.03
Surface/volume	>5	2–4	<1

[a] Values expressed in % dry weight; +, (+), trace amounts
[b] Chromoplast internal membranes of *Narcissus pseudonarcissus* [83]
[c] *Tropaeolum majus* [162], *Rosa rugosa* [168]
[d] *Caltha, Viola, Tulipa* [169]

disregards the pure carotene crystals, then the remaining 'true' carrier structures resemble spheres, cylinders, and planes. The SVRs of these simple geometric forms are quite different (Table 3). If the surfaces of all the different carrier structures were covered with semipolar compounds, whereas the inner cores bear apolar lipids, including the carotenoids, then quite different relative proportions of structural lipids and proteins on the one hand and apolar lipids including the pigments on the other must correspondingly exist. This expectation is well borne out by chemical analysis, carried out during the last two years in the author's laboratory (see Table 3 and Refs. [83, 162, 168, 169]). It is tempting therefore to hypothesize that the plastids, during chromoplast transformation, direct the formation of a particular carrier structure just by producing (or retaining) a definite proportion of apolar and semipolar compounds to be incorporated into pigment carrier structures. Such a hypothesis, which could explain morphogenesis on a subcellular, though supramolecular, level in unexpectedly simple terms as some sort of self-assembly, must now of course to be checked by model experiments that are already underway in our laboratory at Freiburg.

F. The Molecular Architecture of Elementary Membranes

Biomembranes contain not only lipids but also proteins (and up to 10% heterosaccharides, bound covalently to either the protein or the lipid molecules [24, 80, 85]). The lipid-to-protein ratio, expressed in terms of dry weight, varies greatly. In many instances, however, a proportion of roughly 1:1 has been

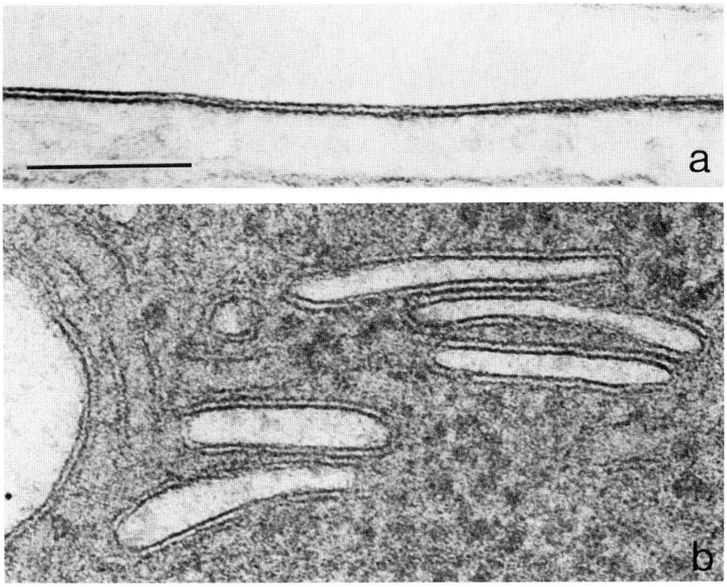

Fig. 14a and b. Elementary membranes in cross section, exhibiting typical trilaminate 'unit membrane look' after fixation with glutaraldehyde and OsO_4. (a), plasma membrane of *Botrydium* spec. (b), fusiform vesicles [36] of *B. becherianum*. Bar [corresponding to both (a) and (b)]: 0.1 µm. (Electron micrographs: Courtesy of Dr. H. Falk)

reported, with the proteins being slightly predominant. What is the spatial arrangement of these main membrane constituents?

The answer to this question has been sought for half a century. Since the days of Overton and his postulation of the 'lipid theory of permeation' [108], it was clear that a coherent and (at least partly) fluid lipid phase must exist in biomembranes. In 1925, Gorter and Grendel published their classic work on the *bi*layer structure of erythrocyte plasma membranes [51]. Around 1960 J.D. Robertson postulated, after extensive EM investigations, a common basic structure for all the otherwise different exo- and endomembranes of eucytes and coined the term unit membrane. This unit membrane, which presents itself as a trilaminate structure in cross section [119] after fixation by OsO_4 or $MnO_4^{\ominus}$ (Fig. 14), was supposed to be essentially a lipid bilayer [119, 120] covered with proteins, according to the old Danielli model.

Meanwhile, however, the investigation of membrane proteins has made great progress, and since the mid-sixties, much attention has been paid to this field, which had in fact been 'underprivileged' until then. It became clear that globular, biologically active proteins with α-helical regions not only adhere to a lipid bilayer by ionic bonds but also tunnel into the lipid phase of membranes where they act, for example, as mediators of specific and active transport. In different ways, the existence of a lipid–protein interaction, based on a hydrophobic effect [149], has been shown.

For a few years, radical alternatives have been under discussion in membrane modeling. Due to certain inconsistencies and insecurities of the Danielli-Robertson model, and mainly due to the increased self-confidence of membrane protein-ologists, several *protein-particle models* appeared on the market [170]. In these models it was assumed that, rather than a lipid bilayer, a two-dimensional lattice of protein molecules forms the backbone of the membrane and that the single lipid molecules become orientated according to the lipid-protein interaction and not according to lipid-lipid interactions.

To cut a long story short, and simplifying, of course, a labyrinthine situation, one may say today that this attack against the bilayer hypothesis resulted not in the collapse of this model but in its reinforcement [13, 172]. The new formulation of the old concept, as provided by Singer and Nicolson in 1972, has been symptomatic of the revival of the lipid bilayer [15, 17, 20, 65, 104, 105, 114, 137, 138, 139, 150, 172), much to the satisfaction of lipidologists and textbook writers. It seems unnecessary to summarize once more all the evidence at hand in favor of the fluid-bilayer model [3]. In addition to the demonstration of thermal phase transitions in biomembranes [17, 107], freeze-etch electron microscopy [12], X-ray studies [38, 59], recent repetitions of the Gortel-Grendel experiment ([172]; see also [74]), and to the notion that protein-dominated membrane regions such as junctional areas behave quite differently when compared with 'normal' membranes [48, 84], one very strong argument in favor of a real bilayer structure of biomembranes comes from the demonstration of the spectacular differences in lipid composition of the outer *vs.* the inner half of erythrocyte membranes ([13, 136, 154, 171]; see also Ref. [1]). The outer half is rich in sphingomyelins, choline phospholipids, glycolipids, and cholesterol, whereas in the inner half amino phospholipids (and PE and PS in particular) prevail. This asymmetry is also reflected by the fatty acid moiety, with the fatty acids of the outer half of the membrane being saturated to a considerably higher degree than those of the inner half. If one combines these data with the well-known experiments of Kornberg and McConnell [76], in which it was shown that in artificial lipid bilayers molecular 'flip-flop' occurs at a frequency that is 10 to 12 orders of magnitude less than that of changes in lateral position, one inevitably arrives at a bilayer model of the erythrocyte plasma membrane.

G. Conventional Electron Microscopy of Biomembranes

One last question remains to be answered: Why is electron microscopy not able to solve the structure problem of biomembranes by direct observation?

Despite all the progress made in bio-electron microscopy during the last few years, we are still left with this rather boring 'railroad track' aspect of membrane cross sections (Fig. 14), and we still have difficulties in interpreting even this. One must not forget that Robertson's postulate, that the 'unit mem-

[3] This popular term is misleading insofar as it could support the wrong impression of a lipid bilayer being in a fluid state throughout. Recent evidence points strongly to the coexistence, in most elementary membranes, of liquid areas with islands of restricted fluidity (see, e.g., Refs. [4, 107]).

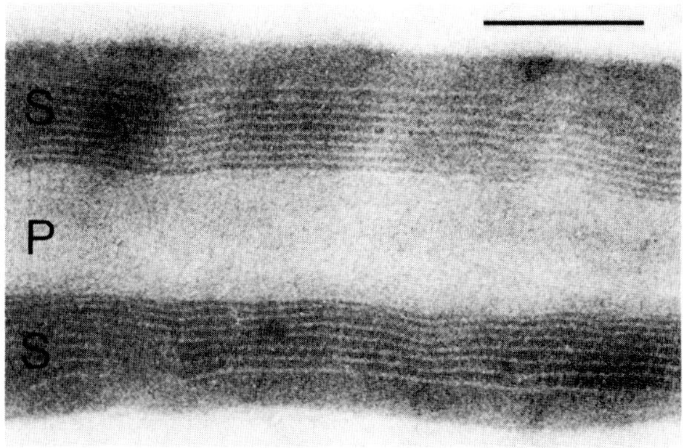

Fig. 15. Finely laminated suberin layers (*S*) of two adjacent cells of potato tuber cork tissue. *P*: primary wall layer. Bar: 0.2 μm. (Electron micrograph: Courtesy of Dr. H. Falk)

brane look' is the silhouette of a bilayer membrane, has been refuted because of the lack of evidence to support it. In fact, one cannot talk reasonably about membrane cross sections as long as one does not understand how the final localization of the contrasting agents, such as the dioxides of osmium or manganese, occurs.

In order to clarify the situation a little, we have carried out some experiments and obtained the following results [145]:

1. OsO_4 exhibits pronounced *lipophilia*. Its partition coefficient has been found to be 17.3 in a diethyl ether-water biphasic system, that means, only 5% OsO_4 is to be found in the aqueous hypophase. It is to be assumed, therefore, that, during the earliest stages of the fixation process, OsO_4 becomes concentrated within the lipid phases of the cell.

2. This assumption has been verified by polarizing microscope studies of concentrically arranged membranes as found in corona chromoplasts of the daffodil (see Fig. 13). Their positive optical character turns negative immediately upon addition of OsO_4 [83].

3. The primary reduction products of osmium tetroxide concentrate at polar/apolar interfaces, that is, they are *amphiphilic*. This could be demonstrated by model experiments and also by osmication of the suberin layer of cork cells (Fig. 15). These lipophilic wall layers are laminated, apolar wax films alternating with semipolar suberin sheets [140]. By osmication, only the latter become positively stained [37].

It can be postulated, therefore, that during the fixation process OsO_4 at first becomes concentrated within phases of low polarity; but afterward (as soon as oxidation of unsaturated compounds by OsO_4 takes place), a displacement occurs to the lipid-water interfaces. Such a secondary shift of contrast-giving fixation products apparently explains all the known images of cellular lipid and lipoprotein structures. It is, furthermore, in perfect agreement with spin-label experiments carried out by Jost and co-workers with lecithin multilayers [69]. These authors explain their electron spin resonance spectra by the assumption that OsO_4, covalently bound to former CC-double bonds, eventually becomes located at the polar surface of the membrane's lipid bilayer.

4. *Permanganate fixation* creates images of lipid structures similar to those created by OsO_4. However, its mechanism of action is totally different, because the permanganate ion is not at all lipophilic. Permanganate ions penetrate into lipid phases very slowly and not at all by diffusion but exclusively according to the oxidative destruction of superficially located lipid molecules under the deposition of MnO_2, which remains located wherever it has been formed [144]. Due to the extremely high packing density of lipid bulk phases, MnO_2 is deposited most heavily at the surface of apolar phases within the cell.

The results just presented, when considered together, strongly support the lipid bilayer model. The 'unit membrane aspect' is in fact (as Robertson postulated more than 15 years ago) a crude, though correct, expression of what we use to visualize a biomembrane's molecular architecture.

Acknowledgements. The author wishes to express his gratitude to Dr. Heinz Falk and Professor Hans Kleinig for many valuable discussions, and to John A. Thompson, B. Sc., for correcting the English manuscript. The experimental work on chromoplasts and on the chemical fixation of lipid structures was supported by the *Deutsche Forschungsgemeinschaft* and *Sonderforschungsbereich* 46 (University of Freiburg).

References

1. Anderson, J.M.: The molecular organization of chloroplast thylakoids. Biochim. Biophys. Acta **416**, 191–235 (1975)
2. Azzone, G.F., Klingenberg, M.E., Quagliarello, E., Siliprandi, N. (eds.): Membrane Proteins in Transport and Phosphorylation. Amsterdam-New York: North-Holland 1974
3. Barber, J.: Cation control in photosynthesis. TIBS **1**, 33–36 (1976)
4. Bardele, C.F.: Particle movement in heliozoan axopods associated with lateral displacement of highly ordered membrane domains. Z. Naturforsch. **31c**, 190–194 (1976)
5. Bennett, H.S.: The concept of membrane flow and membrane vesiculation as mechanisms for active transport and ion pumping. J. Biophys. Biochem. Cytol. **2**, Suppl., 99–103 (1956)
6. Bentrup, F.-W.: Lichtabhängige Membranpotentiale bei Pflanzen. Ber. Deut. Botan. Ges. **87**, 515–528 (1974)
7. Berlin, R.D.: Specificities of transport systems and enzymes. Science **168**, 1539–1545 (1970)
8. Bittar, E.E. (ed.): Membranes and Ion Transport. London: Wiley-Interscience 1970
9. Bolis, L., Katchalsky, A., Keynes, R.D., Loewenstein, W.R., Pethica, B.A. (eds.): Permeability and Function of Biological Membranes. Amsterdam-London: North-Holland 1970
10. Bracker, C.E., Grove, S.N.: Continuity between cytoplasmic endomembranes and outer mitochondrial membranes in fungi. Protoplasma **73**, 15–34 (1971)
11. Branton, D.: Fracture faces of frozen myelin. Exptl. Cell Res. **45**, 703–707 (1967)
12. Branton, D., Deamer, D.W.: Membrane Structure. Wien-New York: Springer 1972
13. Bretcher, M.S.: Membrane structure: Some general principles. Science **181**, 622–629 (1973)
14. Buvat, R.: Origin and continuity of cell vacuoles. In: Origin and Continuity of Cell Organelles. Reinert, J., Ursprung, H. (eds.). Berlin-Heidelberg-New York: Springer, 1971, pp. 127–157
15. Capaldi, R.A.: A dynamic model of cell membranes. Sci. Am. **230**/3, 26–33 (1974)
16. Chapman, D. (ed.): Biological Membranes. Physical Fact and Function. London-New York: Academic Press 1968
17. Chapman, D.: Phase transitions and fluidity characteristics of lipids and cell membranes. Quart. Rev. Biophys. **8**, 185–235 (1975)

18. Chapman, D., Wallach, D.F.H. (eds.): Biological Membranes. London-New York-San Francisco: Academic Press, 1973 and 1976, Vols. 2, 3
19. Chappell, J.B.: Systems used for the transport of substrates into mitochondria. Brit. Med. Bull. **24**, 150–157 (1968)
20. Cherry, R.J.: Protein and lipid mobility in biological and model membranes. In: Biological Membranes. Chapman, D., Wallach, D.F.H. (eds.). London-New York-San Francisco: Academic Press, 1976, Vol. 3., pp. 47–102
21. Clowes, F.A.L., Juniper, B.E.: Plant Cells. Oxford-London: Blackwell 1968
22. Cobb, A.H., Wellburn, A.R.: Changes in plastid envelope polypeptides during chloroplast development. Planta **121**, 273–282 (1974)
23. Coleman, R.: Membrane-bound enzymes and membrane ultrastructure. Biochim. Biophys. Acta **300**, 1–30 (1973)
24. Cook, G.M.W., Stoddart, R.W.: Surface Carbohydrates of the Eukaryotic Cell. London-New York-San Francisco: Academic Press 1973
25. Crotty, W.J., Ledbetter, M.C.: Membrane continuities involving chloroplasts and other organelles in plant cells. Science **182**, 839–841 (1973)
26. Cuatrecasas, P.: Membrane receptors. Ann. Rev. Biochem. **43**, 169–214 (1974)
27. Davis, B.D., Warren, L. (eds.): The Specificity of Cell Surfaces. Englewood Cliffs, N.J.: Prentice-Hall 1967
28. Decker, R.S.: Hormonal regulation of gap junction differentiation. J. Cell Biol. **69**, 669–685 (1976)
29. de Gier, J., Mandersloot, J.G., van Deenen, L.L.M.: The role of cholesterol in lipid membranes. Biochim. Biophys. Acta **173**, 143–145 (1969)
30. de Souza, W.: Associations membrane-microtubules chez *Trypanosoma cruzi*. J. Microscopie Biol. Cell **25**, 189–190 (1976)
31. Drawert, H.: Vitalfärbung und Vitalfluorochromierung pflanzlicher Zellen und Gewebe. Wien-New York: Springer 1968
32. Drawert, H., Bock, U.: Ein Modellversuch zum Ionenfallenprinzip der Farbstoffspeicherung in „leeren" Zellsäften. Ber. Dtsch. Bot. Ges. **74**, 321–323 (1961)
33. Ernster, L., Estabrook, R.W., Slater, E.C. (eds.): Dynamics of Energy-Transducing Membranes. Amsterdam-London-New York: Elsevier 1974
34. Erwin, J.A. (ed.): Lipids and Biomembranes of Eukaryotic Microorganisms. New York-London: Academic Press 1973
35. Estrada-O., S., Gitler, C. (eds.): Perspectives in Membrane Biology. New York-San Francisco-London: Academic Press 1974
36. Falk, H.: Fusiform vesicles in plant cells. J. Cell Biol. **43**, 167–174 (1969)
37. Falk, H., El-Hadidi, M.N.: Der Feinbau der Suberinschichten verkorkter Zellwände. Z. Naturforsch. **16b**, 134–137 (1961)
38. Finean, J.B., Coleman, R., Michell, R.H.: Membranes and their Cellular Functions. Oxford: Blackwell Sci. Publ. 1974
39. Fox, C.F. (ed.): Membrane Research. New York-London: Academic Press 1972
40. Fox, C.F., Keith, A.D. (eds.): Membrane Molecular Biology. Stamford, Conn.: Sinauer 1972
41. Franke, W.W.: Structure, biochemistry, and functions of the nuclear envelope. Internat. Rev. Cytol., Suppl. 4, 71–236 (1974)
42. Franke, W.W., Kartenbeck, J.: Some principles of membrane differentiation. In: Progress in Differentiation Research. Müller-Bérat, N., *et al.* (eds.). Amsterdam: North-Holland, 1976, pp. 213–243
43. Franke, W.W., Scheer, U.: Structures and functions of the nuclear envelope. In: The Cell Nucleus. Busch, H. (ed.). New York-San Francisco-London: Academic Press 1974, Vol. 1, pp. 219–347
44. Franke, W.W., Lüder, M.R., Kartenbeck, J., Zerban, H., Keenan, T.W.: Involvement of vesicle coat material in casein secretion and surface regeneration. J. Cell Biol. **69**, 173–195 (1976)
45. Franke, W.W., Morré, D.J., Deumling, B., Cheetham, R.D., Kartenbeck, J., Jarasch, E.-D., Zentgraf, H.-W.: Synthesis and turnover of membrane proteins in rat liver: An examination of the membrane flow hypothesis. Z. Naturforsch. **26b**, 1031–1039 (1971)
46. Frazier, W.A.: The role of cell surface components in the morphogenesis of the cellular slime molds. TIBS **1**, 130–133 (1976)

47. Getz, G.S.: Organelle biogenesis. In: Membrane Molecular Biology. Fox, C.F., Keith, A.D. (eds.). Stamford, Conn.: Sinauer, 1972, pp. 386–438
48. Gilula, N., Branton, D., Satir, P.: The septate junction: A structural basis for intercellular coupling. Proc. Natl. Acad. Sci. U.S. **67**, 213–220 (1970)
49. Gimmler, A., Schäfer, G., Kraminer, H., Heber, U.: Amino acid permeability of the chloroplast envelope as measured by light scattering, volumetry and amino acid uptake. Planta **120**, 47–61 (1974)
50. Glaser, L.: Cell-cell recognition. TIBS **1**, 84–86 (1976)
51. Gorter, E., Grendel, F.: On bimolecular layers of lipoids on the chromocytes of the blood. J. Exp. Med. **41**, 439–443 (1925)
52. Gräber, P., Witt, H.T.: Relations between the electrical potential, pH gradient, proton flux and phosphorylation in the photosynthetic membrane. Biochim. Biophys. Acta **423**, 141–163 (1976)
53. Greenawalt, J.A., Whiteside, T.L.: Mesosomes: Membranous bacterial organelles. Bacteriol. Rev. **39**, 405–463 (1975)
54. Guidotti, G.: The structure of membrane transport systems. TIBS **1**, 11–13 (1976)
55. Gunning, B.E.S., Robards, A.W. (eds.): Intercellular Communication in Plants: Studies on Plasmodesmata. Berlin-Heidelberg-New York: Springer 1976
56. Gunning, B.E.S., Steer, M.W.: Plant Cell Biology—An Ultrastructural Approach. London: Arnold 1975
57. Harris, E.J. (ed.): Transport and Accumulation in Biological Systems, 3rd ed. London: Butterworth 1972
58. Harrison, R., Lunt, G.G.: Biological Membranes. Their Structure and Function. Glasgow: Blackie & Son 1976
59. Harrison, S.C., David, A., Jumblatt, J., Darnell, J.E.: Lipid and protein organization in Sindbis virus. J. Mol. Biol. **60**, 523–528 (1971)
60. Hatefi, Y., Djavadi-Ohaniance, L. (eds.): The structural Basis of Membrane Function. New York-San Franciso-London: Academic Press 1976
61. Hauska, G.: Vergleichende Untersuchungen zur Topographie der Membranen in Mitochondrien und Chloroplasten. Angew. Chem. **84**, 123–124 (1972)
62. Hauska, G.: Biochemical aspects of electron transport and phosphorylation in chloroplasts. Ber. Dtsch. Bot. Ges. **88**, 303–318 (1975)
63. Heinz, E. (ed.): Coupling between electrolyte and nonelectrolyte transport in cells. Hoppe-Seylers Z. Physiol. Chem. **353**, 1–13 (1972)
64. Heldt, A.W., Sauer, F.: The inner membrane of the chloroplast envelope as the site of specific metabolite transport. Biochim. Biophys. Acta **234**, 83–91 (1971)
65. Höchli, M., Hackenbrock, C.R.: Fluidity in mitochondrial membranes: Thermotropic lateral translational motion of intramembrane proteins. Proc. Natl. Acad. Sci. U.S. **73**, 1636–1640 (1976)
66. Höfler, K.: Progress in permeability. Protoplasma **60**, 150–158 (1965)
67. Hope, A.B.: Ion Transport and Membranes. A Biophysical Outline. London: Butterworth 1971
68. Jost, P., Brooks, U.J., Griffith, O.H.: Fluidity of phospholipid bilayers and membranes after exposure to osmium tetroxide and glutaraldehyde. J. Mol. Biol. **76**, 313–318 (1973)
69. Jost, P., Waggoner, A.S., Griffith, O.H.: Spin labeling and membrane structure. In: Structure and Function of Biological Membranes. Rothfield, L.I. (ed.). New York-London: Academic Press, 1971, pp. 83–144
70. Khera, P.K., Tilney-Bassett, R.A.E.: The origin of vacuoles in young embryos of *Pelargonium* × *Hortorum* Bailey. Planta **130**, 333–338 (1976)
71. Kimmich, G.A.: Coupling between Na^+ and sugar transport in small intestine. Biochim. Biophys. Acta **300**, 31–78 (1973)
72. Kinzel, H.: Theoretische Betrachtungen zur Ionenspeicherung basischer Vitalfarbstoffe in leeren Zellsäften. Protoplasma **44**, 52–72 (1955)
73. Kitamura, A., Hiwatashi, K.: Mating-reactive membrane vesicles from cilia of *Paramaecium caudatum*. J. Cell Biol. **69**, 736–740 (1976)
74. Kleinig, H.: Nuclear membranes from mammalian liver II. Lipid composition. J. Cell Biol. **46**, 396–402 (1970)

75. Klingenberg, M.: Energy transfer in mitochondrial synthesis of ATP; a survey. In: Energy Transformation in Biological Systems. Ciba Found. Symp. **31**, 23–40 (1975)
76. Kornberg, R.D., McConnell, H.M.: Inside-outside transition of phospholipids in vesicle membranes. Biochemistry **10**, 1111–1120 (1971)
77. Kotyk, A.: Mechanism of nonelectrolyte transport. Biochim. Biophys. Acta **300**, 183–210 (1973)
78. Law, J.H., Snyder, W.R.: Membrane lipids. In: Membrane Molecular Biology. Fox, C.F., Keith, A.D. (eds.). Stamford, Conn.: Sinauer, 1972, pp. 3–26
79. Ledbetter, M.C., Porter, K.R.: Introduction to the Fine Structure of Plant Cells. Berlin-Heidelberg-New York: Springer 1970
80. Lee, E.Y.C., Smith, E.E. (eds.): Biology and Chemistry of Eucaryotic Cell Surfaces. New York-London: Academic Press 1974
81. Leibowitz, P.J., Schaechter, M.: The attachment of the bacterial chromosome to the cell membrane. Internat. Rev. Cytol. **41**, 1–28 (1975)
82. Leive, L. (ed.): Bacterial Membranes and Walls. New York: Marcel Dekker 1973
83. Liedvogel, B., Sitte, P., Falk, H.: Chromoplasts in the daffodil: Fine structure and chemistry. Cytobiologie **12**, 155–174 (1976)
84. Loewenstein, W.R.: Intercellular communication. Sci. Am. **222**/5, 78–86 (1970)
85. Luft, J.H.: The structure and properties of the cell surface coat. Internat. Rev. Cytol. **45**, 291–382 (1976)
86. Manson, L.A. (ed.): Biomembranes. New York-London: Plenum Press, 1971–1976, Vols. 1–8
87. Marinos, N.: Vacuolation in plant cells. J. Ultrastruct. Res. **9**, 177–185 (1963)
88. Markham, R., Horne, R.W., Hicks, R.M. (eds.): The Electron Microscopy and Composition of Biological Membranes and Envelopes. London: The Royal Society 1974
89. Martonosi, A.N. (ed.): The Enzymes of Biological Membranes. London: Plenum Publ. Corp., 4 Volumes, appearing from 1976 on
90. Matile, P., Moor, H.: Vacuolation: Origin and development of the lysosomal apparatus in root-tip cells. Planta **80**, 159–175 (1968)
91. Mazliak, P.: Glyco- and phospholipids of biomembranes in higher plants. This volume, pp. 48–74
92. McLean, R.J., Laurendi, C.J., Brown, R.M., Jr.: The relationship of gamone to the mating reaction in *Chlamydomonas moewusii*. Proc. Natl. Acad. Sci. U.S. **71**, 2610–2613 (1974)
93. Mehlman, M.A., Hanson, R.W. (eds.): The Role of Membranes in Metabolic Regulation. New York-London: Academic Press 1972
94. Mesquita, J.F.: Electron microscope study of the origin and development of the vacuoles in root-tip cells of *Lupinus albus* L. J. Ultrastruct. Res. **26**, 242–250 (1969)
95. Mitchell, P.: Coupling of phosphorylation to electron and hydrogen transfer by a chemi-osmotic type of mechanism. Nature (London) **191**, 144–148 (1961)
96. Mitchell, P.: Proton-translocation phosphorylation in mitochondria, chloroplasts and bacteria: Natural fuel cells and solar cells. Federation Proc. **26**, 1370–1379 (1967)
97. Mitchell, P.: Structural and functional organisation of energy-transducing membranes and their ion-conducting properties. Amsterdam: North-Holland/American Elsevier 1972, 8. FEBS Meeting, Vol. 28, pp. 353–370
98. Morré, D.J.: Membrane biogenesis. Ann. Rev. Plant Physiol. **26**, 441–481 (1975)
99. Morré, D.J., Mollenhauer, H.H.: The endomembrane concept: A functional integration of endoplasmic reticulum and Golgi apparatus. In: Dynamic Aspects of Plant Ultrastructure. Robards, A.W. (ed.). London: McGraw-Hill, 1974, pp. 84–137
100. Morré, D.J., Mollenhauer, H.H., Bracker, C.E.: Origin and continuity of Golgi apparatus. In: Origin and Continuity of Cell Organelles. Reinert, J., Ursprung, H. (eds.). Berlin-Heidelberg-New York: Springer, 1971, pp. 82–126
101. Moscona, A.A. (ed.): The Cell Surface in Development. London: John Wiley & Sons 1976
102. Neame, K.D., Richards, T.G.: Elementary Kinetics of Membrane Carrier Transport. Oxford: Blackwell 1972
103. Nicolson, G.L.: Transmembrane control of the receptors on normal and tumor cells I. Cytoplasmic influence over cell surface components. Biochim. Biophys. Acta **457**, 57–108 (1976)
104. Niedermeyer, W.: The elasticity of the yeast cell tonoplast related to its ultrastructure and chemical composition II. Chemical and cytochemical investigations. Cytobiologie **13**, 380–393 (1976)

105. Niedermeyer, W., Parish, G.R., Moor, H.: The elasticity of yeast cell tonoplast related to its ultrastructure and chemical composition I. Induced swelling and shrinking; a freeze-etch membrane study. Cytobiologie **13**, 364–379 (1976)
106. Oelze, J., Drews, G.: Membranes of photosynthetic bacteria. Biochim. Biophys. Acta **265**, 209–239 (1972)
107. Overath, P., Thilo, L., Träuble, H.: Lipid phase transitions and membrane function. TIBS **1**, 186–189 (1976)
108. Overton, E.: Über die allgemeinen osmotischen Eigenschaften der Zelle, ihre vermutlichen Ursachen und ihre Bedeutung für die Physiologie. Vierteljahresschr. Naturforsch. Ges. Zürich **44**, 88–135 (1899)
109. Palade, G.: Intracellular aspects of the process of protein synthesis. Science **189**, 347–358 (1975)
110. Pardee, A.B., Wu, P.C., Zusman, D.R.: Bacterial division and the cell envelope. In: Bacterial Membranes and Walls. Leive, L. (ed.). New York: Marcel Dekker, 1973, pp. 357–412
111. Parsons, D.S. (ed.): Biological Membranes: Twelve Essays on their Organization, Properties, and Functions. London: Clarendon Press, Oxford Univ. Press 1975
112. Pearse, B.M.F.: Clathrin: A unique protein associated with intracellular transfer of membranes by coated vesicles. Proc. Natl. Acad. Sci. U.S. **73**, 1255–1259 (1976)
113. Poste, G., Allison, A.C.: Membrane fusion. Biochim. Biophys. Acta **300**, 421–465 (1973)
114. Poste, G., Papahadjopoulos, D.: Lipid vesicles as carriers for introducing materials into cultured cells: Influence of vesicle lipid composition on mechanism(s) of vesicle incorporation into cells. Proc. Natl. Acad. Sci. U.S. **73**, 1603–1607 (1976)
115. Postma, P.W., van Dam, K.: The ATPase complex from energy-transducing membranes. TIBS **1**, 16–17 (1976)
116. Racker, E. (ed.): Membranes of Mitochondria and Chloroplasts. New York: Van Nostrand Reinhold 1970
117. Robards, A.W.: Plasmodesmata. Ann. Rev. Plant Physiol. **26**, 13–29 (1975)
118. Robards, A.W. (ed.): Dynamic Aspects of Plant Ultrastructure. London: McGraw-Hill 1974
119. Robertson, J.D.: A molecular theory of cell membrane structure. Proc. 4th Internat. Congr. EM. Berlin-Göttingen-Heidelberg: Springer, 1960, pp. 159–171
120. Robertson, J.D.: Current problems of unit membrane structure and substructure. In: Intracellular Membraneous Structure. Seno, S., Cowdry, E.V. (eds.). Symp. Soc. Cell. Chem. **14**, pp. 379–428, Okayama 1964
121. Robertson, R.N.: Protons, Electrons, Phosphorylation and Active Transport. Cambridge Univ. Press 1968
122. Robinson, D.G., Sachs, H.: Cytokinesis in the green alga *Eremosphaera viridis*: Plasmalemma formation from 'open membranes'. Planta **129**, 75–82 (1976)
123. Roseman, S.: The bacterial phosphoenolpyruvate: sugar phosphotransferase system. In: Energy Transformation in Biological Systems. Ciba Found. Symp. **31**, 225–238 (1975)
124. Rothfield, L.I. (ed.): Structure and Function of Biological Membranes. New York-London: Academic Press 1971
125. Sandermann, H., Jr.: A possible correlation between lipid hydration and lipid activation of the C_{55}-isoprenoid alcohol phosphokinase apoprotein. Europ. J. Biochem. **62**, 479–484 (1976)
126. Satir, B.: The final steps in secretion. Sci. Am. **233**/4, 28–37 (1975)
127. Scanu, A.M.: Structural studies on serum lipoproteins. Biochim. Biophys. Acta **265**, 471–508 (1972)
128. Schatz, G.: Biogenesis of mitochondria. In: Membranes of Mitochondria and Chloroplasts. Racker, E. (ed.). New York: Van Nostrand Reinhold, 1970, pp. 251–314
129. Schnepf, E.: Zur Feinstruktur von *Geosiphon pyriforme*. Ein Versuch zur Deutung cytoplasmatischer Membranen und Kompartimente. Arch. Mikrobiol. **49**, 112–131 (1964)
130. Schnepf, E.: Organellen-Reduplikation und Zellkompartimentierung. In: Probleme der biologischen Reduplikation. Sitte, P. (ed.). Berlin-Heidelberg-New York: Springer, 1966, pp. 372–393
131. Schnepf, E.: Die Kompartimentierung der Zelle in morphologischer Sicht. Biol. Rundschau **4**, 259–275 (1966)
132. Schnepf, E.: Membranfluß und Membrantransformation. Ber. Dtsch. Bot. Ges. **82**, 407–413 (1969)

133. Schnepf, E.: Sekretion und Exkretion bei Pflanzen. Wien-New York: Springer 1969
134. Schnepf, E., Brown, R.M., Jr.: On relationships between endosymbiosis and the origin of plastids and mitochondria. In: Origin and Continuity of Cell Organelles. Reinert, J., Ursprung, H. (eds.). Berlin-Heidelberg-New York: Springer, 1971, pp. 299–322
135. Schuldiner, S., Rudnick, G., Weil, R., Kaback, H.R.: Mechanism of β-galactoside transport in *Escherichia coli* membrane vesicles. TIBS **1**, 41–45 (1976)
136. Sheetz, M.P., Painter, R.G., Singer, S.J.: Biological membranes as bilayer couples III. Compensatory shape changes induced in membranes. J. Cell Biol. **70**, 193–203 (1976)
137. Singer, S.J.: The molecular organization of membranes. Ann. Rev. Biochem. **43**, 805–833 (1974)
138. Singer, S.J.: On the fluidity and asymmetry of biological membranes. In: Perspectives in Membrane Biology. Estrada-O., S., Gitler, C. (eds.). New York-San Francisco-London: Academic Press, 1974, pp. 131–147
139. Singer, S.J., Nicolson, G.: The fluid mosaic model of the structure of cell membranes. Science **175**, 720–731 (1972)
140. Sitte, P.: Zum Feinbau der Suberinschichten im Flaschenkork. Protoplasma **54**, 555–559 (1961)
141. Sitte, P.: Biomembranen: Struktur und Funktion. Ber. Dtsch. Bot. Ges. **82**, 329–383 (1969)
142. Sitte, P.: Allgemeine Morphologie der Pflanzenzelle. In: Grundlagen der Cytologie. Hirsch, G.C., Ruska, H., Sitte, P. (Hsg.). 2. Aufl. Jena: G. Fischer, 1974, pp. 391–412
143. Sitte, P.: Plastiden-Metamorphose und Chromoplasten bei *Chrysosplenium*. Z. Pflanzenphysiol. **73**, 243–265 (1974)
144. Sitte, P.: Die Bedeutung der molekularen Lamellenbauweise von Kork-Zellwänden. Biochem. Physiol. Pflanzen **168**, 287–297 (1975)
145. Sitte, P.: Zur chemischen Fixierung von Lipidstrukturen. Mikroskopie **32**, 208–209 (1976)
146. Sitte, P.: Die lebende Zelle als System, Systemelement und Übersystem. Nova Acta Leopoldina, in press
147. Stämpfli, R.: Der Erregungsvorgang und seine Fortleitung. Klin. Wschr. **49**, 777–785 (1971)
148. Steck, T.L.: Membrane isolation. In: Membrane Molecular Biology. Fox, C.F., Keith, A.D. (eds.). Stamford, Conn.: Sinauer, 1972, pp. 76–114 (see in particular pp. 101–102)
149. Tanford, C.: The Hydrophobic Effect: Formation of Micelles and Biological Membranes. New York: John Wiley & Sons 1973
150. Tien, H.T.: Bilayer Lipid Membranes (BLM) – Theory and Practice. New York: Marcel Dekker 1974
151. Trebst, A.: Energy conservation in photosynthetic electron transport of chloroplasts. Ann. Rev. Plant Physiol. **25**, 423–458 (1974)
152. Triggle, D.J.: Neurotransmitter-Receptor Interactions. London-New York: Academic Press 1971
153. Tzagoloff, A. (ed.): Membrane Biogenesis – Mitochondria, Chloroplasts, Bacteria. London: Plenum Publ. Corp. 1975
154. Verkleij, A.J., Zwaal, R.F.A., Roelofsen, B., Comfurius, P., Kastelijn, D., van Deenen, L.L.M.: The asymmetric distribution of phospholipids in the human red cell membrane. A combined study using phospholipases and freeze-etch electron microscopy. Biochim. Biophys. Acta **323**, 178–193 (1973)
155. Vredenberg, W.J.: Energy control of ion transport processes at the membranes of cells and organelles. Ber. Dtsch. Bot. Ges. **87**, 473–481 (1974)
156. Wallach, D.F.H., Fischer, H. (eds.): The Dynamic Structure of Cell Membranes. Berlin-Heidelberg-New York: Springer 1971
157. Wallach, D.F.H., Winzler, R.J.: Evolving Strategies and Tactics in Membrane Research. Berlin-Heidelberg-New York: Springer 1974
158. Wartiovaara, V., Collander, R.: Permeabilitätstheorien. Wien: Springer 1960
159. Weissmann, G., Claiborne, R. (eds.): Cell Membranes. Biochemistry, Cell Biology & Pathology. New York: HP Publ. Co. 1975
160. Whaley, W.G.: The Golgi Apparatus. Wien-New York: Springer 1975
161. Whaley, W.G., Dauwalder, M., Kephart, J.E.: Assembly, continuity, and exchanges in certain cytoplasmic membrane systems. In: Origin and Continuity of Cell Organelles. Reinert, J., Ursprung, H. (eds.). Berlin-Heidelberg-New York: Springer, 1971, pp. 1–45
162. Winkenbach, F., Falk, H., Liedvogel, B., Sitte, P.: Chromoplasts of *Tropaeolum majus* L.: Isolation and characterization of lipoprotein elements. Planta **128**, 23–28 (1976)

163. Witt, H.T.: Biophysikalische Primärvorgänge in der Photosynthesemembran. Naturwissenschaften **63**, 23–27 (1976)
164. Witt, H.T., Rumberg, B., Junge, W.: Electron transfer, field changes, proton translocation and phosphorylation in photosynthesis. Coupling in the thylakoid membrane. In: Biochemie des Sauerstoffs. Hess, B., Staudinger, Hj. (eds.). Berlin-Heidelberg-New York: Springer, 1968, pp. 262–306
165. Woessner, J.F., Jr., Huijing, F. (eds.): The Molecular Basis of Biological Transport. New York-London: Academic Press 1972
166. Wohlfarth-Bottermann, K.E., Stockem, W.: Die Regeneration des Plasmalemmas von *Physarum polycephalum*. Wilhelm Roux' Archiv **164**, 321–340 (1970)
167. Wunderlich, F., Berezney, R., Kleinig, H.: The nuclear envelope: An interdisciplinary analysis of its morphology, composition and functions. In: Biological Membranes. Chapman, D., Wallach, D.F.H. (eds.). London-New York-San Francisco: Academic Press, 1976, Vol. 3, pp. 241–333
168. Wuttke, H.-G.: Chromoplasts in *Rosa rugosa*: Development and chemical characterization of tubular elements. Z. Naturforsch. **31c**, 456–460 (1976)
169. Wuttke, H.-G., Sitte, P.: Results on the chemical analysis of chromoplast globules, to be published
170. Zahler, P.: The structure of the erythrocyte membrane. Experientia **25**, 449–456 (1969)
171. Zwaal, R.F.A., Roelofsen, B., Colley, C.M.: Localization of red cell membrane constituents. Biochim. Biophys. Acta **300**, 159–182 (1973)
172. Zwaal, R.F.A., Demel, R.A., Roelofsen, B., van Deenen, L.L.M.: The lipid bilayer concept of cell membranes. TIBS **1**, 112–114 (1976)

CHAPTER 2

The Prenyllipids of the Membranes of Higher Plants

T.W. GOODWIN

A. Introduction

The prenyllipids found in higher plants are either completely terpenoid in origin (simple), e.g., the carotenoids, or consist of a terpenoid and a non-terpenoid part (mixed), as is the case with ubiquinone. Terpenoids are biosynthesized from a basic branched five carbon unit, $\substack{C \\ C}{>}C{-}C{-}C$. If we formally designate this unit *ip,* then two units join head to tail to give a monoterpene, *ipip;* further sequential head-to-tail additions lead to sesquiterpenes (*ipipip*), diterpenes (*ipipipip*), and polyprenols (8 or more *ip* units). If, however two sesquiterpene units join tail-to-tail, the result is a triterpene (*ipipippipipi*) and if two diterpenes join in this way a tetraterpene is formed. Typical examples of simple prenyllipids are given in Table 1. The mixed terpenoids with which we are mainly concerned are listed in Table 2. A diterpene phytol, or a closely related compound, is a constituent of all these compounds except the ubiquinones and plastoquinones, which contain a polyprenol side chain. The ubiquinone (Q) isoprenologue usually present is Q-9 or Q-10; i.e., the quinone residue with a side chain of nine (45C) or ten (50C) isoprene residues [38].

The major plastoquinone is PQA (plastoquinone A). However, a whole series of two families, PQ-Bs and PQ-Cs, is also widely present in smaller

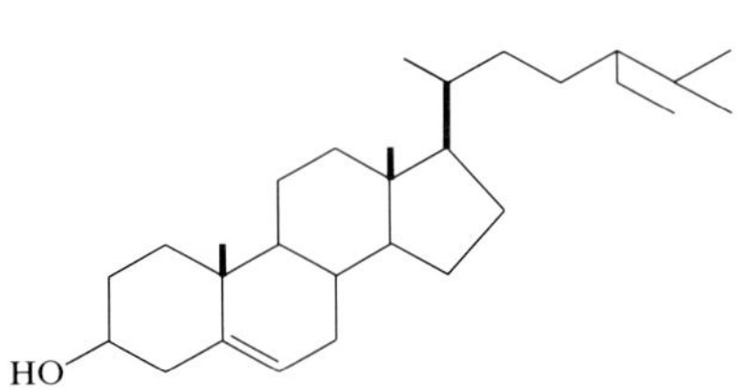

1. Sitosterol

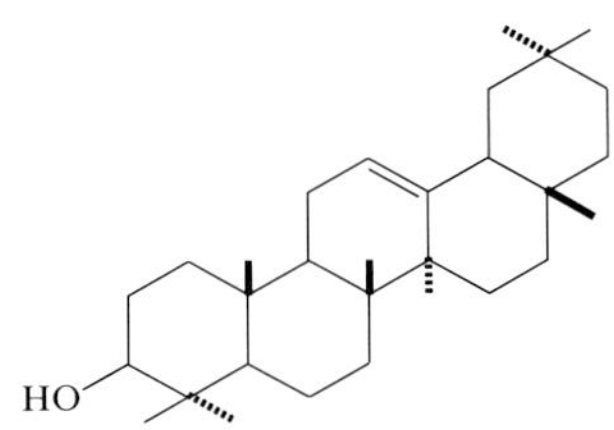

2. β-Amyrin

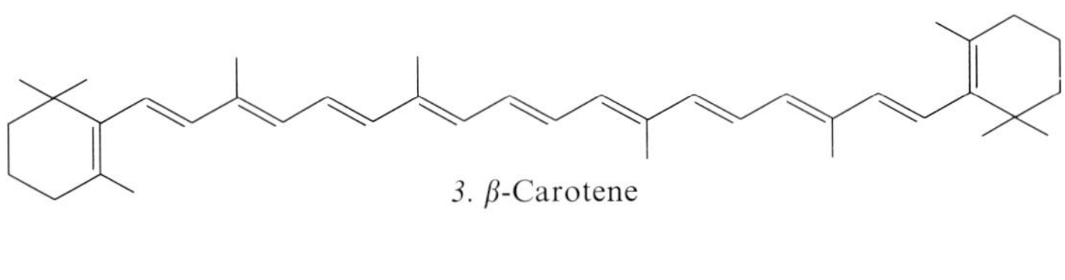

3. β-Carotene

$$CH_3{-}\overset{\displaystyle CH_3}{\overset{|}{C}}{=}CH{-}CH_2{-}\left[{-}CH_2{-}\overset{\displaystyle CH_3}{\overset{|}{C}}{=}CH{-}CH_2{-}\right]_{10}{-}CH_2{-}\overset{\displaystyle CH_3}{\overset{|}{C}}{=}CH{-}CH_2OH$$

cis, trans — 6 cis: 4 trans — trans

4. Castaprenols

amounts; they may be biological degradation products [42]. There also exists plastochromanol-8 (*11*), which is formed by cyclization of reduced plastoquinone-9 [45]. Quinones and chromanols also exist in the tocopherol series (Formulae *8, 10*).

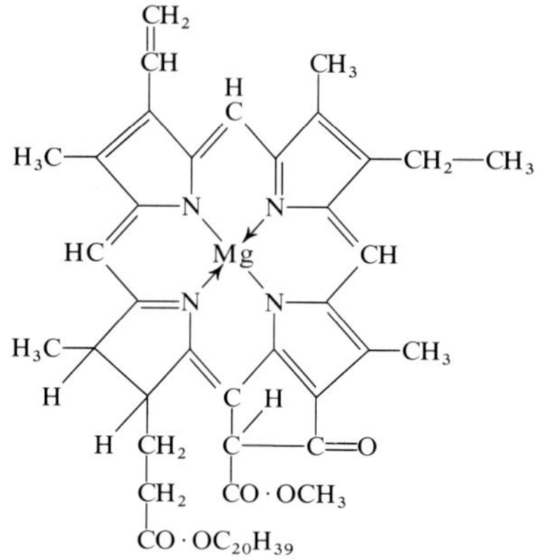

5. Chlorophyll a

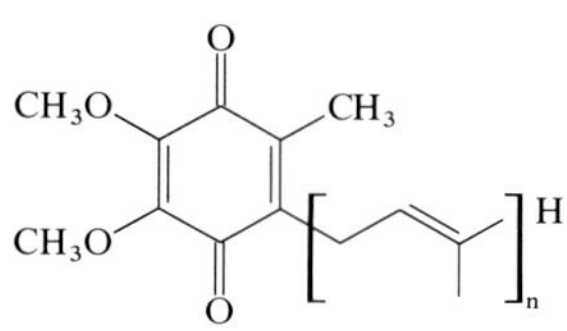

6. Ubiquinone

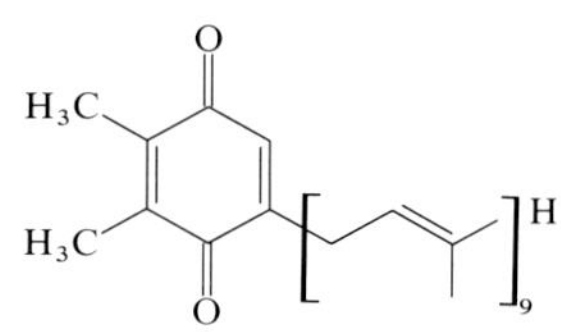

7. Plastoquinone-9

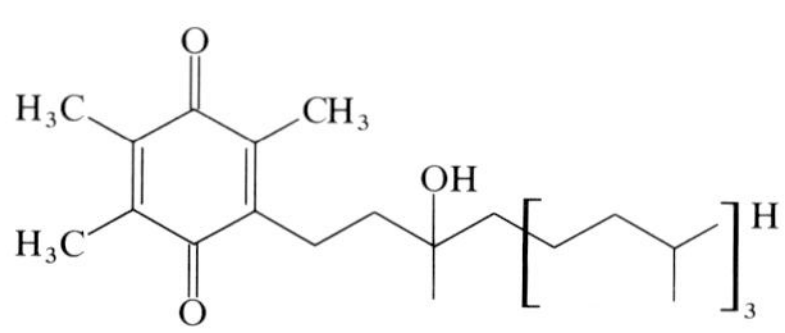

8. α-Tocopheryl quinone

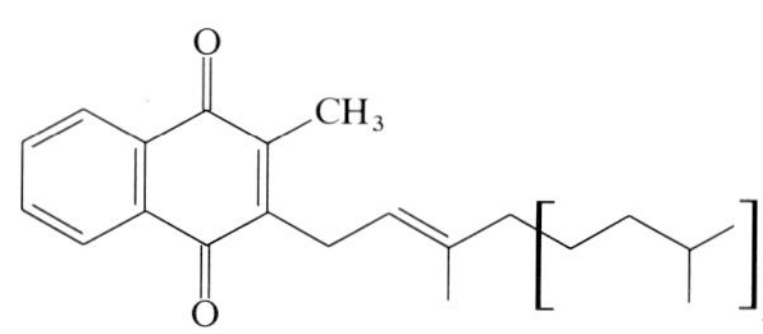

9. Phylloquinone

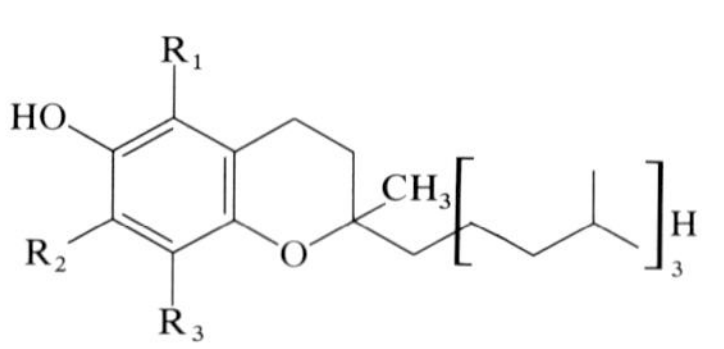

R_1	R_2	R_3	*Tocopherol*
H	H	CH_3	δ(δ-T)
CH_3	H	CH_3	β(β-T)
H	CH_3	CH_3	γ(γ-T)
CH_3	CH_3	CH_3	α(α-T)

10. Tocopherols

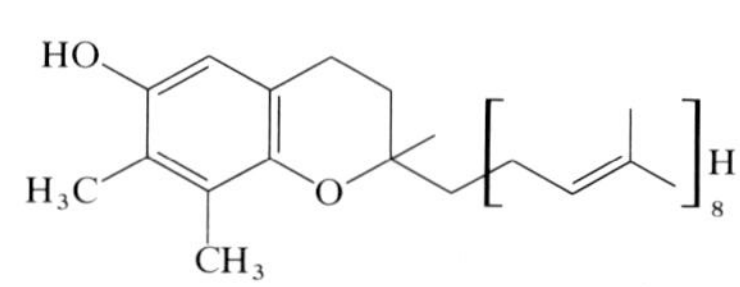

11. Plastochromanol

Table 1. Characteristic simple prenyllipids found in higher plants

Group	Compound	Comment
1. *Triterpenes*		
(a) sterols	Sitosterol (*1*)	Plant sterols are numerous and characterized by supernumerary carbon atoms at C-24
(b) pentacyclics	β-Amyrin (*2*)	Also very numerous
2. *Tetraterpenes*	β-Carotene (*3*)	Numerous: Mandatory components of stable chloroplasts
3. *Polyprenols*	Castaprenols (*4*)	$(ip)_n$ where *n* varies from 10 to 16

Table 2. Characteristic mixed terpenoids in higher plants

Compound	Structure	Terpenoid component
(i) *Porphyrins*		
Chlorophyll a	(*5*)	Phytol
(ii) *Benzoquinones*		
Ubiquinone	(*6*)	C_{45} or C_{50} polyprenol
Plastoquinone-9	(*7*)	C_{45}-polyprenol
α-Tocopheryl quinone	(*8*)	C_{20} (phytol derivative)
(iii) *Naphthoquinones*		
Phylloquinone (vitamin K)	(*9*)	Phytol
(iv) *Chromanols*		
Tocopherols	(*10*)	C_{20}-(phytol derivative)
Plastochromanol	(*11*)	C_{45}-polyprenol

B. Localization

I. Intracellular Distribution

The general distribution pattern of prenyllipids among various intracellular organelles of the leaves of higher plants is outlined in Table 3. Most of the information is unequivocal in that the compounds generally are located specifically in one organelle. Only the tocopherols raise any serious problems, and these have been discussed recently by Pennock and Janiszowska [29]. The major part of α-tocopherol is present in the chloroplast, but a small fraction probably exists in the mitochondria. On the other hand, in peas for example, γ-tocopherol (*10*) is chloroplastic and microsomal while most of the δ-tocopherol (*10*) occurs mainly in the chloroplast and in the supernatant fraction. A recent report suggests that phylloquinone is also present in the mitochondria as well as in the chloroplasts [16]. Sterols, on the other hand, are present in all membranes, regardless of where they are situated [17].

Table 3. Intracellular distribution of prenyllipids in leaves of higher plants

Compounds	Distribution
Chlorophylls	Chloroplasts
Carotenoids	Chloroplasts, chromoplasts
Plastoquinone	Chloroplasts, chromoplasts
Ubiquinone	Mitochondria
α-Tocopherol	Chloroplasts, (mitochondria), chromoplasts
β-Tocopherol	Chloroplasts (?)
γ-Tocopherol	Chloroplasts, microsomes
δ-Tocopherol	Chloroplasts
Plastochromanol	Chloroplasts (?)
Tocopheryl quinones	Chloroplasts
Phylloquinone	Chloroplasts, chromoplasts, mitochondria (?)
Polyprenols	Chloroplasts, cell walls (Chloroplasts, mitochondria)
Sterols	(Microsomes, nuclei, chloroplasts)

II. Intraorganelle Distribution

When one considers the chloroplast, which is our main concern in this essay, one can distinguish three structural components: the thylakoids, the chloroplast envelope, and the osmiophilic plastoglobuli. It is not yet possible to provide a direct comparison of all the prenyl components present in these structures, but it is clear that the chlorophylls are present exclusively in the thylakoids. This situation was considered also to be the case with carotenoids until a recent report appeared indicating that they may also be present in the envelope membrane. Nothing is known about the distribution of the other prenyllipids in the membranes, but they do accumulate in the thylakoids and osmiophilic plastoglobuli.

Let us first consider a typical, overall quantitative distribution of prenyllipids in chloroplasts and then look for evidence of differential distribution in sub-chloroplast particles. Table 4 contains results for *Nicotiana tabaca* leaves; in

Table 4. Distribution of prenyllipids in leaf tissue of *Nicotiana tabacum* (recalculated from Ref. [25])

Lipid	Amount[a]	Molar ratio (chlorophyll = 100)
Chlorophylls	11.7	100
Carotenoids	2.43	21.8
Plastoquinones	2.20	19
Tocopherols	1.0	9
Phylloquinone	0.06	0.5
Sterols[b]	1.2	10

[a] μmol/g dry tissue

[b] Figure extrapolated from average values for beans and maize [27]

Table 5. Carotenoid distribution in sub-chloroplast particles obtained by detergent treatment (Triton X-100) of spinach chloroplasts [41]

Pigment	Light particle (photosystem I)	Heavy particle (photosystem II)	Intact thylakoids
β-Carotene	16	6	6.3
Lutein	6	16	9.6
Violaxanthin	1	3	2.8
Neoxanthin	1	2	2.5

Concentration μmol/100 mmol total chlorophyll

effect it gives a good indication of the quantitative distribution within the chloroplasts because the components we are considering are to a great extent confined to these organelles. Recent work has, however, been concerned with the preparation of biologically active sub-chloroplast particles. For example, when chloroplasts are treated with detergent, the dispersed chromoproteins separated into two main fractions corresponding roughly to particles with photosystem I and photosystem II activity. A typical distribution of carotenoids in early preparations of such particles is given in Table 5. It is clear that the most important quantitative difference is that β-carotene accumulates specifically in the light particles (photosystem I) and the xanthophylls lutein (*12*), violaxanthin

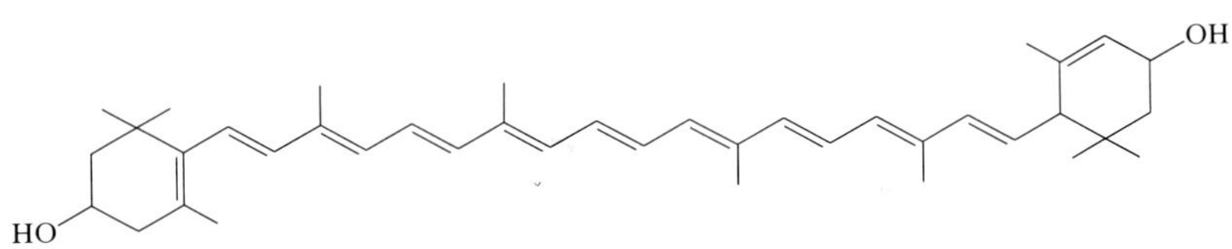

12. Lutein

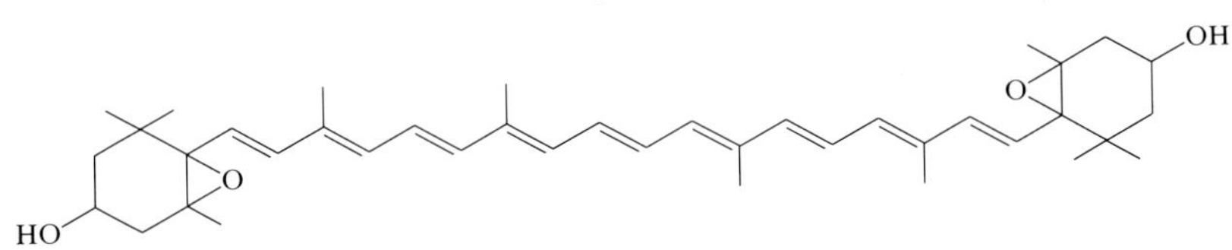

13. Violaxanthin

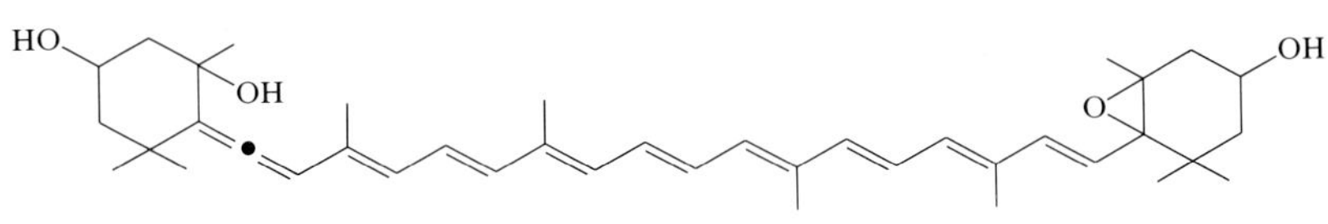

14. Neoxanthin

(*13*), and neoxanthin (*14*) in the heavy particles (photosystem II). Plastoquinone also accumulates in the photosystem II particles and is considered to be on the acceptor side of the system while lutein is on the donor side [30]. Tocopherols appear to be distributed equally between photosystems I and II while phylloqui-

Table 6. Prenyllipid content of photosystems I and II from spinach chloroplasts

	Photosystem I	Photosystem II
Plastoquinones	2.0	8.0
α-Tocopherol	2.0	3.0
α-Tocopherylquinone	0.3	0.5
Phylloquinone	1.3	0.1
Total	5.6	11.6

mol/100 mol chlorophyll a

Table 7. Comparison of carotenoids and chlorophylls in the lamellae envelope fraction of chloroplasts from spinach [7]

Pigment	Concentration (μg/mg protein)	
	Lamellae Conc.% of total carotenoids	Envelope Conc.% of total carotenoids
Chlorophylls	147	1.2
β-Carotene	7.6	1.2
Lutein + zeaxanthin	11.6	2.5
Violaxanthin	5.1	4.1
Neoxanthin	2.7	0.4

none accumulates in photosystem I (Table 6) [22]. Particles prepared by polyacrylamide gel electrophoresis, *P-700 chlorophyll a protein* and the *light-harvesting chlorophyll a/b protein,* respectively, have recently been obtained in a relatively pure form [37]. The former contains chlorophyll a and β-carotene in a molar ratio of 20–30 : 1 [37] while the distribution of prenylquinones, although present, is unknown [28]. The second system contains all the various carotenoids found in the chloroplast although not in the same stoicheiometric ratios as in the chloroplast; lutein and β-carotene are the major pigments but there are also considerable amounts of neoxanthin. Nothing is known of other prenyllipids in these complexes, and investigations in this area are clearly required.

A preparation of chloroplast envelopes has recently been obtained [7]. This membrane is presumably concerned with permeability to metabolities rather than with photosynthesis and is characterised by the virtual absence of chlorophylls and a total carotenoid concentration of about one-third that of the thylakoids. Violaxanthin, however, appears to be selectively accumulated in the envelope because its concentration is about the same as in the thylakoids (Table 7). Criticism has, however, been levelled at the purity of membranes prepared by the method used. The pigments are probably present in trapped osmiophilic globules and are probably absent from the outer envelope membrane

Table 8. Distribution of prenyllipids in young and old spinach leaves [24]

Component	Age of Leaf (days)	
	10	42
Chlorophyll a	100[a]	100[a]
Chlorophyll b	36	36
Carotenoids	39	44
Plastoquinone	8	18
α-Tocopherol	7	17
α-Tocopherol quinone	0.7	1.8
Phylloquinone	0.8	2.0

Amounts (mol) relative to chlorophyll a = 100

Table 9. Ratio of amount of prenyllipids in thylakoids and in plastoglobuli in 28-week-old leaves of *A. hippocastanum* [44]

Component	Ratio (thylakoids : plastoglobuli)	Total amount (mg/mmol chlorophyll)
Solanesol	4.9 : 1	2.0
Castaprenols	0.23 : 1	22.4
Plastoquinone	0.27 : 1	15.3

[32, 33; Sprey, 1976, personal communication]. The general pattern of carotenoid distribution in chloroplasts outlined in Table 5 is reproduced in many higher plants and varies only slightly with age and cultural conditions of the plant [13]. However, this is not so for the mixed terpenoids, particularly those of C-3 plants, in which considerable increase in amount is observed as the chloroplasts age (Table 8). The effect is much less marked in C-4 plants such as *Zea mais* [46]. This accumulation is the result of the development within the chloroplast of lipid droplets, plastoglobuli [21, 24], separate from the thylakoids in which these lipophilic compounds accumulate. The greater part of the plastoquinone in the plastoglobuli is in the reduced form, plastohydroquinone, while the reverse situation exists in the thylakoids [15]. It must be emphasized that carotenoids and chlorophylls are not components of the plastoglobuli. According to experiments with *Aesculus hippocastanum,* the castaprenols (tritrans-poly cisprenols) also tend to accumulate in the plastoglobuli of old leaves while solanesol (C_{45}-all trans) does not (Table 9) [43]. Furthermore, as the leaves age the polyprenols move out of the chloroplast and appear in the cell wall. For up to 20 weeks after terminal bud unfolding, 80 to 90% of the castaprenols are in the chloroplasts of *A. hippocastanum;* after 24 weeks only 50% remains in the chloroplast with about 40% appearing in the cell wall preparation [44].

C. Localization of Biosynthesis

I. General Pattern of Synthesis

The first biosynthetic steps are the same for all the terpenoids discussed in the previous sections, i.e., the production of isopentenyl pyrophosphate from acetyl-CoA via mevalonic acid (Fig. 1). Furthermore, the elongation process involving (1) the isomerization of isopentenyl pyrophosphate to dimethylallyl pyrophosphate (Fig. 2) and (2) the elongation by successive additions of two isopentenyl pyrophosphate molecules to form a C_{15} precursor (farnesyl pyrophosphate) is also common to all compounds (Fig. 3). Thereafter the pathways separate and the bifurcated pathway is indicated in Figure 4 (2). When one considers the multiplicity of terpenoids and the diversity of their intercellular location, then it is evident that a complicated and subtle control mechanism must exist for the synthesis of these compounds in order to produce a perfectly functioning leaf tissue. The details of how this control is exercised are still unknown, but it appears that localization of synthesis within the organelles

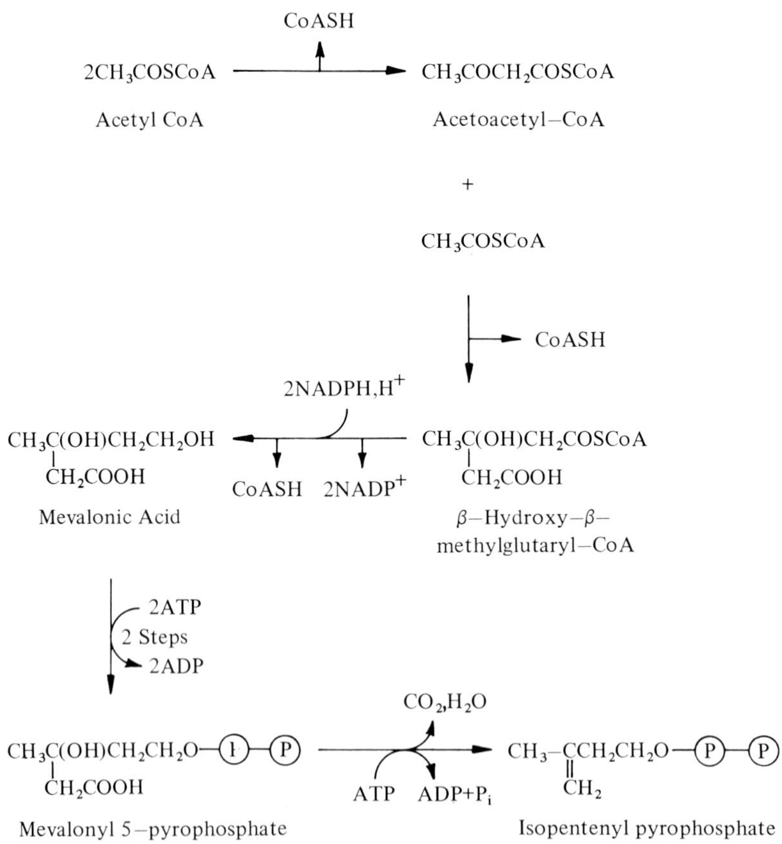

Fig. 1. Formation of isopentenyl pyrophosphate from acetyl-CoA

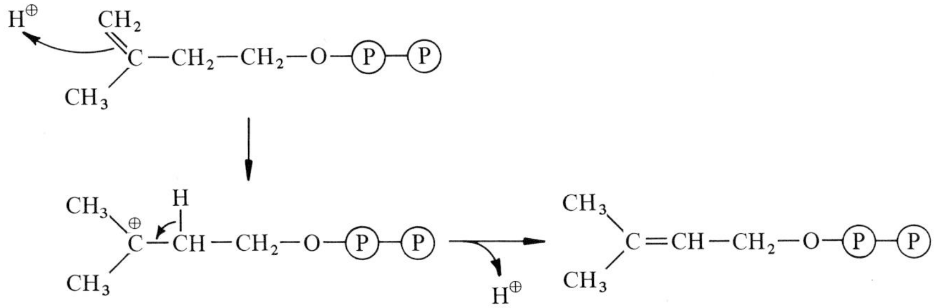

Fig. 2. Isomerization of isopentyl pyrophosphate to dimethylallyl pyrophosphate

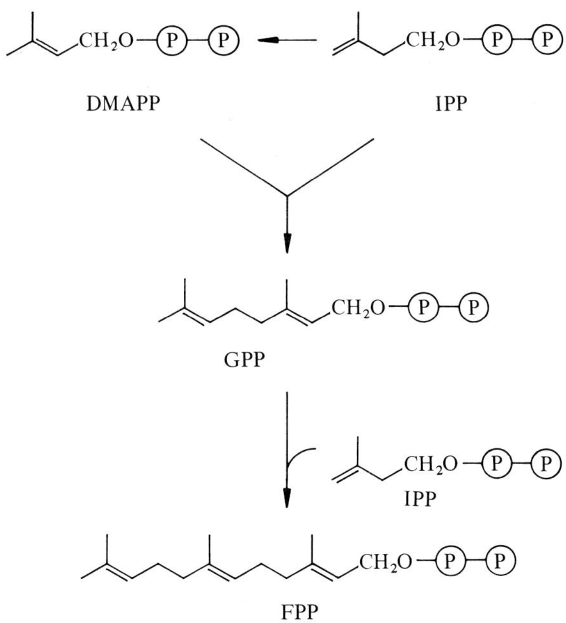

Fig. 3. Formation of farnesyl pyrophosphate from isopentenyl pyrophosphate

is probably an important factor. In the present review we shall confine ourselves to the chloroplast.

II. Synthesis in the Chloroplast

The first point to be considered is whether the chloroplast is able alone (1) to synthesize the fundamental precursor, acetyl-CoA, (2) whether it can convert acetyl-CoA into mevalonic acid (Fig. 1); (3) whether it can carry out the chain elongation process (Figs. 2 and 3) and (4) whether it can carry out the later stages to produce the various terpenoids found in chloroplast.

If the chloroplast is autonomous it must be able to synthesize de novo the enzymes necessary to carry out these various biosynthetic steps.

There is no doubt that chloroplasts can synthesize proteins (see Refs. [11, 14]), but it is also clear that the number of proteins it synthesizes is small

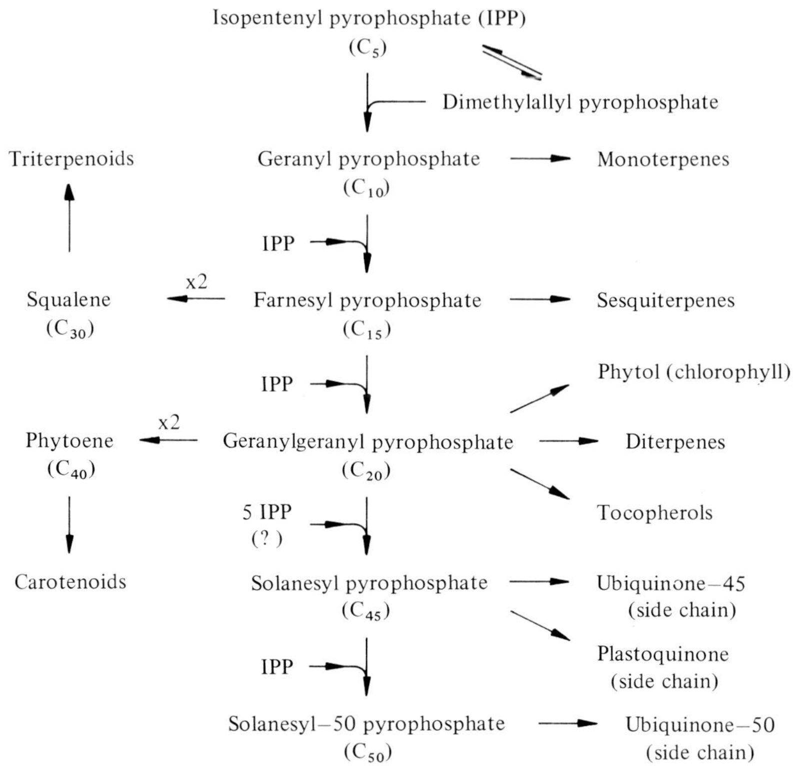

Fig. 4. General pattern of synthesis of terpenoids in higher plants

and that many chloroplast proteins must be assembled outside the organelle and transported there at some time during its development (see Ref. [8]). Having established that chloroplasts can synthesize proteins, we have to decide whether these include those enzymes involved in terpenoid synthesis.

1. Formation of Acetyl-CoA

Acetyl-CoA is normally synthesized either from pyruvate by the action of pyruvate dehydrogenase, pyruvate itself being formed by glycolysis, or by β-oxidation of fatty acids. Neither glycolysis, oxidative decarboxylation of pyruvate, nor β-oxidation has yet been shown to occur in chloroplasts. One is thus led to the view that acetyl-CoA is formed in the chloroplast by an unusual route or that it is formed outside the chloroplast. If one assumes that acetyl-CoA is formed in the leaf mitochondria then the problem of transport into the chloroplast must be solved. Four possibilities exist: (1) direct entry as acetyl-CoA; (2) entry as acetate followed by activation within the chloroplast; (3) transport as acetyl-*L*-carnitine as in animals; (4) conversion into citrate by condensation with oxaloacetate in the mitochondria and transfer of citrate, which could then be cleaved in the chloroplast by ATP-citrate lyase (Eq. 1).

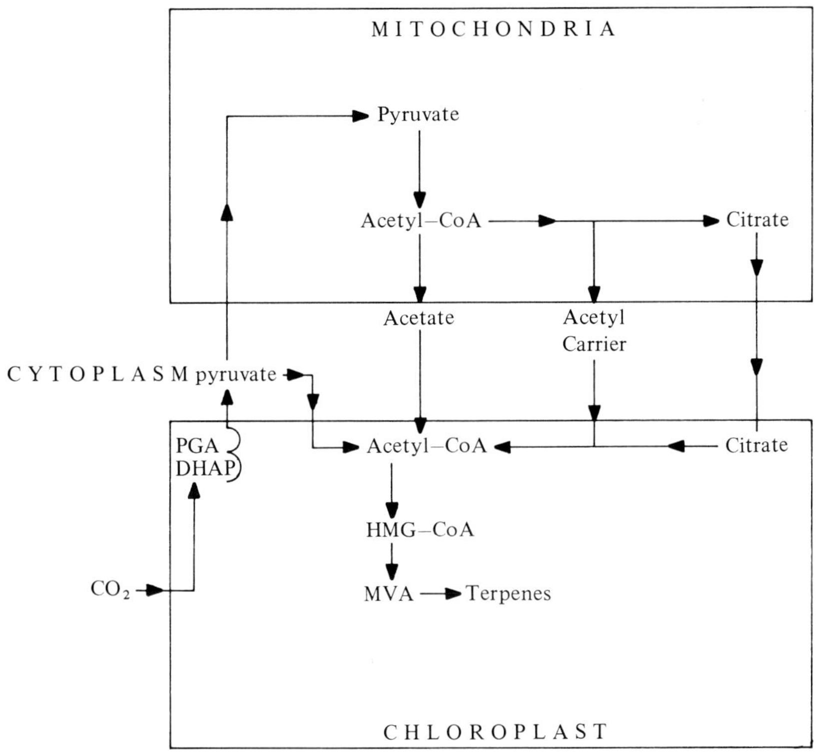

Fig. 5. Metabolic collaboration between mitochondria and chloroplasts to produce terpenes within the chloroplast

$$\text{ATP} + \text{citrate} + \text{CoASH} \rightleftharpoons \text{Acetyl-CoA} + \text{oxaloacetate} + \text{ATP} + \text{Pi}. \quad (1)$$

If we consider these (1) would appear to be eliminated because in experiments with intact chloroplasts, acetyl-CoA was not utilized for fatty acid synthesis [4] although acetate in the presence of the appropriate co-factors was [34]. The hydrolysis of acetyl-CoA to acetate (possibility 2) by acetyl-CoA acetylase, which is found in plants, transfer of acetate into the chloroplast, and conversion back to acetyl-CoA by acetyl-CoA synthetase, also found in plants, is an attractive idea. However, free acetate has not yet been detected in plants although acetyl-CoA synthetase must be very active in chloroplast preparations [34]. Acetyl-*L*-carnitine (possibility 3) is only spasmodically present in plants [34], and for (4) to have any standing, citrate lyase must be clearly located in chloroplasts; this has not yet been demonstrated unequivocally although slight activity is always associated with carefully prepared chloroplasts (P.D.J. Weitzman, personal communication). So the problem of the source of chloroplast acetyl-CoA is still not solved; but the likely possibilities are summarized in Figure 5. There is, however, the additional possibility that an as yet unidentified mechanism of chloroplast synthesis of acetyl-CoA is operating or that a novel transport system exists between chloroplast and mitochondria.

2. Formation of Mevalonic Acid

The next question to arise is whether acetyl-CoA, however formed, can be converted into mevalonic acid (MVA) inside the chloroplast via the conventional pathway indicated in Figure 1. The only evidence we have for this pathway in chloroplasts is the demonstration of the key enzyme HMG-CoA reductase in pea chloroplasts [3]. An alternative route to MVA in the chloroplast could involve a carbon skeleton of an amino acid formed outside the chloroplast and transported into the chloroplast, because amino acids can pass freely across the chloroplast membrane. An analogous situation is already known to exist in chlorophyll biosynthesis; δ-aminolaevulinic acid is formed in higher plant chloroplasts not from succinyl-CoA and glycine, as in animals, but from α-keto-glutarate, which can arise by transamination of glutamic acid [1]. The amino acid that could be involved in terpenoid biosynthesis is leucine. It is certainly converted into carotene in the fungus *Phycomyces blakesleeanus* by a pathway (Fig. 6) that involves a CO_2-fixing step (see Ref. [12]). It must be emphasized that this pathway does not eliminate the requirement for acetyl-CoA; the carbon skeleton of leucine arises from three molecules of acetyl-CoA (see Ref. [9]).

If MVA is formed in the chloroplast from acetyl-CoA produced in the mitochondria, there must be a brisk traffic between chloroplasts and mitochondria because in a greening tissue, $^{14}CO_2$ is much more effectively incorporated into chloroplast terpenoids than is exogenously applied [2-^{14}C] mevalonic acid. The reverse is true for sterols and ubiquinone that are formed outside the chloroplast (see Refs. [10, 11]).

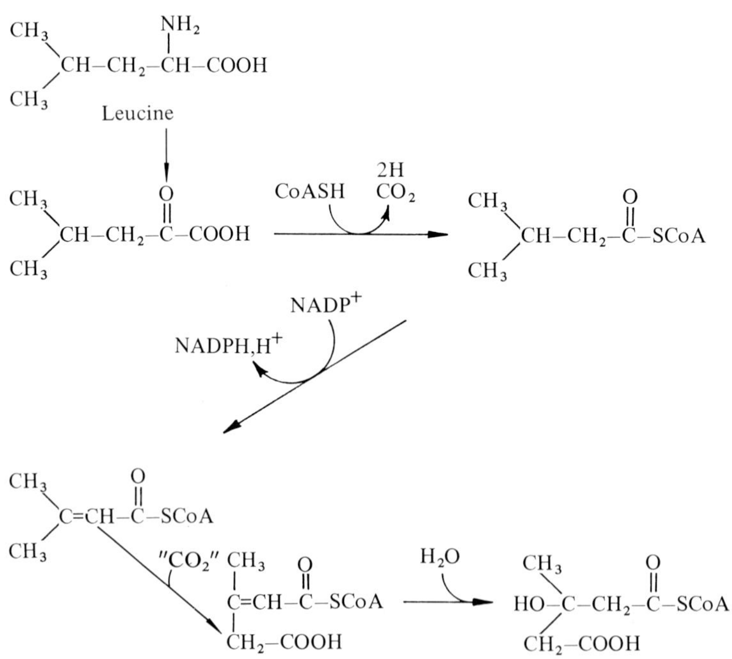

Fig. 6. Formation of β-hydroxy-β-methyl glutaryl-CoA (HMG-CoA) from leucine

3. *Conversion of MVA into Carotenoids*

The synthesis of phytoene from mevalonic acid has been clearly demonstrated in isolated freeze-dried chloroplasts [5], and its conversion into coloured carotenoids has been achieved in partly purified chloroplast preparations [35]. In spite of this, a number of the enzymes concerned with the desaturation of phytoene are under nuclear control (see Ref. [18]) (Fig. 7) and therefore must have been synthesized outside the chloroplast and transported thereto.

It is important to emphasize here that isolated chloroplasts will not synthesize squalene, the precursor of sterols and the compound corresponding to phytoene in the carotenoid pathway. Squalene and phytosterols are synthesized in the supernatant and microsomal fractions, respectively [20]. This means that there must be transport of sterols from the site of synthesis to the chloroplast, and it has been suggested that sterol esters are involved [10], although current views favour sterol glycosides for this role.

It must be concluded that the chloroplast probably cannot make its own acetyl-CoA, but can convert acetyl-CoA into MVA and can convert MVA into various terpenes, in particular carotenoids, but not into sterols. At least some of the enzymes concerned with carotenoid biosynthesis are not themselves synthesized within the chloroplast.

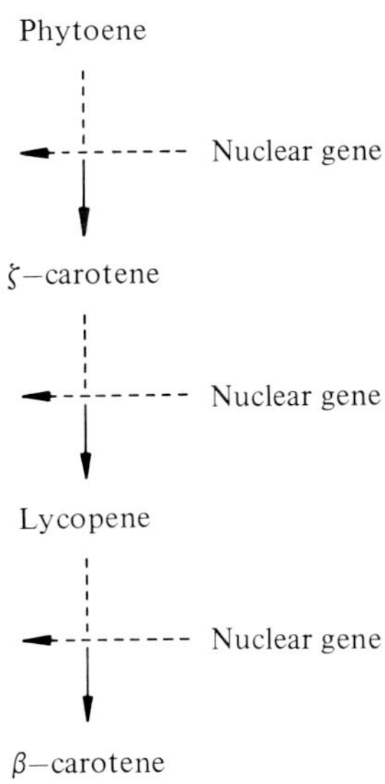

Fig. 7. Genetic control of carotenoid synthesis in higher plants

4. *Formation of Mixed Terpenoids*

a) General. Existing evidence points convincingly to the view that plastoquinone, some tocopherols and tocopherol quinones, and phylloquinone can be synthesized in chloroplasts. For example: (1) synthesis of plastoquinone and α-tocopherol quinone parallels chloroplast formation; (2) as with carotenoids $^{14}CO_2$ is more effective incorporated into plastoquinone than is [2-^{14}C]MVA; (3) with $^{14}CO_2$ as substrate the specific activity of the quinone residue of plastoquinone is the same as that of the terpenoid residues; (4) chloroplast preparations will condense homogentistic acid, the putative precursor of the non-prenol part, with protein-bound polyphenyl pyrophosphate to form 2-demethylplastoquinol, a probable precursor of plastoquinone (Eq. 2) [36]. Similar preparations will form demethylphylloquinone from phytyl pyrophosphate and 4-(2-carboxy-phenyl)-4-oxobutrate (Eq. 3). It appears well authenticated that the prenyl side chain of phylloquinone, plastoquinone, and tocopherol is synthesized in the chloroplast [38], but there is one claim that it is formed outside the chloroplast [16].

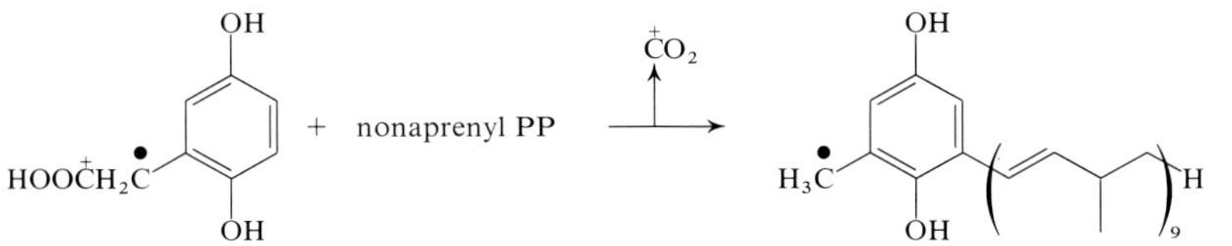

(2)

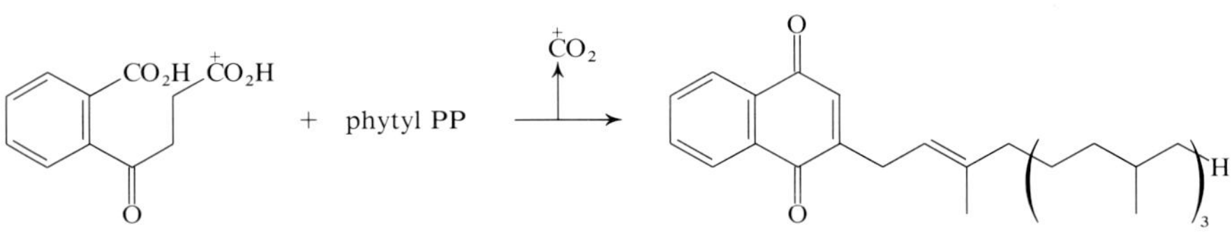

(3)

b) Tocopherols. In this last section we shall consider some recent information regarding tocopherol biosynthesis in plants. The generally accepted pathway is that outlined in Figure 8, in which the key intermediates are 6-phytyltoluquinone (6PTQ) (*15*) and 8-methyl tocol (δ-tocopherol). Recent work by Pennock and Janiszowska [29] has revealed a much more complex situation. They found that 5PTQ (*16*) and 3PTQ (*7*) are present together with 6PTQ (*15*) in the leaves of a number of higher plants and that when these are re-presented to the plant, all are converted into 5,7,8-trimethyltocol (α-tocopherol), with 5PTQ

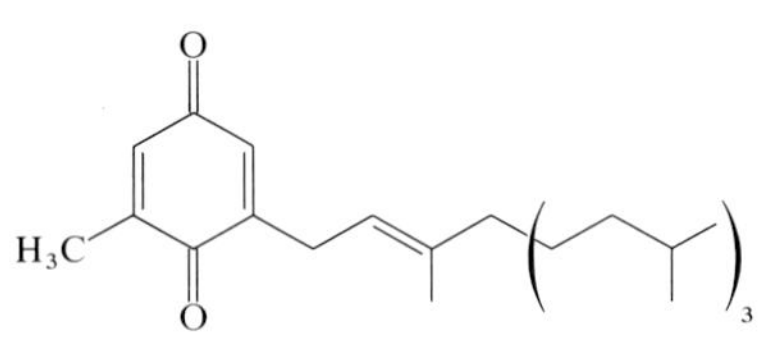

15. 6-Phytyltoluquinone

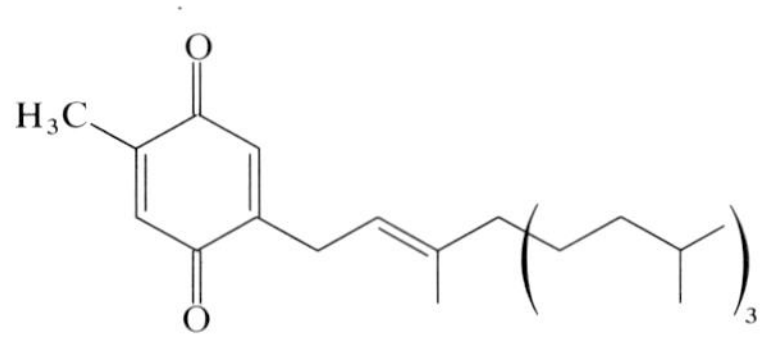

16. 5-Phytyltoluquinone

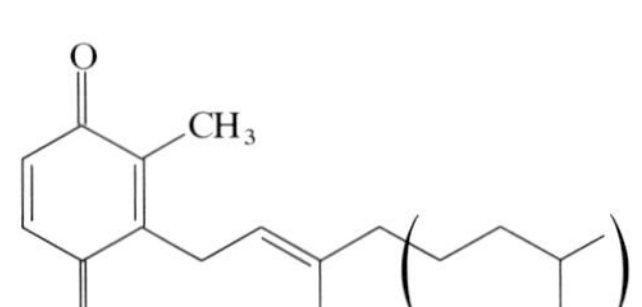

17. 3-Phytyltoluquinone

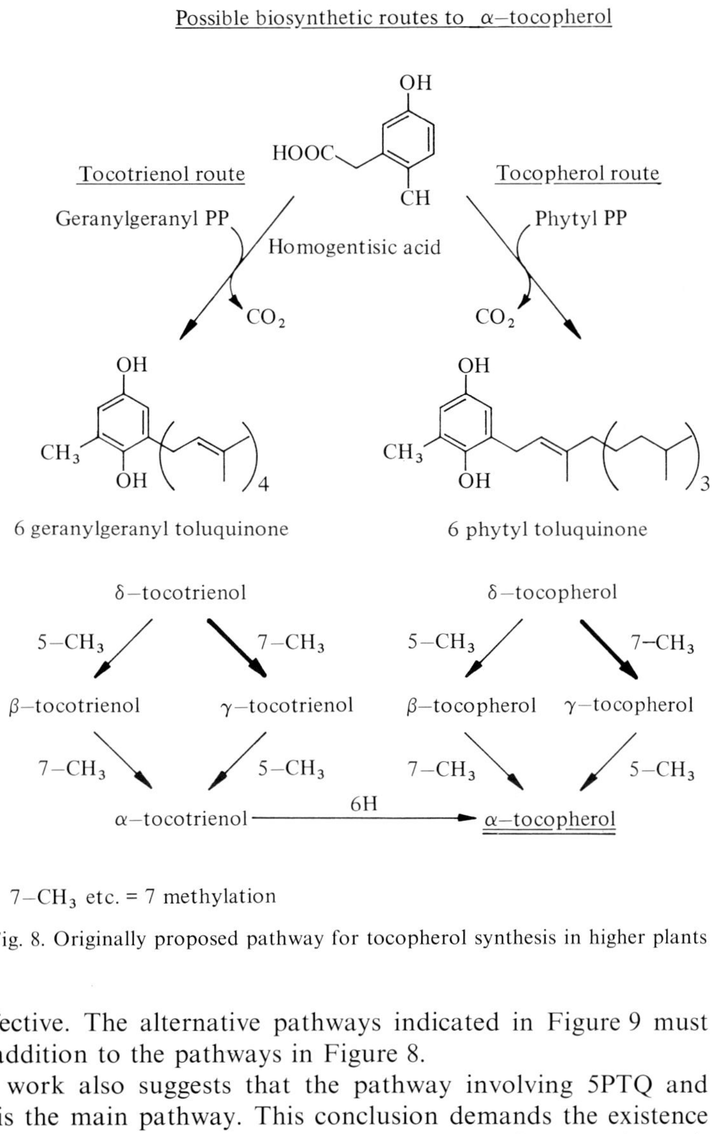

Fig. 8. Originally proposed pathway for tocopherol synthesis in higher plants

being the most effective. The alternative pathways indicated in Figure 9 must be functioning in addition to the pathways in Figure 8.

The Liverpool work also suggests that the pathway involving 5PTQ and 7,8-dimethyltocol is the main pathway. This conclusion demands the existence of a very active methyl transferase that inserts a methyl group from S-adenosyl-methionine at C-8. Earlier work with maize indicated that the methyl group at C-8 was derived from the side chain of homogentisic acid (Fig. 8) [39]. This discrepancy needs to be sorted out; it may be that different pathways exist in different plants or that the same pathways exist side by side in all plants but that their individual contributions differ widely from organism to organism. Another possibility is that the pathways are compartmentalized, one being located inside and one outside the chloroplast. This is a real possibility because in early experiments in which the incorporations of $^{14}CO_2$ and [2-^{14}C] MVA into tocopherols were compared, the results with α-tocopherol and α-tocopherol quinone fell neither into the 'chloroplastidic' nor into the 'extra-chloro-

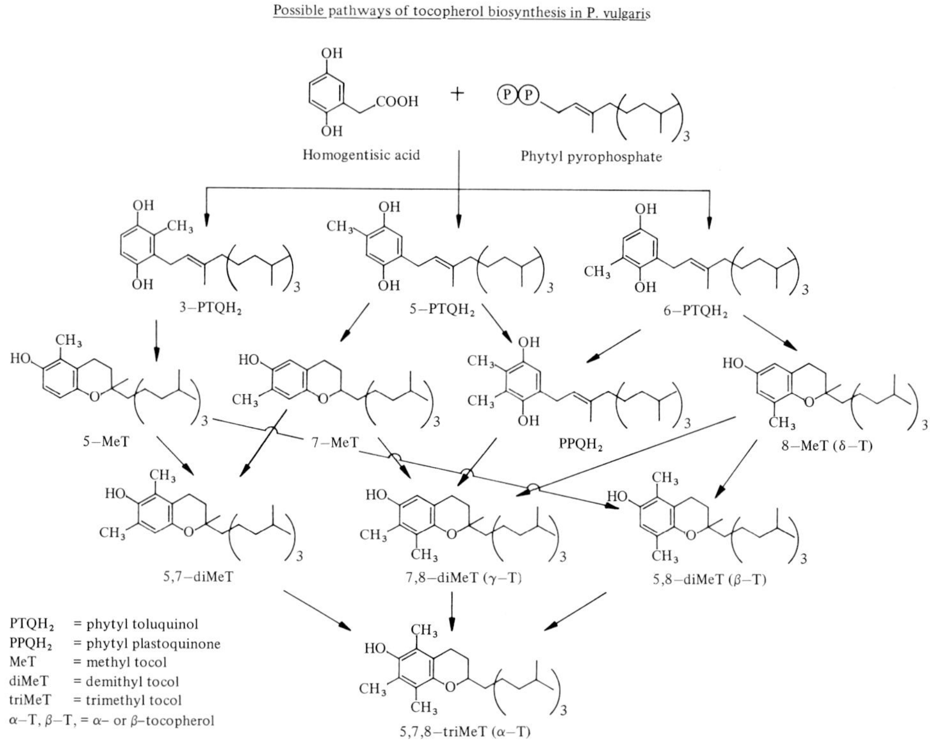

Fig. 9. Alternative pathways of tocopherol synthesis in higher plants

plastidic' pattern of labelling; they were labelled equally from both substrates. In addition, when [^{14}C] tyrosine is added to *Pisum* leaves, the radioactivity found after 24 h in the tocopherols broadly mirrors their cellular distribution. The label is mainly in the chloroplastidic α-tocopherol and in the extraplastidic γ-tocopherol, although some activity was also noted in the chloroplastidic γ-tocopherol and extraplastidic α-tocopherol. Furthermore [^{14}C] γ-tocopherol applied to the leaf surface yielded labelled α-tocopherol in both the chloroplast and supernatant fractions, thus indicating synthesis of α-tocopherol in two sites (Pennock and colleagues, unpublished observations). Finally, a chloroplast-rich fraction of sugar beet will convert homogentisate and phytyl pyrophosphate into a phytyl toluquinol, although the position of the phytyl group has not yet been determined [36].

D. Regulation of Prenyllipid Synthesis

I. Terpenes

We still do not know how the synthesis of prenyllipids in the chloroplast is regulated. The stimulation of synthesis on illumination is clearly an all-or-nothing response that produces a fully functional chloroplast containing every component in just the right amount; that is, the accumulation of prenyllipids parallels the formation of the thylakoids. The synthesis of prenyllipids is under phytochrome control as first demonstrated for carotenoids by Cohen and Goodwin [6], who showed that the lag period normally observed on illuminating etiolated maize seedlings is abolished by prior brief treatment with red light. This effect is nullified by a subsequent treatment with far-red light. The elimination of lag was also demonstrated for phytylation of chlorophyllide [26]. The phytochrome-controlled stimulation of carotenoid synthesis was confirmed and extended by Schnarrenberger and Mohr [31] and Lichtenthaler [23], and the effect was also shown to apply to the lipoquinone group [19]. The close association of carotenoid synthesis with thylakoid development has been indicated from studies of mutants.

When a chloroplast is functionally mature, synthesis of carotenoids and chlorophylls is virtually switched off and the turnover of the molecules is very low. The incorporation of $^{14}CO_2$ into these pigments in mature chloroplasts drops almost to zero [40]. However, this cannot be the case for the other terpenoids because, as indicated previously (Section B.II), they continue to increase as the chloroplast ages and accumulate in considerable amounts in the plastoglobuli.

References

1. Beale, S.I., Castelfranco, P.A.: Biosynthesis of δ-aminolevulinic acid in higher plants. II. Formation of ^{14}C-δ-aminolevulinic acid from labelled precursors in greening plant tissues. Plant Physiol. **53**, 297–303 (1974)
2. Britton, G.: Biosynthesis of carotenoids. In: Chemistry and Biochemistry of Plant Pigments. Goodwin, T.W. (ed.) London: Academic Press, 1976, 2nd ed., Vol. 1., pp. 262–327
3. Brooker, J.D., Russell, D.W.: Subcellular localization of 3-hydroxy-3-methyl glutaryl coenzyme A reductase in *Pisum sativum* seedlings. Arch. Biochem. Biophys. **167**, 730–737 (1975)
4. Brooks, J.L., Stumpf, P.K.: Fat metabolism in higher plants XXXIX. Properties of a soluble fatty acid biosynthesizing system from lettuce chloroplasts. Arch. Biochem. Biophys. **116**, 108–116 (1966)
5. Buggy, M.J., Britton, G., Goodwin, T.W.: Stereochemistry of phytoene biosynthesis by isolated chloroplasts. Biochem. J. **114**, 641–643 (1969)
6. Cohen, R.Z., Goodwin, T.W.: The effect of red and far-red light on carotenoid synthesis by etiolated maize seedlings. Phytochemistry **1**, 67–72 (1962)
7. Douce, R., Holtz, R.B., Benson, A.A.: Isolation and properties of the envelope of spinach chloroplasts. J. Biol. Chem. **248**, 7215–7222 (1975)
8. Ellis, R.J., Blair, E.E., Hartley, M.R.: The nature and function of chloroplast protein synthesis. In: Nitrogen Metabolism in Plants. Goodwin, T.W., Smellie, R.M.S. (eds.) Biochem. Soc. Symp. **38**, 137–162 (1973)
9. Goodwin, T.W.: Biosynthesis of Vitamins. London: Academic Press 1964
10. Goodwin, T.W.: Terpenoids and chloroplast development. In: Biochemistry of Chloroplasts. Goodwin, T.W. (ed.) London: Academic Press, 1967, Vol. 2, pp. 721–733
11. Goodwin, T.W.: Biosynthesis by chloroplasts. In: Structure and Function of Chloroplasts. Gibbs, M. (ed.) Berlin: Springer, 1971 a, pp. 215–276
12. Goodwin, T.W.: Biosynthesis of carotenoids. In: Carotenoids. Isler, O. (ed.) Basel: Birkhäuser, 1971 b, pp. 577–636

13. Goodwin, T.W.: Comparative Biochemistry of the Carotenoids. London: Chapman and Hall, in press, 1978, 2nd ed.
14. Goodwin, T.W., Mercer, E.I.: Introduction to Plant Biochemistry. London: Pergamon Press, 1972, 359 pp.
15. Grumbach, K.H., Lichtenthaler, H.K.: Photo-oxidation of the plastohydroquinone pool during onset of photosynthesis. Proc. 3rd Int. Cong. Photosynthesis, 1974, p. 516
16. Janiszowska, W., Michalski, W., Kasprzyk, S.: Polyprenol quinones and α-tocopherol in *Calendula officinalis*. Phytochemistry **15**, 125–127 (1976)
17. Kemp, R.J., Mercer, E.I.: The sterol esters of maize seedlings. Biochem. J. **110**, 111–118 (1968)
18. Kirk, J.T.O., Tilney-Bassett, R.A.E.: The Plastids. London: W.H. Freeman, 1967, 608 pp.
19. Kleudgen, H.K., Lichtenthaler, H.K.: Induction and reversion of prenyl-lipid synthesis in etiolated barley seedlings by red and far-red light. Proc. 3rd Int. Cong. Photosynthesis, 1974, pp. 2017–2020
20. Knapp, F.F., Aexel, R.T., Nicholas, H.J.: Sterol biosynthesis in subcellular particles of higher plants. Plant Physiol. **44**, 442–446 (1969)
21. Lichtenthaler, H.K.: Plastoglobuli und Plastidenstruktur. Ber. Dtsch. Bot. Ges. **79**, 82–88 (1966)
22. Lichtenthaler, H.K.: Localization and functional concentrations of lipoquinones in chloroplasts. In: Progress in Photosynthetic Research. Metzner, H. (ed.) 1969, Vol. 1, pp. 304–314
23. Lichtenthaler, H.K.: Control of light induced carotenoid synthesis in *Raphanus* seedlings by phytochrome. Physiol. Plantarum **34**, 357–358 (1975)
24. Lichtenthaler, H.K., Sprey, B.: Über die osmiophilen globularen Lipideinschlüsse der Chloroplasten. Z. Naturforsch. **21**, 690–697 (1966)
25. Lichtenthaler, H.K., Straub, V., Grumbach, K.H.: Unequal formation of prenyl-lipids in a plant tissue culture and in leaves of *Nicotiana Tabacum* L. Plant Sci. Lett. **4**, 61–65 (1975)
26. Liljenberg, C.: The effect of light on the phytolization of chlorophyllide a and the spectral dependence of the process. Physiol. Plantarum **19**, 848–853 (1966)
27. Mercer, E.I., Treharne, K.J.: Occurrence of sterols in chloroplasts. In: Biochemistry of Chloroplasts. Goodwin, T.W. (ed.) London: Academic Press 1966, Vol. 1, pp. 181–186
28. Ogawa, T., Obata, F., Shibata, K.: The pigment proteins in spinach chloroplasts. Biochim. Biophys. Acta **112**, 223–234 (1966)
29. Pennock, J.F., Janiszowska, W.: Biochemistry of vitamin E in plants. Vitamins and Hormones (in press) (1976)
30. Radunz, A., Schmid, G.H.: Reactions of antisera to lutein and plastoquinone with chloroplast preparations and their effects on photosynthetic electron transport. Z. Naturforsch. **28C**, 36–44 (1973)
31. Schnarrenberger, L., Mohr, H.: Carotenoid synthesis in mustard seedlings as controlled by phytochrome and inhibitors. Planta **94**, 296–307 (1970)
32. Sprey, B., Laetsch, W.M.: Chloroplast envelopes of *Spinach oleracea* L. II. Ultrastructure of chloroplast envelopes and lamellae. Z. Pflanzenphysiol. **78**, 146–163 (1976a)
33. Sprey, B., Laetsch, W.M.: Chloroplast envelopes of *Spinach oleracea* L. III. Freeze-fracturing of chloroplast envelopes. Z. Pflanzenphysiol. **78**, 360–371 (1976b)
34. Stumpf, P.K.: Lipid biosynthesis in higher plants. In: Biochemistry of Lipids. Goodwin, T.W. (ed.) London: Butterworth, 1976, in press
35. Subbarayan, C., Kushwaha, S.C., Suzu, G., Porter, J.W.: Enzymic conversion of isopentenyl pyrophosphate-4-^{14}C and phytoene-^{14}C to acyclic carotenes by an ammonium sulphate-precipitated spinach enzyme system. Arch. Biochem. Biophys. **137**, 547–557 (1970)
36. Thomas, G., Threlfall, D.R.: Synthesis of polyprenyltoluquinols from homogentisate and polyprenyl pyrophosphates in particulate fractions of *Euglena* and sugar beet. Biochem. J. **142**, 437–440 (1974)
37. Thornber, J.P.: Chlorophyll-proteins: light-harvesting and reaction center components of plants. Ann. Rev. Plant Physiol. **26**, 127–158 (1975)
38. Threlfall, D.R., Griffiths, W.T.: Biosynthesis of terpenoid quinones. In: Biochemistry of Chloroplasts. Goodwin, T.W. (ed.) London: Academic Press, 1966, Vol. 2, pp. 257–282
39. Threlfall, D.R., Whistance, G.R.: Biosynthesis of isoprenoid quinones and chromanols. In: Aspects of Terpenoid Chemistry and Biochemistry. Goodwin, T.W. (ed.) London: Academic Press, 1971, pp. 357–404

40. Treharne, K.J., Mercer, E.I., Goodwin, T.W.: Incorporation of [^{14}C] carbon dioxide and [2-^{14}C]mevalonic acid into terpenoids of higher plants during chloroplast development. Biochem. J. **99**, 239–245 (1966)
41. Vernon, L.P., Ke, B., Mollenhauser, H., Shaw, E.R.: Composition and structure of spinach sub-chloroplast fragments obtained by the action of Triton X-100. In: Progress in Photosynthesis Research. Metzner, H. (ed.) Tübingen: IUBS, 1969, Vol. 1, 137–148
42. Wallwork, J.C., Crane, F.L.: The nature, distribution, function and biosynthesis of prenyl phytoquinones and related compounds. In: Progress in Phytochemistry. Reinhold, L., Liwschitz, Y. (eds.), New York: John Wiley, 1970, Vol. 2, pp. 267–341
43. Wellburn, A.R., Hemming, F.W.: Long chain polyisoprenoid alcohols in chloroplasts. In: The Biochemistry of Chloroplasts. Goodwin, T.W. (ed.). London: Academic Press, 1966, Vol. 1, pp. 173–188
44. Wellburn, A.R., Hemming, F.W.: The subcellular distribution and biosynthesis of castaprenols and plastoquinone in the leaves of *Aesculus hippocastanum*. Biochem. J. **104**, 173–177 (1967)
45. Whittle, K.J., Dunphy, P.J., Pennock, J.F.: Plastochromanol in the leaves of *Hevea brasiliensis*. Biochem. J. **96**, 17C–19C (1965)
46. Wild, A., Conrad, G., Zickler, H.-O.: Der Gehalt an Plastidenchinonen und Pigmenten von *Zea mays* und *Atriplex rosea* in Abhängigkeit von Lichtintensität und dem Alter der Pflanzen. Planta **103**, 181–187 (1972)

CHAPTER 3

Glyco- and Phospholipids of Biomembranes in Higher Plants

P. MAZLIAK

A. Introduction

Lipids are important constituents of all biological membranes and play a central role in the molecular organization of these membranes. Structural lipids in plant biomembranes are essentially phospholipids, galactolipids, and sterols. In this chapter we shall first review the lipid composition of plant biomembranes; then we shall examine where these lipids come from in plant cells, how they are built, and how they integrate the various biomembranes or cell organelles. However, membrane lipids have not only a structural role. The activities of many membrane-bound enzymes are controlled by the extent and the quality of the lipid environment within the membranes. The last part of this chapter will deal with the results of some recent studies concerning the effect of lipids upon the activities of some membrane-bound enzymes.

B. Lipid Composition of Plant Biomembranes

Techniques to purify plant cell organelles have progressed significantly during the last ten years, primarily because of the systematic use, in many laboratories, of sucrose density gradients. The identity of prepared subcellular fractions was generally controlled by combining the use of marker enzymes and electron microscopy. Excellent preparations of plant plasmalemma, purified mitochondria, purified inner and outer membranes of plant mitochondria, chloroplasts and chloroplast envelopes, peroxisomes, glyoxysomes, nuclei, Golgi apparatus elements, microsomes, and lysosomes have been obtained.

Detailed lipid analyses of these various cellular fractions have been published, and useful conclusions may be drawn from a comparison of the results collected from different plants by many different laboratories.

I. Plasmalemma

Lipid analyses of plant plasmalemma are not very numerous because this subcellular fraction remains difficult to prepare from plant tissues. During grinding and homogenization of plant tissues, the plasmalemma is ruptured either into *large fragments* that cosediment in the gradients with nuclei and mitochondria or into *small vesicles* that cosediment with microsomes. Consequently, one can

Table 1. Lipids of plant plasmalemma

Plant organs	Initial pellet for sedimentation of plasmalemma	mg Lipid/ mg Protein	% Total lipids				mol Sterol/ mol Phospholipid	References
			Phospho-lipids	Galacto-lipids	Neutral lipids	Sterols		
Oat roots	13,000–80,000 *g*	–	–	–	–	–	1.074–1.213	[27]
Mushroom sporophores	2,600 *g*	0.91	–	–	–	–	0.77	[28]
Oat root tips	13,000–80,000 *g*	1.17	28.60	+	+	+	–	[35]
Maize coleoptiles	100,000 *g*	–	–	–	–	116–135[a]	–	[22]
Maize coleoptiles	100,000 *g*	–	–	–	–	140–160[a]	–	[61]
Potato tubers	3,000 *g*	0.71	31	21	48 (neutral lipids + sterols)		–	[10]
Lupin roots	2,000 *g*	–	41	24	35 (neutral lipids + sterols)		–	[37]
Broadbean roots	2,000 *g*	–	65	9	26 (neutral lipids + sterols)		–	[37]

Plant organs	% Total Phospholipids								References
	PC	PE	PGP	PA	PI	PS	PG	Lyso compounds and unknown	
Phaseolus limensis root tips	++	++	–	–	+	+	+	+	[4]
Oat root tips	18.2	10.5	–	13.9	3.7	–	4.0	49.7	[35]
Potato tubers	32	46	3	traces	19	–	–	–	[11]

Plant organs	% Total fatty acids						mg Saturated/ mg Unsaturated	References
	$C_{16:0}$	$C_{16:1}$	$C_{18:0}$	$C_{18:1}$	$C_{18:2}$	$C_{18:3}$		
Mushroom sporophores	15.1	traces	15.6	13.6	55.1	traces	0.44	[28]
Oat root tips	47.8	2.0	7.3	8.6	15.9	0.3	1.9	[35]
Potato tubers	28.5	–	7.5	4.5	41	17.5	0.56	[10]

[a] μg/mg protein

find in the literature two types of methods for preparing plant plasmalemma: (1) methods that start from crude 3,000 *g* pellets and (2) methods that permit the purification of plant plasmalemma from the microsomal pellet sedimenting at 100,000 *g*. Published lipid analyses of plant plasmalemma prepared by both types of methods have been grouped in Table 1. It must be clearly indicated that most authors could not distinguish between tonoplasts and plasmalemma in their preparations and thus, the results presented in Table 1 have, in fact, been obtained from mixtures of plasmalemma and tonoplast fragments.

One can draw some general conclusions from these data, despite the limited number of complete analyses:

1. The lipids of plant plasmalemma constitute 40 to 54% of total the membrane weight, which is rather high compared to other cell membranes.

2. Plasmalemma lipids comprise essentially four molecular types: phospholipids, glycolipids, sterols, and neutral lipids (i.e., hydrocarbons, di- and triglycerides, free fatty acids). Neutral lipids and sterols represent 30 to 50% of total lipid weight, which again is high compared to other cell membranes.

3. The sterol:phospholipid ratio, when calculated, has been found to be much higher in plasma membranes than in any other cell membrane [27]. Hartmann et al. [22] have identified at least four sterols in the plasmalemma from maize coleoptiles: sitosterol, stigmasterol, campesterol, and cholesterol. Because it is known that sterol molecules can complex phospholipids in monolayers and thus decrease the fluidity and the permeability of the lipid phase, it has been suggested that the high concentration of sterols in the plasmalemma helps restrict the permeability of this membrane, which is directly exposed to the external medium. However, treatments of *Avena* coleoptile protoplasts with polyene antibiotics indicated that little sterol was present at the external surface of these protoplasts [66].

4. Results from the phospholipid and fatty acid analyses of plant plasmalemma are not uniform. Plasmalemma prepared from microsomal pellets by long centrifugation procedures contain a relatively large amount of lysocompounds, or of an unknown compound or phosphatidic acid; phosphatidylcholine and phosphatidylethanolamine together do not constitute more than 30% of the total phospholipids. The unsaturated fatty acid content of the same membranes is very low. In contrast, plasma membranes prepared from an initial 3,000 *g* pellet (from mushroom or potato tuber homogenates) contain the usual large amount of phosphatidylcholine, phosphatidylethanolamine, or polyunsaturated fatty acids. Discrepancies between these analyses may result from the different origins of the plasmalemma, but the degradation of lipids by phospholipases, which could easily have occurred during the long centrifugation times implied in the preparation method starting from the microsomal pellet, should also be considered.

II. Mitochondria

Purified mitochondria have been prepared from numerous plant tissues in recent years; the outer and inner membranes of these organelles have been separated

Table 2. Lipids of plant mitochondria

Plant organs (origin of mitochondria)	% Total phospholipids							References
	PC	PE	PGP	PI	PG	PS	PA	
Scorsonere tissue culture	45	21	3	15	2	–	10	[17]
Potato tubers								
Whole organelles	43	30	8	7	3	3	5	[52]
Outer membrane	52.6	25	12.1		10.3		–	
Inner membrane	27	29	19.5		24.5		–	
Cauliflower buds								
Whole organelles	44	34	11	7	4	–	–	[53]
Outer membrane	42	24	3	21	10	–	–	
Inner membrane	41	37	14	5	3	–	–	
Broadbean root	47.2	20.3	3.2	13.2	15.1	–	1	[64]
Lupin root	32.8	23.7	9	12.7	12.8	9		
Castor bean endosperm	46.3	29.5	5.4	9.4	2.3	–	–	[62]

Sources	% Total fatty acids					
	$C_{16:0}$	$C_{18:0}$	$C_{18:1}$	$C_{18:2}$	$C_{18:3}$	
Potato tuber mitochondria						
Outer membrane	24	3.5	1.8	58.0	12.7	[52]
Inner membrane	12.5	2.3	0.8	67.4	16.8	
Cauliflower mitochondria						
Outer membrane	50	4	20	8	18	[53]
Inner membrane	14	1	9	13	62	
Mung bean mitochondria						
Outer membrane	25	3.5	5.5	28.5	37	[43]
Inner membrane	14.5	1.5	4	35	45	

in several laboratories by different procedures (digitonin treatment or osmotic shock). Detailed analyses of the lipids of both mitochondrial membranes have been performed, and some representative data are grouped in Table 2. Phospholipids are the major lipids of mitochondria, forming 70% or more of the total lipids on a weight basis. The presence of galactolipids in mitochondrial preparations from avocado or cauliflower buds has been reported [69], but several groups [5, 46, 62] have shown that these glycolipids found in mitochondrial preparations could be attributed to contamination by membranes of other organelles such as proplastids of endoplasmic reticulum membranes. This is not the opinion of Mackender and Leech [42], who have analyzed various subcellular fractions from *Vicia faba* leaves and found that the mitochondrial fraction does contain galactosyldiglycerides. These authors have found striking differences in the fatty acid compositions, of the galactolipids from chloroplasts or mitochondria, and so they have been able to exclude cross-contamination as a possible source of galactolipids in mitochondria.

All mitochondria from all plant tissues examined thus far present the same phospholipid composition, with two major molecular species: phosphatidylcholine and phosphatidylethanolamine; the diphosphatidylglycerol (or cardiolipin) may be used as a biochemical marker of the inner mitochondrial membrane. Differences may be noticed between the phospholipid compositions of the two mitochondrial membranes: the outer membrane resembles plasmalemma or endoplasmic reticulum membranes in having twice as much phosphatidylcholine as phosphatidylethanolamine, while both phospholipids are present in equal amounts within the inner membrane. The major fatty acid found in these lipids varies from tissue to tissue, but in all species, the inner mitochondrial membrane contains many more unsaturated fatty acids than the outer.

III. Chloroplasts

The lipids of these characteristic plant cell organelles have been analyzed in great detail so many times that we found it sufficient, in order to illustrate the results, to include only some representative and well-established data in Table 3 (see also Chapter 7).

In chloroplasts, the main lipids are three galactolipids: monogalactosyldiglyceride (MGDG), digalactosyldiglyceride (DGDG), and sulfoquinovosyldiglycer-

Table 3. Lipids of higher plant chloroplasts

Source (plant leaves)	% Total lipids						References
	MGDG	DGDG	SL	PG	PC	PI	
Beta vulgaris							
Whole chloroplasts	36	20	5.0	7.0	7.0	2	[60]
Spinacia oleracea							
Chloroplast envelope	22	32	5.0	8.4	27	1.3	[21]
Chloroplast envelope	8.4	29.3	5.5	13.2	27.5	–	[24]
Chloroplast lamellae	51	26	7.1	9.1	3.2	1.4	[21]
Chloroplast lamellae	39.1	20.1	7.3	16.5	10.1	–	[24]
Vicia faba							
Chloroplast envelope	29.1	32.4	–	8.9	29.6	–	[42]
Chloroplast lamellae	65.4	26.2	–	5.5	2.8	–	[42]

Source (plant leaves)	% Total fatty acids						References
	16:0	16:1Δ^3 trans	16:3	18:1	18:2	18:3	
Spinacia oleracea							
Whole chloroplasts							
MGDG	–	–	25	–	2	72	[1]
DGDG	3	–	5	2	2	87	[1]
PG	11	32	2	2	4	47	[1]
SL	39	–			6	52	[1]
PC	12	–	4	9	16	58	[1]

ide (SL). These lipids are not uniformly distributed in chloroplast membranes: MGDG is concentrated in the internal thylakoid membranes whereas DGDG is dominant in the chloroplast envelopes. Besides galactolipids, phospholipids have been found in chloroplasts, namely, phosphatidylglycerol and phosphatidylcholine. The latter molecular species is strikingly concentrated in the chloroplast envelope. Phosphatidylethanolamine has never been found (in significant amount) within the chloroplasts: this phospholipid is now generally considered to be a negative lipid marker for highly purified plant chloroplasts. The fatty acid composition of chloroplast lipids has been studied many times. Recently, for instance, Rullkötter et al. [67] analyzed the composition and fatty acid distribution of the DGDG from 29 plants. Some general characteristics of chloroplast lipids may be briefly indicated:

1. Galactolipids and phospholipids from chloroplasts both have a high content of trienoïc fatty acids (linolenic acid, predominantly, accompanied in some species by hexadecatrienoïc acid).

2. Phosphatidylglycerol from chloroplast is unique in that it contains a high amount of Δ^3-trans-hexadecenoïc acid in the C_2-position of glycerol.

3. Sulfoquinovosyldiglyceride is more saturated than other chloroplast lipids and presents a high palmitic acid content.

IV. Glyoxysomes and Peroxisomes

These organelles have been isolated from plants only very recently and very few lipid analyses are presently available. Nevertheless, one can see, from the published data collected in Table 4, that phosphatidylcholine and phosphatidylethanolamine are the major phospholipid components of plant glyoxysomes and peroxisomes.

Table 4. Phospholipid composition of plant glyoxysomes and peroxisomes

Organelles	% Total phospholipids								References
	PC	PE	PI	PGP	PG	PS	PA	Unidentified	
Castor bean endosperm glyoxysome	49.0	31.4	6.1	2.4	–	0.0	–	11.4	[15]
Potato tuber peroxisomes	61	19.6	4.3	–	15.1	–	–	–	[73]

V. Microsomes

This subcellular fraction is defined only as the membrane pellet that is obtained by centrifuging a 15,000 *g* supernatant at $60 \times 100{,}000$ *g* mn. In this pellet, one can find, all together, membranes from endoplasmic reticulum, Golgi apparatus,

Table 5. Lipid composition of plant microsomes

Source	% Total glycerolipids			References
	Galactolipids	Phospholipids	Neutral lipids	
Potato tubers	16	47	36.5	[5]
Vicia faba roots	6.6	82.3	11.1	[64]
Lupin roots	16.2	71.5	12.3	[64]

Source	% Total polar lipids									References
	MGDG	DGDG	SL	PC	PE	PI	PG	PS	PA	
Spinach leaves	12	8	2	35	30	7	3	–	–	[19]
Vicia faba leaves	8.2	20.4	–	47.6	14.2	–	9.6	–	–	[42]
Vicia faba roots	3.6	3.8	–	39	25	9.5	12.2		2.0	[64]
Lupin roots	9.4	9.1	–	28.5	23	11.5	7.7		4.7	[64]

Source	% Total phospholipids						References
	PC	PE	PI	PGP	PS+PG	PA	
Potato tubers	44.8	32.5	16	1	2	3.5	[7]
Cauliflower buds	50	36	6	tr.	8	–	[7]
Vicia faba roots	42.1	26.6	10.3	5.3	13.5	2.1	[64]
Lupin roots	34.9	28.2	14.1	7.5	9.5	5.8	[64]
Castor bean endosperm	50.0	26.6	18.9	2.7	1.8	–	[15]

Source	% Total fatty acids						References
	$C_{16:0}$	$C_{16:1}$	$C_{18:0}$	$C_{18:1}$	$C_{18:2}$	$C_{18:3}$	
Potato tubers	31	–	7	0.5	49.5	12.0	[7]
Cauliflower buds	24.3	–	2.2	9.7	13.2	50.0	[7]
Vicia faba roots	19.5	traces	5.9	2.0	49.5	4.4	[64]
Vicia faba leaves	19.8	1.1	4.8	6.1	27.6	40.7	[42]
Lupin roots	27.4	traces	5.0	2.7	6.4	49.5	[64]

tonoplast and plasmalemma fragments, etc. Table 5 shows some representative lipid analyses of the membranes found in this pellet. It appears that polar lipids are predominant in microsomal membranes and that, although phospholipids are the more abundant among them, the percentage of galactolipids found in microsomes is far from negligible. However, it is difficult to exclude the presence of chloroplast, proplastid, plastid, or amyloplast envelopes in the microsomal pellet; if present, these organelle envelopes could contribute to the galactolipid content of microsomes.

Among the microsomal phospholipids, three molecular types are clearly dominant, namely, phosphatidylcholine, phosphatidylethanolamine, and phosphatidylinositol. The fatty acids engaged in these lipids may vary from tissue to tissue, e.g., linolenic acid is dominant in microsomal lipids from broadbean leaves whereas linoleic acid is dominant in microsomal lipids from broadbean roots.

A very interesting point has been raised by Oursel and collaborators [64] and Lamant and Heller [37]. These authors have shown that microsomal membranes as well as plasmalemma of plants growing on a medium enriched in calcium have lower phospholipid contents than plants growing on a normal medium. Also Donaire and Oursel [14] have observed the inhibition by calcium ions, in vitro and in vivo, of the CDP-choline transferase from broadbean roots: this could explain the differences in phospholipid composition of microsomal membranes from plants growing on mediums differently enriched in calcium.

VI. Nuclei

Pure preparations of isolated plant nuclei are still very difficult to obtain. Very few lipid analyses have been realized up to now on these plant organelles. Table 6 shows that in the nuclear membranes from two different plant organs

Table 6. Lipid composition of plant nuclei

Tissue	% Total lipids			References
	Neutral lipids	Glycolipids	Phospholipids	
Onion stem	46 (neutral lipids + glycolipids)		54	[65]
Onion root tip	37 (neutral lipids + glycolipids)		63	[65]
Sunflower hypocotyl	29.0	13.0	58.0	Mazliak (1976) unpublished

Tissue	% Total phospholipids								References
	PC	PE	PI	PGP	PG	PS	LYSO	PA	
Onion root tip	27.1	20.4	8.0	1.2	5.6	3.1	10.1	24.6	[65]
Sunflower hypocotyl	23.0	24.6	10.3	15.6 (PGP + PG)		19.1 (PS + LYSO)		7.4	Mazliak (1976) unpublished

Tissue	% Total fatty acids						References
	$C_{16:0}$	$C_{16:1}$	$C_{18:0}$	$C_{18:1}$	$C_{18:2}$	$C_{18:3}$	
Onion stem	22	traces	2	20	52	traces	[65]
Onion root tip	33	traces	3	8	51	5	[65]
Sunflower hypocotyl	21.4	traces	3.9	8.3	59.5	6.8	Mazliak (1976) unpublished

(onion root tip and sunflower hypocotyl), phospholipids have been found of the major lipid types; among these, phosphatidylcholine and phosphatidylethanolamine were the predominant species, as in all other cellular membranes except chloroplast membranes.

VII. Comparisons Between Different Membranes

In a few cases, simultaneous analyses of the lipid compositions of different biomembranes from the same plant tissue have been carried out. In our own laboratory, for instance, we have compared the phospholipid compositions of different cell organelles prepared from potato tubers or cauliflower florets. Tables 7 and 8 show that there is a striking uniformity in the phospholipid compositions, of the different biomembranes present in these two nongreen organs. Phosphatidylcholine and phosphatidylethanolamine are the most abundant molecular types of membrane phospholipids, constituting together at least 60% of the total phospholipids. There are, however, some minor variations between biomembranes, e.g., diphosphatidylglycerol is clearly concentrated in the inner membrane of mitochondria. Nevertheless, despite these minor variations, the overwhelming impression is one of great uniformity of lipid composition among the different cell membranes.

Table 7. Percentage composition of the membrane phospholipids of potato tubers

Lipid	Cytoplasmic membranes (plasma-lemma)	Mitochondria			Peroxisomes	Microsomes
		Whole organelles	Inner membranes	Outer membranes		
PC	38	43	27	52.6	61	44.8
PE	47	30	29	25	19.6	32.5
PGP	–	8	19.5	12.1	–	1
PA	–	5	–	–	–	3.5
PI	15	7 }			4.3	16
PS	–	3 }	24.5	10.3	–	1
PG	–	3 }			15.1	1

Table 8. Percentage composition of the membrane phospholipids of cauliflower florets

Lipid	Cytoplasmic membranes (plasma-lemma)	Mitochondria			Peroxisomes	Microsomes
		Whole organelles	Inner membranes	Outer membranes		
PC	66	43	41	42	48.0	50
PE	21	32	37	24	45.6	36
PGP	–	7	14	3	–	tr
PA	–	1.9	–	–	–	–
PI	13	11	5	21	6.4	6
PS	–	1	–	–	–	–
PG	–	3	3	10	–	8

Table 9. Percentage composition of the fatty acids of potato tuber membranes

Fatty acid	Cytoplasmic membranes (plasmalemma)	Mitochondria			Peroxisomes	Microsomes
		Whole organelles	Inner membranes	Outer membranes		
14:0	—	—	—	—	2	—
16:0	27.8	14.7	12.5	24	22	31
18:0	8.4	2.7	2.3	3.5	2.1	7
18:1	4.3	1.4	0.8	1.8	4.3	0.5
18:2	40	63.5	67.4	58.0	56	49.5
18:3	8.5	16.9	16.8	12.7	13.3	12.0

Table 10. Percentage composition of the fatty acids of cauliflower bud membranes

Fatty acid	Plasmalemma	Mitochondria			Peroxisomes	Microsomes
		Whole organelles	Inner[a] membranes	Outer[a] membranes		
16:0	26	17.6	14	50	17.3	24.3
18:0	4	2.4	1	4	1.7	2.2
18:1	13	10	9	20	16.1	9.7
18:2	12	12.2	13	8	11.5	13.2
18:3	43	45.3	62	18	53.3	50.5

[a] Data from Ref. [53]

The same is true of the fatty acid composition of the membrane lipids present in different cellular organelles. Tables 9 and 10, showing the fatty acid compositions of the different membranes from potato tubers or cauliflower florets, exemplify this. There are again, however, some minor differences. For example, the fatty acids of the outer mitochondrial membrane are more saturated than those of the inner, and the fatty acids of the plasmalemma are more saturated than those of the intracellular organelles. Nevertheless the same fatty acids appear in the different membranes, in roughly similar proportions.

In conclusion one can say that in all the cell membranes of potato tubers or cauliflower florets—two types of plant organs devoid of chloroplasts—phosphatidylcholine and phosphatidylethanolamine are the major phospholipids and that the same polyunsaturated fatty acids—linoleic in one case or linolenic in the other—are dominant everywhere.

The situation is different in a green organ such as a leaf (Table 11). The chloroplast is unique in having a very high galactolipid content (more than 50% of total lipids are galactolipids) and unique also in its phospholipid and fatty acid composition: no phosphatidylethanolamine can be found in chloroplasts and phosphatidylglycerol represents more than 20% the of total phospholipids; linolenic acid forms more than 60% of the total fatty acids in chloroplasts. However, the chloroplast envelope shares some characteristics with microsomal or mitochondrial membrane, *i.e.*, a high phosphatidylcholine content and a

Table 11. Lipid composition of subcellular fractions isolated from leaves of *Vicia faba* (42)

Subcellular fractions	% Total lipids				
	MGDG	DGDG	PC	PE	PG
Chloroplast envelope	29.1	32.4	29.6	0	8.9
Chloroplast lamellae	65.4	26.2	2.8	0	5.5
Mitochondria	10.2	12.8	43.5	25.7	7.8
Microsomes	8.2	20.4	47.6	14.2	9.6

Subcellular fractions	% Total fatty acids						
	16:0	16:1	16:1 Δ^3 trans	18:0	18:1	18:2	18:3
Chloroplast envelope	13.3	1.8	0.0	4.3	5.8	11.5	63.3
Chloroplast lamellae	5.9	0.0	1.2	1.6	3.2	5.2	82.9
Mitochondria	19.3	1.6	0.0	3.9	5.8	31.1	37.3
Microsomes	19.8	1.1	0.0	4.8	6.1	27.6	40.7

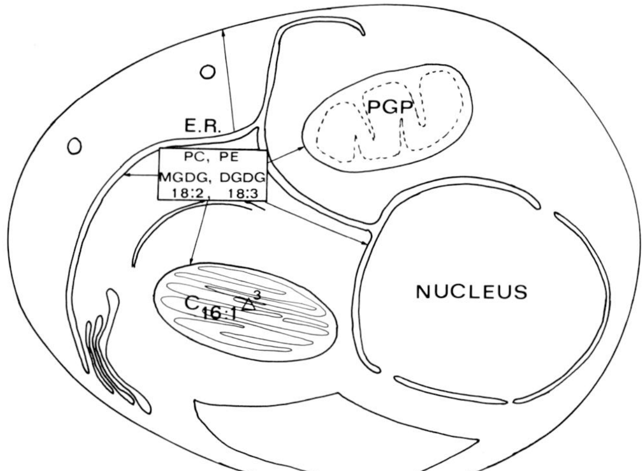

Fig. 1. Schematic division of a typical plant cell into several compartments differing according to lipid composition of biomembranes: (1) matrix and internal membrane of mitochondrion; (2) stroma and internal lamellae of the chloroplast; (3) the remaining lamellar and cytoplasmic systems of the cell

high palmitic acid content. The phosphatidylglycerol of chloroplast lamellae is unique in having trans-3-hexadecenoic acid in the C_2 position of glycerol [77].

To conclude this brief review on the lipids of biomembranes in plant cells, we should like to point out that all published results allow one to distinguish between three compartments in a typical plant cell when considering the membrane lipids (Fig. 1): (1) the inner mitochondrial membrane forming cristae and surrounding the matrix of the mitochondrion, with cardiolipin as a lipid marker; (2) the chloroplast lamellae dividing the internal stroma of the organelle, with trans-3-hexadecenoic acid as a lipid marker; and (3) the remainder of the cell, including all other biomembranes with the outer mitochondrial membrane and the chloroplast envelope that present roughly the same phospholipid, galactolipid, and sterol components. Within the framework of the endosymbiotic theory of the origin of mitochondria and chloroplasts, the inner parts of both organelles would have retained some prokaryotic characteristics expressed at the level of their component lipids by the presence of cardiolipin in mitochondria and trans-3-hexadecenoic acid in chloroplasts. All other parts of the cell would have belonged to the primitive host of endosymbiotic bacteria (mitochondria) and cyanophyceae (chloroplast).

C. Biosynthesis of Membrane Lipids

Where are the membrane lipids synthesized in the plant cell? To answer this question, one can isolate the intracellular organelles (mitochondria, chloroplasts, microsomes, etc.) and see which fatty acids and phospholipids they are capable of synthesizing from radioactively labeled precursors. It is, however, important to ensure that the isolation procedure has not caused any loss of enzymes from the organelles or any adsorption of enzymes from some other part of the cell. Similarly, it is essential that the isolated organelles retain their normal physiologic activity, which means, for instance, that mitochondria must exhibit a good respiratory control or that chloroplasts used for the biosynthetic studies must be of class I.

Table 12 shows the synthetic capabilities of plant mitochondria. One can see that:

1. Isolated mitochondria are only capable of synthesizing saturated and monounsaturated fatty acids, despite the fact that polyunsaturated fatty acids are the main fatty acids present in mitochondrial membranes.

Table 12. Lipid synthesis by isolated plant mitochondria

Tissue	Precursor administered or enzyme assayed	Fatty acids synthesized	Phospholipids synthesized	References
Cauliflower floret Potato tuber *Vicia faba* root *Lupinus albus* root	[1-^{14}C]- and [2-^{14}C] acetate [1-^{14}C] acetyl-CoA [2-^{14}C] malonate [2-^{14}C] malonyl-CoA	14:0, 16:0, 18:0, 18:1		[50]
Peanut cotyledon	AT^{32}P		PA	[8]
Cauliflower floret	[^{32}P]glycerophosphate		lyso PA, PA	[3]
	CT^{32}P		PG	[18]
	[^{14}C]CTP		PG, DPG, PI	[72]

Table 13. Lipid synthesis by isolated plastids or chloroplasts

Organelle	Tissue	Precursor administered or enzyme assayed	Molecules synthesized	References
			Fatty acids	
Proplastids	Castor bean	malonyl CoA	18:1	[84]
Proplastids	Castor bean	sucrose, glucose, G1P, G6P, pyruvate	long=chain fatty acids	[83]
Plastids	Avocado pear	acetate	16:0, 18:0, 18:1	[57]
Plastid stroma	Avocado pear	malonyl CoA	18:0	[23]
Chloroplasts	Barley leaves	acetate	16:0, 18:0, 18:1	[2, 26]
Chloroplasts	Lettuce leaves	acetate	16:0, 18:1	[9]
Chloroplasts	Spinach leaves (immature)	acetate	16:0, 18:1, (18:2, 18:3)	[34]
Chloroplasts	Spinach leaves	acetate	16:0, 18:1, (18:2, 18:3)	[58]
Chloroplasts	Maize leaves			
	mesophyll	acetate	16:0, 18:1, (18:2, 18:3)	[25]
	bundle sheath	acetate	16:0, 18:1, (18:2, 18:3)	
Chloroplast stroma	Lettuce leaves	malonyl CoA	18:0	[9]
Chloroplasts	Spinach leaves	stearyl-ACP	oleyl-ACP	[29]
Chloroplasts	Spinach leaves	? 12:3→16:3	18:3	[71]
	Pea leaves	18:2	18:3	[45]
			Glycerolipids	
Chloroplasts	Spinach leaves	UDP-galactose	MGDG, DGDG	[59]
Chloroplasts	Spinach leaves	UDP-galactose	MGDG, DGDG	[63]
Etioplasts	Maize leaves	glycerol-3-phosphate	PA, lyso PA, diglyceride, MGDG	[20]
Chloroplast envelopes	Spinach leaves	UDP-galactose	MGDG, DGDG, TGDG, tetra GDG	[19] [32] [78]

2. Isolated mitochondria are only capable of synthesizing acidic phospholipids such as phosphatidic acid, phosphatidylglycerol, or diphosphatidylglycerol, despite the fact that the main phospholipids of mitochondrial membranes are phosphatidylcholine and phosphatidylethanolamine.

Table 13 shows the synthetic capabilities of isolated plastids or chloroplasts. One can see that:

1. Isolated plastids or chloroplasts do possess the enzymatic machinery necessary for synthesizing most or all component fatty acids. If, however, starting from a simple precursor such as ^{14}C-acetate, it is very easy to obtain with isolated chloroplasts the biosynthesis of palmitic, stearic, and oleic acids, then the biosynthesis of polyunsaturated fatty acids, and particularly of linolenic acid (the major fatty acid of chloroplast lipids), is still a matter for further research, since the rate of synthesis of this peculiar fatty acid is very low in isolated organelles. Stumpf [71] has hypothesized that linoleic acid is synthesized

by elongation of a dodecatrienoic acid resulting from the progressive desaturation of a dodecanoic acid, probably formed outside the chloroplast; Tremolieres and Mazliak [75] are of the opinion that linolenic acid is synthesized by the direct desaturation of linoleic acid and that this acid in turn is formed by an oleyl-CoA desaturase located outside the chloroplast, in the microsomes. Even when the direct desaturation of linoleic acid to linolenic acid has been readily shown with entire leaves, up to now all assays to demonstrate this desaturation directly, with isolated chloroplasts, have failed [75]. All the results obtained thus far by the various groups working on fatty acid synthesis in chloroplasts seem to indicate that the eventual precursors of linoleic acid come from outside the chloroplast.

2. Isolated chloroplasts and etioplasts, and particularly isolated chloroplast envelopes, have a high capacity for galactosylation of diglycerides, thus easily forming monogalactosyl-diglycerides and digalactosyldiglycerides. Douce has suggested that the chloroplast envelope may indeed be the real site of galactosylation of diglycerides in chloroplasts.

3. The capacity for phospholipid biosynthesis in isolated chloroplasts seems to be much more restricted. Only the synthesis of phosphatidic acid has been observed, up to now, in isolated organelles. Marshall and Kates [44] have suggested that the phosphatidylglycerol of chloroplasts (*i.e.*, the major phospholipid of these organelles) is synthesized in the microsomes.

Table 14 shows that the microsomes of plant cells contain all the enzymes necessary for the biosynthesis of either polyunsaturated fatty acids (at least

Table 14. Lipid synthesis by isolated plant microsomes

				References
Spinach leaves	[^{32}P] glycerophosphate		lyso PA, PA	[68]
	[1,3-^{14}C] glycerol		PG	[44]
	[U-^{14}C-serine] ^{14}C-ethanolamine, ^{14}C serine		PS, PE, PC	[45]
	[1,2-^{14}C] choline, CDP-choline [1,2-^{14}C]		PC	[13]
	^{14}C-CDP-ethanolamine		PE	[41]
Pea seedlings	[1,2-^{14}C] ethanolamine		PE	[76]
Safflower seeds	oleyl-CoA desaturase	18:2		[80]
Potato	oleyl-CoA desaturase	18:2		[6]
Castor bean endosperm	PA-phosphatase CDP-diglyceride: *myo*-inositol-phosphatidyltransferase PE: L-serine-phosphatidyltransferase ^{14}C-serine		PE, PS, PI	[55] [54]
Golgi vesicles Onion stem	CTP: choline phosphate cytodyltransferase		PC	[56]

linoleic acid) or the major phospholipids of biomembranes, phosphatidylcholine, phosphatidylethanolamine, etc. Slack and collaborators [70] have suggested recently that oleyl-phosphatidylcholine is the true substrate of the microsomal oleate desaturase from pea leaves.

Therefore, in order to correlate the synthetic capabilities of the various organelles with the presence of the same major phospholipids and fatty acids in all of them, one is forced to conclude that there must be some transfer of lipids from microsomes to other intracellular organelles. To account for this transfer of lipids, we have proposed a continuous intermembrane lipid exchange within plant cells [51]; since 1968, we have devoted a great deal of time in our laboratory to study this phenomenon in vivo and in vitro.

D. Lipid Exchanges Between Cell Organelles

First, one can use model systems to show the lipid exchange phenomenon in vitro. Such model systems may consist simply of mixtures of labeled liposomes and unlabeled biomembranes or organelles. In such systems, one can observe that radioactive lipids pass progressively from the liposomes toward the membranes. For instance, Figure 2a demonstrates the transfer of radioactive lipids

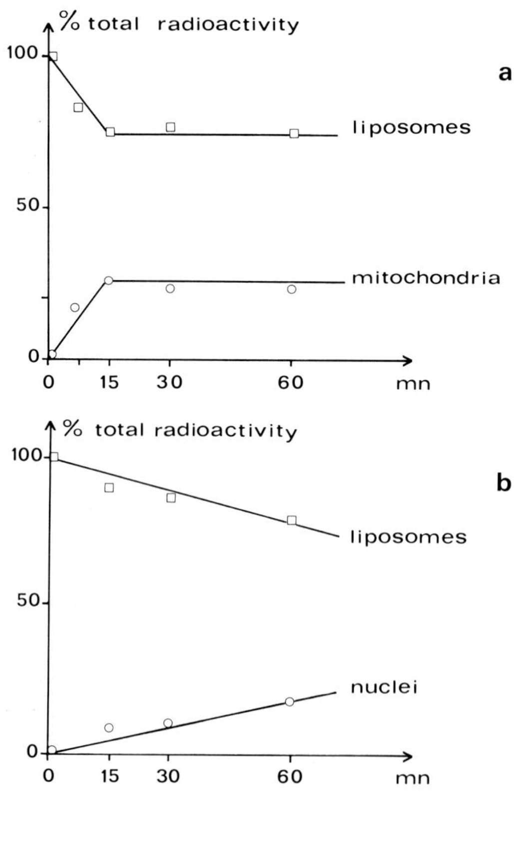

Fig. 2a and b. In vitro lipid transfers from ^{14}C-liposomes towards (a) cauliflower mitochondria and (b) sunflower nuclei

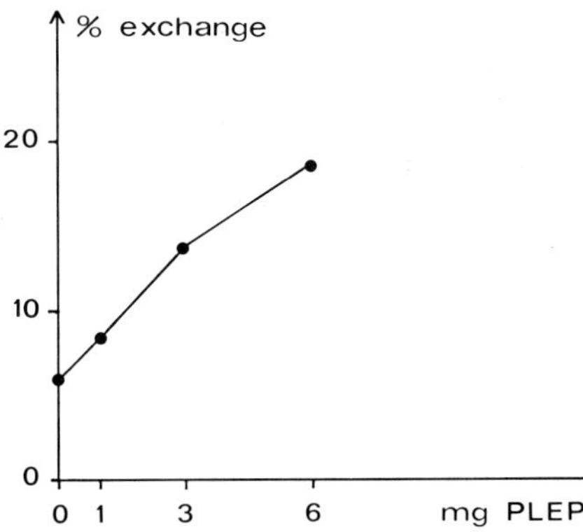

Fig. 3. Effects of increasing quantities of potato lipid exchange protein (*PLEP*) on lipid transfers from ^{14}C-liposomes toward potato mitochondria. ^{14}C-choline=containing liposomes were incubated at 30° C for 15 min with unlabeled mitochondria (1.2 mg protein) in 0.05 M sodium phosphate buffer (pH=7.5) containing various amounts of PLEP, as indicated in abscissa. The tubes were then centrifuged at 12,500*g* for 15 min. The pellets were washed. Lipids of supernatants and pellets were then extracted

from ^{14}C-liposomes toward cauliflower mitochondria [16] and Figure 2b demonstrates the transfer of radioactive lipids from ^{14}C-liposomes toward sunflower nuclei (Mazliak, unpublished observations).

The lipid transfers from liposomes to biomembranes are greatly stimulated in vitro by some proteins (the phospholipid exchange proteins, PLEP) present in the cytoplasmic supernatant of cells. The first PLEP was isolated from a rat liver supernatant by Wirtz et al. [82]. Kader [33] succeeded in our laboratory in isolating from the cytoplasmic supernatant of potato cells a plant phospholipid exchange protein that stimulated the lipid transfers between labeled liposomes and potato mitochondria (Fig. 3). The method of isolating this phospholipid exchange protein has been described by Kader [33]. A phospholipid exchange protein stimulating lipid transfers from labeled liposomes toward cauliflower mitochondria has been also isolated by Douady and Kader from cauliflower cytoplasmic supernatant (Douady and Kader, unpublished results).

Lipid transfers have also been demonstrated in vitro between two different kinds of biomembranes: e.g., mitochondria and microsomes [7]. Microsomes and mitochondria containing radioactively labeled lipids were isolated from potato tuber or cauliflower floret slices previously incubated with ^{14}C-acetate, ^{14}C-glycerol, ^{14}C-choline, or $Na_2H^{32}PO_4$. Labeled microsomes were then mixed with unlabeled mitochondria and vice versa. The resulting mixtures were suspended in a small volume of cytoplasmic supernatant, incubated for varying lengths of time, and then separated by density gradient centrifugation into microsomal and mitochondrial fractions, which were then analyzed. This experiment yielded the following results:

1. Radioactive lipids leave the initially labeled (donor) organelle and are transferred to the initially unlabeled (acceptor) organelle. Transfers of up to 40% have been recorded (Fig. 4).

2. The exchange process operates equally well in either direction, microsomes to mitochondria or vice versa. An examination of the specific radioactivity (Fig. 5) of the phospholipid in the "donor" and "acceptor" fractions shows that it decreases with time in the former fraction and increases with time in the latter. This indicates not only that labeled phospholipid is continuously transferred from the donor organelle to the acceptor organelle, but also that

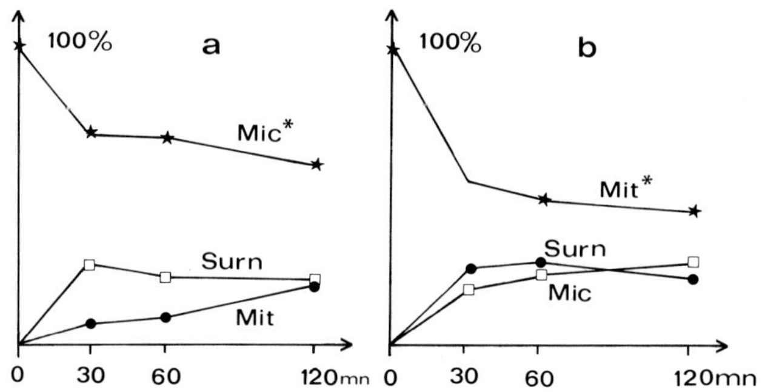

Fig. 4a and b. Lipid exchange between microsomes (*Mic*) and mitochondria (*Mit*) in vitro. (a) Incubation of microsomes containing radioactively labeled lipids with unlabeled mitochondria, followed by reisolation of each organelle and determination of content or radioactivity. Latter is expressed as % of total radioactivity present. (b) Incubation of mitochondria containing radioactively labeled lipids with unlabeled microsomes, handled as in (a). *Surn*: cytoplasmic supernatant. *: initial location of radioactive lipid

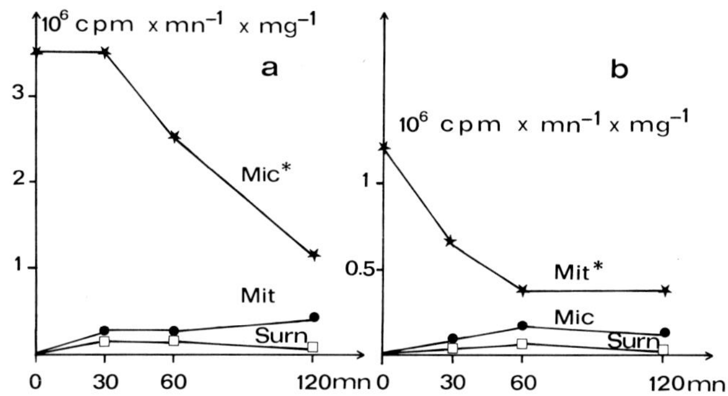

Fig. 5a and b. Lipid exchange between microsomes and mitochondria in vitro. (a) Incubation of microsomes containing radioactively labeled lipids with unlabeled mitochondria, followed by reisolation of each organelle and determination of the specific radioactivity of its lipid. (b) Incubation of mitochondria containing radioactively labeled lipids with unlabeled microsomes, then handled as in (a). *Mic*: microsomes; *Mit*: mitochondria; *Surn*: cytoplasmic supernatant; *: initial location of radioactive lipid

the donor organelle is continuously receiving unlabeled phospholipid from the acceptor organelle.

3. Calculation of the quantities of lipid exchanged between the organelles shows (i) that it is twice as high when starting with lipids labeled with [^{14}C]-acetate as with lipids labeled with [^{14}C] glycerol or [^{32}P] phosphate, indicating that fatty acids are exchanged more easily than entire phospholipid molecules, and (ii) that phosphatidylcholine and phosphatidylethanolamine are exchanged to a greater extent, regardless of the radioactive precursor used, than any of the other phospholipids, indicating that the more abundant a phospholipid is in the membrane the more actively it will be exchanged. This suggests that the exchange process is a physical rather than an enzymic phenomenon. A

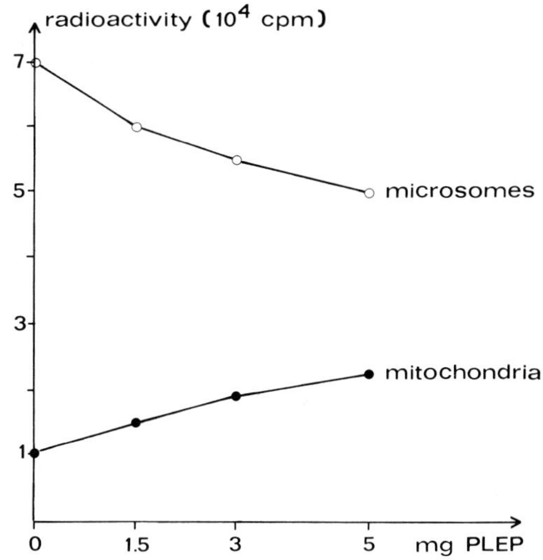

Fig. 6. Effects of increasing quantities of potato lipid exchange protein (PLEP) on lipid transfers from radioactively labeled potato microsomes toward unlabeled potato mitochondria. Unlabeled mitochondria (1.2 mg protein) and ^{32}P-labeled microsomes (0.2 mg protein) were incubated at 30° C for 15 min in 0.05 M sodium phosphate buffer solutions (pH 7.5) containing various amounts of PLEP, as indicated in abscissa. The tubes were then centrifuged at 12,500 *g* for 15 min. The pellets were washed twice. Lipids of supernatants and pellets were then extracted [33]

study of the exchange of labeled fatty acids between microsomes and mitochondria [48] has led to the same conclusion.

Kader [33] then showed that the phospholipid exchange protein isolated from potato tuber markedly stimulated the in vitro lipid transfers between microsomes and mitochondria. Adding increasing quantities of PLEP to the mixture of labeled microsomes and unlabeled mitochondria resulted in increasing percentages of radioactive lipid transfer (Fig. 6). The intensities of lipid exchanges stimulated by PLEP could be 80% greater than those observed in the controls.

Finally, these lipid transfers between different biomembranes have been observed in vivo using pulse-chase experiments done with aged potato slices [5]. It was found that an initial 2-h pulse of [1-^{14}C] acetate labeled the phospholipids of the microsomal fraction more rapidly and more intensively than those of the mitochondria. During the 40-h chase in a medium containing nonradioactive acetate, it was found that the total and specific radioactivity of the microsomal phospholipid decreases continuously while that of the mitochondrial phospholipid is still increasing (Fig. 7a). A similar result was also obtained in the same tissue with ^{14}C-choline (Fig. 7b) and leads to the conclusion that the newly synthesized, radioactive lipids are transferred from the microsomes to the mitochondria in the chase period. We have also recently shown (Mazliak, unpublished results) that in etiolated sunflower germinations, the newly synthesized, radioactive lipids (labeled from ^{14}C-acetate) accumulate first in the microsomes, secondly in the mitochondria, and much more slowly in the nuclei, which also suggests a lipid transfer from microsomes toward other cell organelles such as mitochondria and nuclei (Fig. 8).

All of these results indicate the role played by intracellular exchange processes in the distribution of lipids from the sites of biosynthesis to all cellular mem-

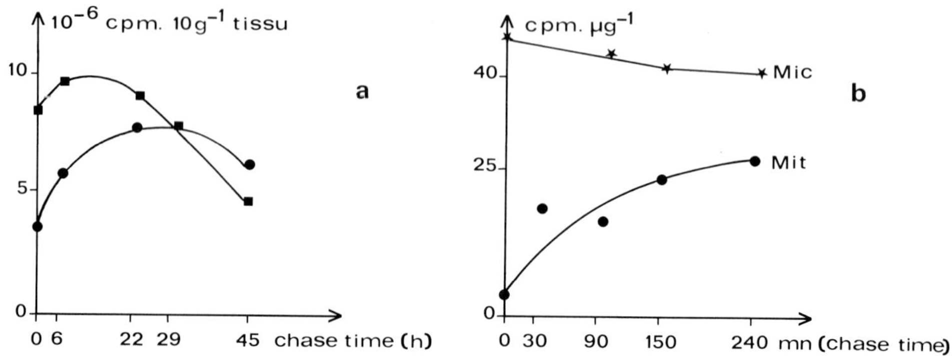

Fig. 7a and b. Lipid exchange between microsomes and mitochondria in vivo. (a) Potato tuber slices were incubated for 2 h in solution of [1-^{14}C] acetate. They were then rinsed and transferred to solution containing unlabeled acetate. Samples were removed after various incubation times in the unlabeled acetate, and total radioactivity/10 g of tissue in microsomal lipid (■) and mitochondrial lipid (●) was determined. (b) The same experiment as (a) was carried out using a 30=min pulse of [^{14}C] choline instead of a pulse of [1-^{14}C] acetate. Specific radioactivities (cpm/µg) of microsomal lipid (*) and mitochondrial lipid (●) were determined after chasing with unlabeled choline for various times

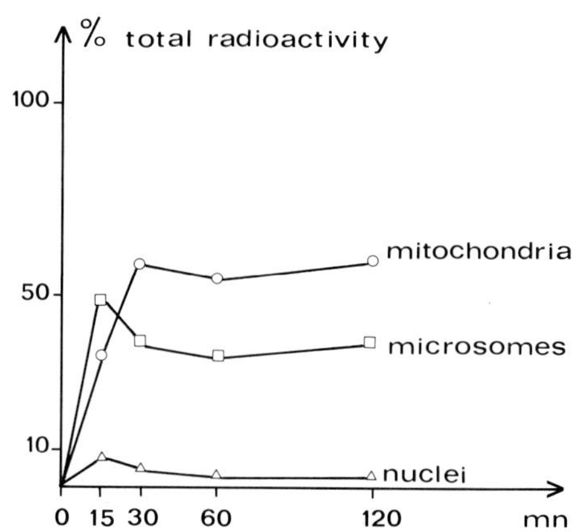

Fig. 8. Lipid exchange in vivo between microsomes, mitochondria, and nuclei. Etiolated sunflower fragments were incubated for various times (indicated in abscissa) in a solution of ^{14}C-acetate. They were then rinsed and nuclei, mitochondria, and microsomes were prepared from the fragments. Total radioactivity of lipids was determined in each subcellular fractions

branes and organelles. However, from similar in vivo pulse-chase experiments in castor bean endosperm, Lord [38] has recently proposed a slightly different interpretation. The data obtained from castor bean endosperm are consistent with the concept that the endoplasmic reticulum is a major site of membrane proliferation where structural lipids, and possibly proteins, are inserted into, and thus expand, a pre-existing membrane fraction. Other organelles and cellular membranes could therefore originate from the proliferating endoplasmic reticulum by a process of membrane flow and differentiation, which could be an alternative, in growing tissues, to the process of phospholipid exchange between cellular membranes.

E. Role of the Lipid Environment in the Activities of Membrane-Bound Enzymes

Many recent studies on the behavior of electron spin resonance probes in artificial and natural membranes indicate that the lipids of biomembranes are in a dynamic state. The hydrocarbon chains of phospholipids appear to flex at rapid rates, and individual molecules can shift laterally within the plane of the membrane at fast rates [12]. Most of the phospholipids bound to membranes appear to be arranged in the form of a bilayer, and studies with electron spin resonance probes have shown that phospholipid molecules in one half of the bilayer probably do not shift over to the other side (the so-called flip-flop movement) or do so at very low rates [36]. This result is consistent with the molecular asymmetry found in the distribution of phospholipids, e.g., in red cell membranes [79].

The tridimensional structure of integral proteins included within the membranes may rely, at least partly, on their lipid environment. It has been shown by Warren et al. [81], for instance, that a sarcoplasmic ATPase from muscle must keep around itself a lipid annulus of 30 phospholipid molecules to remain active in calcium transport; if this lipid annulus was lost, or if phospholipid molecules were substituted by cholesterol molecules, a conformational change occurred in the membrane protein and it became inactive. Similarly it has been shown by Jost et al. [31] that the mitochondrial cytochrome oxidase interacted strongly with a shell of lipid molecules surrounding the protein in a relatively rigid state. According to these results, membrane lipids can be divided into two classes: (1) boundary lipids existing in shells immediately surrounding membrane proteins, and (2) mobile bilayer lipids, diffusing laterally in the plane of the membrane. Short-range interactions, involving lipid molecules that are in contact with the intercalated proteins, may be specific and involve a close association of given lipid types with particular proteins. Direct experimental evidences for these lipid-protein associations in plant membranes are beginning to appear in the literature.

Lyons and Raison [39] pointed out that the inhibition of mitochondrial respiration below 10°C in chill-sensitive plants such as tomato, cucumber, and sweet potato appears to be the result of a transition of membrane lipids from a liquid to a gel-like state. Lyons et al. [39, 40] observed that mitochondrial membrane flexibility and the degree of fatty acid unsaturation in membranes were greater in chill-insensitive than in sensitive plants. It can be concluded that the chilling sensitivity of plants depends on the quality of the lipids synthesized by plant cells, because these lipids are partly integrated in biomembranes, where they surround membrane proteins and subsequently modify the activities of membrane enzymes according to their physical state.

Evidence was also presented by Jolliot and Mazliak [30] that lipids played a role in the electron-transfer chain of mitochondria from cauliflower inflorescences. Extraction of mitochondrial lipids with aqueous acetone caused an important decrease of NADH-cytochrome c reductase, NADH-dehydrogenase, succinate-cytochrome reductase, and succinate dehydrogenase activities. The

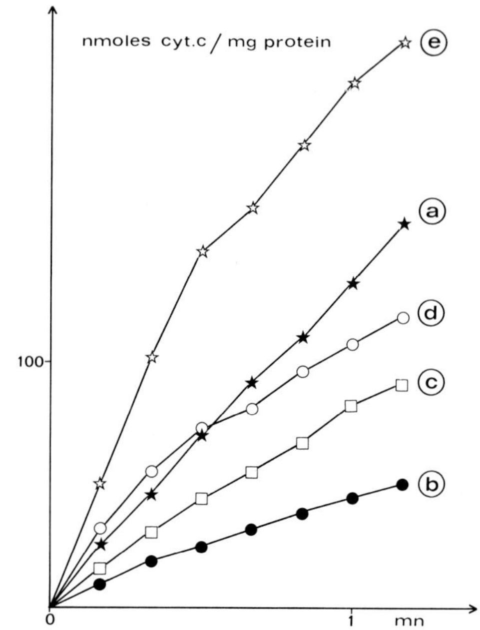

Fig. 9a–e. Effects of increasing quantities of mitochondrial lipids upon mitochondrial NADH-cytochrome c reductase activity. (a) cauliflower mitochondria (28 μg or protein) incubated 10 min in Tris buffer (pH 7.2); (b) mitochondria extracted for 10 min by an aqueous 90% acetone solution; (c–e), lipid-extracted mitochondria incubated with (c), 150 μg, (d), 300 μg and (e), 600 μg of mitochondrial phosphilipids

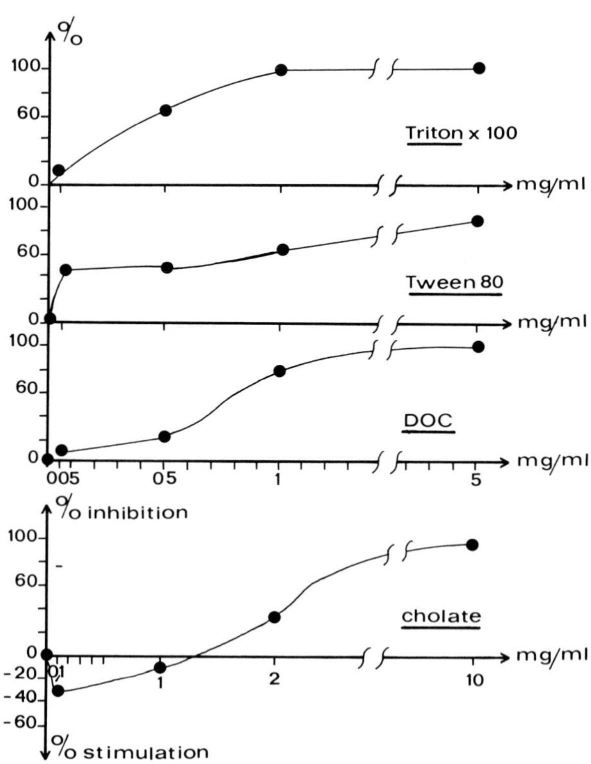

Fig. 10. Effects of four detergents upon the oleyl-CoA desaturase activity of aged potato microsomes

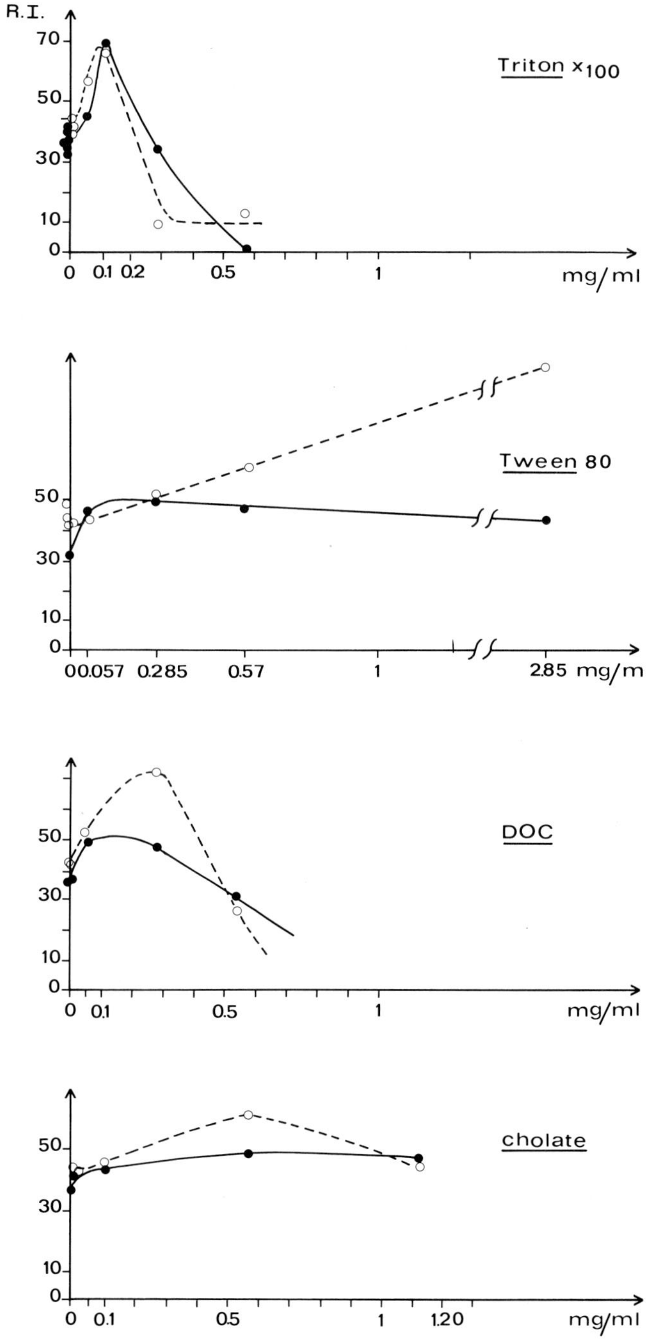

Fig. 11. Effects of four detergents upon respiration of potato mitochondria, followed use of an oxygen electrode. Abscissas: detergent concentrations (mg/ml). Ordinates: respiratory intensity (R.I.) in mμmol O_2/min/mg protein. ●—● Series of experiments No. 1. ○—○ Series of experiments No. 2

activities of the first two enzyme complexes could be restored by adding back extracted lipids as a suspension of liposomes (Fig. 9), which implied lipid transfers from liposomes to the lipid-depleted mitochondrial membranes.

Another approach recently used in our laboratory to demonstrate the role of the lipid environment in the activity of membrane enzymes consisted of studying the effects of some detergents on enzyme activities.

Figure 10 shows that the oleyl-CoA desaturase from aged potato microsomes [47] is very sensitive to the presence of detergents in the reaction medium. Nonionic detergents (Triton X-100 and Tween 80) progressively inhibit the oleyl-CoA desaturation at increasing doses. These detergents progressively extract the lipids from the natural environment of the oleyl-CoA desaturase in the membrane, and thus it appears that the lipids of the microsomes are necessary for oleyl-CoA desaturase activity. Ionic detergents (sodium deoxycholate and above all cholate) have curious effects on oleyl-CoA desaturase activity at low concentrations (less than 0.1 mg/ml): deoxycholate is barely inhibitory, and cholate even stimulates (by more than 30%) the desaturation activity at these low concentrations. The same phenomenon (stimulation by low concentrations of detergents) can be found when studying the effects of the same detergents on the respiration of plant mitochondria [49]: it can be seen from Figure 11 that low concentrations of any of the four assayed detergents (ionic and nonionic) do stimulate mitochondrial respiration. It can be concluded from these results that the low detergent concentrations disturb the lipid environment of membrane enzymes only slightly, thus conferring upon the proteins a greater mobility in the membrane, which would reinforce the enzyme activities.

F. Conclusions

The present picture of biomembrane organization allows the differentiation of membrane lipids into several pools, differing according to their respective mobilities (Fig. 12):

1. In the membrane bilayer, some lipids immediately surrounding the membrane proteins form various lipid annuli, probably with fixed compositions; the respective compositions must remain unchanged for optimal enzyme activities.

2. In each layer, the other lipids are continuously diffusing laterally, but the so-called flip-flop movements from the external layer toward the internal (or vice versa) seem to be impossible (at least frequently).

3. Finally, some renewal of membrane lipids may occur via transfers from cytoplasmic phospholipid exchange proteins to membrane bilayers and vice versa. Supplementary restrictions of the mobilities of lipid molecules are added by various physical states of the lipids in the membranes, according to external temperature.

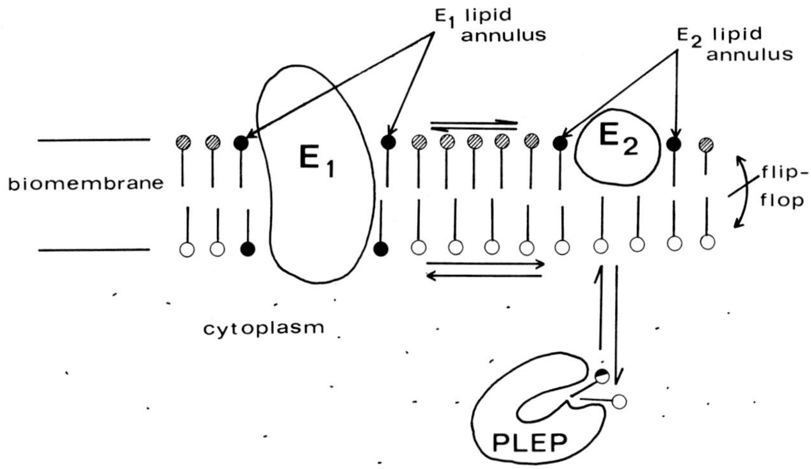

Fig. 12. Schematic distinction between various lipid pools differing in their respective mobilites in biomembranes

References

1. Allen, C.F., Hirayama, O., Good, P.: Lipid composition of photosynthetic systems. In: Biochemistry of Chloroplasts. Goodwin, T.W. (ed.). London: Academic Press, 1966, Vol. 1, pp. 195–200
2. Appleqvist, L.A., Stumpf, P.K., Von Wettstein, D.: Lipid synthesis and ultrastructure of isolated barley chloroplasts. Plant Physiol. **43**, 163–187 (1968)
3. Bahl, J., Guillot-Salomon, T., Douce, R.: Synthèse enzymatique du cytidine diphosphate dyglycéride dans le végétal supérieur. Physiol. Vég. **8**, 55–74 (1970)
4. Bartholomew, L., Mace, K.D.: Isolation and identification of phospholipids from root tip cell plasmalemma of *Phaseolus limensis*. Cytobios **5**, 241–247 (1972)
5. Ben Abdelkader, A.: Biogenèse des lipides membranaires pendant la "survie" de tranches de tubercules de Pomme de terre. Thèse Fac. Sciences, Paris, 1972, 184 pp.
6. Ben Abdelkader, A., Cherif, A., Demandre, C., Mazliak, P.: The oleyl-CoA desaturase of potato tubers: enzymatic properties, intracellular localization and induction during aging of tuber slices. Europ. J. Biochem. **32**, 155–165 (1973)
7. Ben Abdelkader, A., Mazliak, P.: Echange de lipides entre mitochondries, microsomes et surnageant cytoplasmique de cellules de Pomme de terre ou de Chou-fleur. Europ. J. Biochem. **15**, 250–262 (1970)
8. Bradbeer, C., Stumpf, P.K.: Fat metabolism in higher plants XIII. Phosphatidic acid synthesis and diglyceride phosphokinase activity in mitochondria from peanut cotyledons. J. Lipid Res. **1**, 214–220 (1960)
9. Brooks, J.L., Stumpf, P.K.: Properties of a soluble fatty acid synthesizing system from lettuce chloroplasts. Arch. Biochem. Biophys. **116**, 108–116 (1966)
10. Demandre, C.: Activités enzymatiques et composition lipidique de fractions enrichies en membranes cytoplasmiques isolées de tubercules de Pomme de terre. Physiol. Vég. **13**, 103–114 (1975)
11. Demandre, C.: Les membranes cytoplasmiques de tubercules de Pomme de terre: isolement, composition lipidique, biogenèse des lipides associés à ces membranes au cours de la survie de tranches de tubercule. Thèse Doctorat 3è Cycle, Paris, 1976, 75 pp.
12. Devaux, P., Mc Connel, H.M.: Equality of the rates of lateral diffusion of phosphatidylethanolamine and phosphatidylcholine spin labels in rabbit sarcoplasmic reticulum. Ann. N.Y. Acad. Sci. **222**, 489–498 (1973)

13. Devor, K.A., Mudd, J.B.: Biosynthesis of phosphatidylcholine by enzyme preparations of spinach leaves. J. Lipid Res. **12**, 390–419 (1971)
14. Donaire, J.P., Oursel, A.: Influence du calcium sur la biosynthèse de la phosphatidylcholine par les microsomes isolés de racines de Féverole. C. R. Acad. Sci. **281**, 787–789 (1975)
15. Donaldson, R.P., Tolbert, N.E., Schnarrenberger, C.: A comparison of microbody membranes with microsomes and mitochondria from plant and animal tissue. Arch. Biochem. Biophys. **152**, 199–215 (1972)
16. Douady, D., Mazliak, P.: Etude des transferts d'acides gras entre liposomes et mitochondries isolées de Chou-fleur. Physiol. Vég. **13**, 383–392 (1975)
17. Douce, R.: Identification et dosage de quelques glycerophosphatides dans des souches normales et tumorales de Scorsonère cultivées *in vitro*. C. R. Acad. Sci. **259**, 2167–2170 (1964)
18. Douce, R.: Mise en évidence du cytidine diphosphate diglycéride dans les mitochondries végétales isolées. C. R. Acad. Sci. **267**, 534–537 (1968)
19. Douce, R.: Site of biosynthesis of galactolipids in spinach chloroplasts. Science **183**, 852–853 (1974)
20. Douce, R., Guillot-Salomon, T.: Sur l'incorporation de la radioactivité du sn-glycerol-3-phosphate-^{14}C dans le monogalactosyldiglyceride des plastes isolés. FEBS Lett. **11**, 121–126 (1970)
21. Douce, R., Holz, R.B., Benson, A.A.: Isolation and properties of the envelope of spinach chloroplasts. J. Biol. Chem. **248**, 7215–7222 (1973)
22. Hartmann, M.A., Normand, G., Benveniste, P.: Sterol composition of plasma membrane enriched fractions from maize coleoptiles. Plant Sci. Lett. **5**, 287–292 (1975)
23. Harwood, J.L.: Fatty acid biosynthesis by avocado pear. Lipids **9**, 850–854 (1974)
24. Hashimoto, H., Murakami, S.: Dual character of lipid composition of the envelope membrane of spinach chloroplasts. Plant Cell Physiol. (Tokyo) **16**, 895–902 (1975)
25. Hawke, J.C., Leese, B.M., Leech, R.M.: Lipid biosynthesis by intact mesophyll and bundle sheat chloroplasts from maize. Phytochemistry **14**, 1733–1737 (1975)
26. Hawke, J.C., Stumpf, P.K.: The biosynthesis of saturated and unsaturated fatty acids by preparations from barley seedlings. J. Biol. Chem. **240**, 4746–4752 (1965)
27. Hodges, T.K., Leonard, R.T., Bracker, C.E., Keenan, T.W.: Purification of an ion-stimulated adenosine triphosphatase from plant roots: association with plasma membranes. Proc. Natl. Acad. Sci. U.S. **69**, 3307–3311 (1972)
28. Holtz, R.B., Stewart, P.S., Patton, S., Schisler, L.C.: Isolation and characterization of membranes from the cultivated mushroom. Plant Physiol. **50**, 541–546 (1972)
29. Jacobson, B.S., Jaworski, J.G., Stumpf, P.K.: Stearyl-acyl carrier protein desaturase from spinach chloroplasts. Plant Physiol. **54**, 484–486 (1974)
30. Jolliot, A., Mazliak, P.: Rôle des lipides dans diverses activités enzymatiques de la chaîne de transport des électrons d'une fraction mitochondriale isolée d'inflorescences de Chou-fleur. Plant Sci. Lett. **1**, 21–29 (1973)
31. Jost, P.C., Griffith, O.H., Capaldi, R.A., Vanderkooi, G.: Evidence for boundary lipid in membrane. Proc. Natl. Acad. Sci. **70**, 480–484 (1973)
32. Joyard, J., Douce, R.: Mise en évidence et rôle des diacylglycerols de l'enveloppe des chloroplastes. Biochim. Biophys. Acta **424**, 125–131 (1976)
33. Kader, J.C.: Proteins and intracellular exchange of lipids. I. Stimulation of phospholipid exchange between mitochondria and microsomal fractions by proteins isolated from potato tuber. Biochim. Biophys. Acta **380**, 31–44 (1975)
34. Kannangara, C.G., Stumpf, P.K.: The biosynthesis of polyunsaturated fatty acids by isolated chloroplasts. Arch. Biochem. Biophys. **148**, 414–424 (1972)
35. Keenan, T.W., Leonard, R.T., Hodges, T.K.: Lipid composition of plasma membranes from Avena sativa roots. Cytobios **7**, 103–112 (1973)
36. Kornberg, R.D., Mc Connel, H.M.: Inside-outside transitions of phospholipids in vesicle membranes. Biochemistry **10**, 1111–1120 (1971)
37. Lamant, A., Heller, R.: Intervention des systèmes membranaires dans l'absorption du Calcium par les racines de Féverole (calcicole) et de Lupin (Calcifuge). Physiol. Vég. **13**, 685–700 (1975)
38. Lord, J.M.: Phospholipid synthesis and exchange in castor bean endosperm homogenates. Plant Physiol. **57**, 218–223 (1976)
39. Lyons, J.M., Raison, J.K.: Oxidative activity of mitochondria isolated from plant tissues sensitive and resistant to chilling injury. Plant Physiol. **45**, 386–389 (1970)

40. Lyons, J.M., Wheaton, T.A., Pratt, H.K.: Relationship between physical nature of mitochondrial membranes and chilling sensitivity in plants. Plant Physiol. **39**, 262–268 (1964)
41. Macher, B.A., Mudd, J.B.: Synthesis of phosphatidylethanolamine by enzyme preparations from plant tissues. Plant Physiol. **53**, 171–175 (1974)
42. Mackender, L.R.O., Leech, R.M.: The galactolipid, phospholipid and fatty acid composition of the chloroplast envelope membranes of *Vicia faba*. Plant Physiol. **53**, 496–502 (1974)
43. Mannella, C.A., Bonner, W.D.: Biochemical characteristics of the outer membranes of plant mitochondria. Biochim. Biophys. Acta **413**, 213–225 (1975)
44. Marshall, M.O., Kates, M.: Biosynthesis of phosphatidylglycerol by cell-free preparations from spinach leaves. Biochim. Biophys. Acta **260**, 558–570 (1972)
45. Marshall, M.O., Kates, M.: Biosynthesis of nitrogenous phospholipids in spinach leaves. Can. J. Biochem. **52**, 469–482 (1974)
46. McCarty, R.E., Douce, R., Benson, A.A.: The acyl lipids of highly purified plant mitochondria. Biochim. Biophys. Acta **316**, 266–270 (1973)
47. Mazliak, P.: Effet de quatre détergents sur l'oleyl-CoA désaturase des microsomes de Pomme de terre. C. R. Acad. Sci. **281**, 1471–1474 (1975)
48. Mazliak, P., Ben Abdelkader, A.: Exchanges d'acides gras *in vitro* entre mitochondries, microsomes et surnageants cytoplasmiques de cellules de Pomme de terre ou de Chou-fleur. Phytochemistry **10**, 2879–2890 (1971)
49. Mazliak, P., Decotte, A.M.: Effects of four detergents on the oxidative and phosphorylating capacities of potato mitochondria. J. Exp. Botany **27**, 769–777 (1976)
50. Mazliak, P., Oursel, A., Ben Abdelkader, A., Grosbois, M.: Biosynthèse des acides gras dans les mitochondries végétales isolées. Europ. J. Biochem. **28**, 399–411 (1972)
51. Mazliak, P., Stoll, U., Ben Abdelkader, A.: Coopération entre les mitochondries et le reste de la cellule pour le renouvellement des lipides mitochondriaux du parenchyme de Pomme. Biochim. Biophys. Acta **152**, 414–417 (1968)
52. Meunier, D., Mazliak, P.: Différences de composition lipidique entre les deux membranes des mitochondries de Pomme de terre. C. R. Acad. Sci. **275**, 213–216 (1972)
53. Moreau, F., Dupont, J., Lance, C.: Phospholipid and fatty acid composition of outer and inner membranes of plant mitochondria. Biochim. Biophys. Acta **345**, 294–304 (1974)
54. Moore, T.S.: Phosphatidylserine synthesis in castor bean endosperm. Plant Physiol. **56**, 177–180 (1975)
55. Moore, T.S., Lord, J.M., Kagawa, T., Beevers, H.: Enzymes of phospholipid metabolism in the endoplasmic reticulum of castor bean endosperm. Plant Physiol. **52**, 50–53 (1973)
56. Morre, O.J., Nyquist, S., Rivera, E.: Lecithin biosynthetic enzymes of onion stem and the distribution of phorylcholine-cytidyl transferase among cell fractions. Plant Physiol. **45**, 800–804 (1970)
57. Mudd, J.B., Stumpf, P.K.: Factors affecting the synthesis of oleic acid by particulate preparations from avocado mesocarp. J. Biol. Chem. **236**, 2602–2609 (1961)
58. Nakamura, Y., Yamada, M.: Fatty acid synthesis by spinach chloroplasts. I. Property of fatty acid synthesis from acetate. Plant Cell Physiol. (Tokyo) **16**, 650–661 (1975)
59. Neufeld, E.F., Hall, C.W.: Formation of galactolipids by chloroplasts. Biochem. Biophys. Res. Commun. **14**, 503–508 (1964)
60. Nichols, B.W., James, A.T.: Acyl lipids and fatty acids in photosynthetic tissues. In: Progress in Phytochemistry. Reinhold, L., Liwschitz, Y. (eds.). London: Interscience Publishers, 1968, Vol. 1, pp. 1–49
61. Normand, G., Hartmann, M.A., Schuber, F., Benveniste, P.: Caractérisation de membranes de coléoptiles de Maïs fixant l'auxine et l'acide N-naphtylphtalamique. Physiol. Vég. **13**, 743–761 (1975)
62. Ohmori, M., Yamada, M.: Composition of phosphoglycerides and glycosylglycerides in castor bean mitochondria. Plant Cell Physiol. (Tokyo) **15**, 1129–1132 (1974)
63. Ongun, A., Mudd, J.B.: Biosynthesis of galactolipids in plants. J. Biol. Chem. **243**, 1558–1566 (1968)
64. Oursel, A., Lamant, A., Salsac, L., Mazliak, P.: Etude comparée des lipides et de la fixation passive du calcium dans les racines et les fractions subcellulaires du *Lupinus luteus* et de la *Vicia faba*. Phytochemistry **12**, 1865–1874 (1973)
65. Philipp, E.I., Franke, W.W., Keenan, T.W., Stadler, J., Jarasch, E.E.: Characterization of

nuclear membranes and endoplasmic reticulum isolated from plant tissues. J. Cell Biol. **68**, 11–29 (1976)

66. Ruesink, A.W.: The plasma membrane of *Avena* coleoptile protoplasts. Plant Physiol. **47**, 192–195 (1971)
67. Rullkötter, J., Heinz, E., Tulloch, A.P.: Combination and positional distribution of fatty acids in plant digalactosyldiglycerides. Z. Pflanzenphysiologie **76**, 163–175 (1975)
68. Sastry, P.S., Kates, M.: Biosynthesis of lipids in plants II. Incorporation of glycerophosphate-^{32}P into phosphatides by cell-free preparations from spinach leaves. Can. J. Biochem. **44**, 459–467 (1966)
69. Schwertener, H.A., Biale, V.B.: Lipid composition of plant mitochondria and chloroplasts. J. Lipid Res. **14**, 235–241 (1973)
70. Slack, C.R., Roughan, P.G., Terpstra, J.: Some properties of a microsomal oleate desaturase from leaves. Biochem. J. **155**, 71–80 (1976)
71. Stumpf, P.K.: Biosynthesis of fatty acids in spinach chloroplast. In: Recent Advances in the Chemistry and Biochemistry of Plant Lipids. Galliard, T., Mercer, E.T. (eds.). New York: Academic Press, 1975, pp. 95–113
72. Sumida, S., Mudd, J.B.: Biosynthesis of cytidine diphosphate diglyceride by cauliflower mitochondria. Plant Physiol. **43**, 1162–1164 (1968)
73. Tchang, F.: unpublished results, 1974
74. Tremolieres, A., Dubacq, J.P.: Formation of a Δ^9-dodecenoic dibasic acid from linoleic acid by young pea leaves. Phytochemistry **15**, 1123–1124 (1976)
75. Tremolieres, A., Mazliak, P.: Biosynthetic pathway of α-linolenic acid in developing pea leaves. Plant Sci. Lett. **2**, 193–201 (1974)
76. Vandor, S.L., Richardson, K.E.: Incorporation of ethanolamine 1,2^{14}C into plant microsomal phospholipids. Can. J. Biochem. **46**, 1309–1315 (1968)
77. Van Deenen, L.L.M., Haverkate, F.: Chemical characterization of phosphatidylglycerol from photosynthetic tissues. In: Biochemistry of Chloroplasts. Goodwin, T.W. (ed.). London: Academic Press, 1966, Vol. 1, pp. 117–131
78. Van Hummel, H.C., Hulsebos, J.J.M., Wintermand, J.F.G.M.: Biosynthesis of galactosyldiglycerides by non-green fractions from chloroplasts. Biochim. Biophys. Acta **380**, 219–226 (1975)
79. Verkleij, A.J., Zwaal, R.F.A., Roelofsen, B., Comfurius, P., Kastelijn, D., van Deenen, L.L.M.: The asymmetric distribution of phospholipids in the human red cell membrane. Biochim. Biophys. Acta **323**, 178–193 (1973)
80. Vijay, I.K., Stumpf, P.K.: Fat metabolism in higher plants XLXI. Properties of oleyl-CoA desaturase of *Carthamus tinctorius*. J. Biol. Chem. **247**, 360–372 (1971)
81. Warren, G.B., Houslay, M.D., Metcalfe, J.C., Birdsall, N.J.M.: Cholesterol is excluded from the phospholipid annulus surrounding an active calcium transport protein. Nature (London) **255**, 684–687 (1975)
82. Wirtz, K.W.A., Kamp, H.H., van Deenen, L.L.M.: Isolation of a protein from beef liver which specifically stimulates the exchange of phosphatidylcholine. Biochim. Biophysica Acta **274**, 606–617 (1972)
83. Yamada, M., Usami, Q.: Long chain fatty acids in developing castor bean seeds IV. The synthetic system in proplastids. Plant Cell Physiol. (Tokyo) **16**, 879–884 (1975)
84. Zilkey, B., Canvin, D.T.: Subcellular localization of oleic acid biosynthesis enzymes in the developing castor bean endosperm. Biochem. Biophys. Res. Commun. **34**, 646–653 (1969)

CHAPTER 4

Lipid Biosynthesis in Developing Seeds

P.K. STUMPF

A. Introduction

Of major importance in the food economy of the world is the ever-increasing production of oil-rich seeds. With a world production in 1975–76 of over 130,000,000 tons of oilseeds, from which over 27,000,000 tons of oil were extracted, a fuller understanding of the biochemical mechanisms involved in the formation of oils in these seeds has considerable relevance.

The major sources of plant oils are soybean, cotton seed, groundnut, sunflower, and rapeseed, with other sources such as sesame seed, copra, palm kernel, and castor bean being unique but minor sources characteristic of special geographic regions of the world.

It is only in recent years that the biochemist has begun investigations into the biosynthetic reactions involved in the formation of these oils. With the advent of extremely sensitive analytical techniques coupled with the rapid advances made in the lipid biochemistry of bacteria, yeast, animal, and plant systems during the past decade, the plant lipid biochemist today is in a position to define with considerable precision the many reactions involved in the developing seed in the conversion of a number of precursors to the final product, namely, the triacylglycerols.

The results derived from these studies may have a direct effect on increasing the conversion efficiency of precursors to products and, equally important, greatly increase our basic knowledge of the biochemistry of plant plasma membranes, of mitochondrial and chloroplastic membranes, and of the role of lipid polymers associated with outer surfaces of plant tissues, to cite a few examples.

Of considerable interest to lipid biochemists is the dynamics of the process by which a developing oil-rich seed channels the flow of carbon into the formation of large amounts of triacylglycerols whereas seeds low in oil accumulate as their major storage component polysaccharides such as starch. Studies have clearly shown that the machinery for lipid biosynthesis exists in both types of tissues; however, it is not known at present what control systems have been introduced into these developing tissues, making possible in one type the accumulation of large amounts of triacylglycerols with a characteristic fatty acid composition, and in the other, the synthesis of only small amounts of specific fatty acids that are inserted into membrane lipids, resulting in the formation of a minimal amount of triacylglycerols [8].

To understand these and other problems more fully demands a thorough exploration into the basic biosynthetic mechanisms in both types of tissues. This chapter will review briefly recent investigations into several oil-rich seeds pertaining in part to these problems. Because of limited space, the biosynthesis of polyunsaturated fatty acids in developing seeds will not be discussed.

B. General Considerations

A number of experimental approaches are feasible in defining the biochemical events that occur during the development of a seed:

1. Examination of oil-rich seeds for developmental changes that occur in the fatty acid composition of these seeds. A useful recent review on this subject has been written by Appelqvist [1]. Specifically, studies with *Crambe* by Gurr et al. [6, 7]; and Appleby et al. [2]; with safflower by Sims et al. [23] and McMahon and Stumpf [15]; with castor bean by Canvin [4]; with rapeseed by Appelqvist [1] and by Norton and Harris [17]; and with soybean by Privett et al. [18], by Reubel et al. [21] and by Porra and Stumpf [19] provide much information concerning the patterns of fatty acids that emerge as a seed proceeds through several phases of development.

While this approach has considerable value and can be readily conducted with present-day gas chromatographic techniques, the data reveal only the accumulation of fatty acids in complex lipids but give little insight into the dynamics of precursor-product relationships. For example, while it is well known that stearic acid is only a minor fatty acid in most seed tissues [1], in actual fact this acid, as a stearyl acyl carrier protein (ACP) derivative, plays a key role in the formation of the unsaturated fatty acids in the C_{18} series [13].

2. Morphologic examination, either by light microscopy or by electron microscopy, of developing seed tissue to correlate structural changes with the biogenesis of oil droplets [7, 10]. While this approach is useful in that the biochemist can examine directly the changes that are occurring in seed cells, correlation of structure and function must be carried out by the employment of modern cell fractionation techniques [3]. The results of these studies are sometimes difficult to interprete because of the possibility of introducing artifacts from absorption and/or flotation phenomena as a result of the high lipid content of the tissue.

3. The changing capacity of seeds to incorporate ^{14}C-substrates into lipids. The tissue employed may be intact, sliced, or chopped, and the substrate may be introduced by injection, spraying, or imbibition. The results obtained by these techniques, while useful, may be open to criticism. However, the flow of ^{14}C-substrate into ^{14}C-fatty acids under well-defined conditions provides valuable information to the experimentalist.

For example, this approach proved very useful in determining at what period in the development of the castor bean seed cell-free extracts should be made to study the hydroxylating system involved in the biosynthesis of ricinoleic acid from oleic acid [5].

4. The direct analysis of changing patterns of enzymes directly or indirectly related to the biosynthesis of lipids. These studies are limited to the availability of the appropriate substrates for enzyme analysis.

5. The localization of these enzymes in the cell structure leading to a possible correlation with data obtained from electron microscopy.

6. Possible alterations of the environment, i.e., oxygen tension, temperature, growth nutrients, or, growth hormones that may modify the final lipid composition of the seed.

7. Because of the difficulty of obtaining uniform material from developing seeds, future investigations may involve suspension cells of a specific plant tissue, of known genetic background, grown under completely defined conditions, available in small or large amounts at any time of the year, and exposed to a number of environmental and nutrient variables. Although suspension cells could serve very well as model systems, interpretation of the data as relates to developmental changes of seed tissue may lead to conclusions remote from what is actually occurring in the developing seed. Radwan and Mangold [20] have recently reviewed the use of tissue culture techniques as relates to lipid metabolism.

C. The Role of the Leaf Compartment

The leaf compartment is the site for the conversion of CO_2 to 3-phosphoglyceric acid and for the further conversion of this compound in the cytosol compartment of the leaf cell to sucrose [11].

Although the chloroplast contains a full complement of enzymes that can synthesize C_{16} and C_{18} fatty acids, there is no evidence to suggest that any of these fatty acids are transported as such through the vascular system to the seed compartment. The biosynthetic machinery in the chloroplast functions presumably to maintain the lamellar membranes with the full complement of lipids essential for the proper function of these membranes, i.e., photophosphorylation. Since the biosynthesis of fatty acids in chloroplasts has been thoroughly reviewed recently [24], there is no need to discuss this area in this chapter.

D. The Developing Soybean Seed

I. General Considerations

Since this seed is of major importance as a source of vegetable oil, it would be of value to examine in some detail the biosynthesis of fatty acids by various preparations of this seed.

Mature soybean seeds [*Glycine max.* (L.) *Merrill*] contain approximately 20% of their total dry weight as neutral lipids, 40% as protein, 22% as cellulose and hemicellulose, 7% as sugars, and the remaining amount as crude fiber and mineral matter. A number of observers have divided lipid formation in the developing seed into three approximate phases [1]. With reference to the soybean, Reubel [21] and Privett et al. [18] have carried out careful studies on the lipid composition of the developing soybean seed. As summarized in Tables 1 and 2, in the first phase, i.e., the first 24 days after flowering (DAF), less than 1% of the total lipid and only about 2% of the total protein of the mature seed had been synthesized. As would be expected at this early

Table 1. Fatty acid weights in Harosoy 63 (soy-bean) seed during three different phases of development [21]

Phase	DAF	mg/seed						
		Dry weight	Oil	16:0	18:0	18:1	18:2	18:3
1	24	0.9	0.3	0.006	0.002	0.002	0.011	0.009
2	40	70.3	14.1	1.52	0.45	3.63	7.24	1.28
3	64	187	44.9	4.67	1.16	11.80	24.60	2.65
Mature seed	72	186	44.0	4.57	1.27	11.61	23.80	2.55

Table 2A. Relative amounts of accumulating fatty acids as related to phase of development[a]

Ratio of phases	mg Oil	16:0	18:0	18:1	18:2	18:3
2/1	47	254	225	182	659	142
3/2	3.2	3.07	2.58	3.24	3.40	0.96

Table 2B. Relative amounts of accumulated fatty acids at a given phase of development

Phases	16:0	18:0	18:1	18:2	18:3
1	3	1	1	5.5	4.5
2	3.4	1	8.1	16.1	2.8
3	4.01	1	10.1	21.1	2.3

[a] Calculated from data in Table 1

stage of development, the major fatty acids are the polyunsaturated acids and the major lipid classes are the typical membrane lipids, namely, phospholipids and glycolipids with triacylglycerols as a minor class. The second phase includes the period from 24 to 40 DAF. During this phase 30% of the total lipid of the mature seed is synthesized, and equally important, the fatty acid composition undergoes rapid changes so that at the termination of this phase the composition of the fatty acids approaches the composition found in the mature seed. The final phase extends from the 40th to the 64th DAF. During this period the fatty acid percentage values are essentially constant, although during this period 70% of the total lipid as well as protein found in the mature seed is synthesized.

Thus, the final composition of seed oil is achieved through rigidly controlled differential rates of synthesis of the various fatty acids. In Phase 1, membrane lipids are the major products of biosynthesis. The major activities for the synthesis of storage lipids occur during Phases 2 and 3, and final adjustments of fatty acid composition programmed by the informational molecules of the seed are made in Phase 3.

II. Comparison of Capacities of Developing Tissues to Incorporate ^{14}C-Acetate into Long-Chain Fatty Acids

There is good correlation between the accumulation of endogenous fatty acids and the rate of incorporation of ^{14}C-acetate into ^{14}C-fatty acids as a function of the developmental phase of the seed. Thus, most studies are carried out with tissues in the latter part of Phase 2, namely, about the 30th–35th DAF. ^{14}C-Acetate incorporation is high, and the entire tissue is geared for optimal synthesis of fatty acids.

However, as summarized in Table 3, the endogenous fatty acid composition and the ^{14}C-fatty acid pattern that results from ^{14}C-acetate incorporation by that same tissue do not correlate well. In addition, cell-free extracts prepared from these same tissues catalyze only the synthesis of ^{14}C-palmitic and ^{14}C-stearic acids. These results are, however, not surprising when one considers the many factors involved in the biosynthesis of fatty acids. A brief analysis of these factors is pertinent here:

1. The time intervals for fatty acid biosynthesis from ^{14}C-acetate and for the accumulation of fatty acids being formed during the various phases of development are vastly different, i.e., 4 h incubation with ^{14}C-acetate and 20, 30, or 50 days for the accumulation of the endogenously formed fatty acids. The short time period provides a clue to the rate of synthesis by intact tissue at a specific period of its development. However, quantitative data are difficult to obtain since the experimentalist is never certain as to the percentage of the tissue that is actually participating in the biosynthetic reactions.

2. Moreover, incorrect interpretation of the data is possible. For example, since no stearic acid accumulates in the developing seed or since no ^{14}C-stearic acid is detected when ^{14}C-acetate is employed as the substrate, or since ^{14}C-stearic is not converted to ^{14}C-oleic acid, one could conclude that stearic acid does not serve as a precursor of oleic, linoleic, or linolenic acids. This conclusion would be quite incorrect, since these types of experiments only reflect the end

Table 3. Comparison of fatty acid composition with capacity of different preparations to incorporate ^{14}C-acetate into long-chain fatty acids [19]

DAF	Type of preparation	% Fatty acid				
		16:0	18:0	18:1	18:2	18:3
20	Endogenous	25	t	17	44	14
	Tissue slice	21	t	56	23	0
	Extract	33	67	0	0	0
30	Endogenous	20	t	14	55	10
	Tissue slice	17	t	53	30	0
	Extract	38	62	0	0	0
50	Endogenous	24	1	15	55	5
	Tissue slice	20	t	56	24	0
	Extract	57	43	0	0	0

Table 4. Incorporation of [^{14}C] substrates into C_{16} and C_{18} fatty acids by developing soybean cotyledons [19]

Substrate	% of ^{14}C in total fatty acids				
	16:0	18:0	18:1	18:2	18:3
[1-^{14}C] acetate	19	3	36	39	3
[1-^{14}C] palmitate	100	0	0	0	0
[1-^{14}C] stearate	0	100	0	0	0
[1-^{14}C] oleate	0	0	75	25	0

result of a biosynthetic route and obscure the dynamic flow of activated precursors into intermediates and finally into products.

3. Table 4 emphasizes another aspect of this problem. This Table shows clearly that whereas ^{14}C-acetate was readily incorporated into ^{14}C-palmitic, ^{14}C-stearic, ^{14}C-oleic, ^{14}C-linoleic, and ^{14}C-linolenic acids, added ^{14}C-palmitic and ^{14}C-stearic acids were not elongated or modified, although they entered the tissue and labeled complex lipids. In contrast, ^{14}C-oleic acid was rapidly desaturated to linoleic acid, as seen in tissue slice preparations. Similar observations have been made with avocado, spinach leaf, and safflower seed preparations.

An explanation for these anomalous results has recently been proposed by Shine et al. [22]. Evidence has been accumulating over a number of years that there are at least two systems concerned with the biosynthesis of fatty acids [9, 12, 14].

The first system, called the de novo system, catalyzes the specific set of reactions leading to the formations of palmityl ACP:

$$C_2 + 7C_3 \xrightarrow{ACP} C_{16:0}ACP + 7CO_2.$$

The second system, called the elongation system, catalyzes the specific reaction:

$$16{:}0\ ACP + \text{malonyl ACP} \xrightarrow{NADPH} 18{:}0\ ACP + ACP.$$

There is considerable evidence to support the conclusion that different enzymes are involved in the de novo system and in the elongation system [14].

The final product of the coupled de novo and elongation systems is stearyl ACP, and only this product serves as the specific substrate for the first desaturation step, i.e., the introduction of the first *cis* double bond in the 9,10 position of the hydrocarbon chain:

$$\text{stearyl ACP} + O_2 + NADPH + H^+ \rightarrow \text{oleyl ACP} + 2\ H_2O + NADP^+.$$

In contrast to animal and yeast desaturases, plant stearyl ACP desaturase is completely soluble. Ferredoxin is the electron carrier between NADPH and molecular oxygen. Finally, this enzyme is completely inactive with palmityl ACP, palmityl CoA, and stearyl CoA.

Since acyl CoA:ACP transacylase, which catalyzes the reaction;

acyl CoA + ACP ⇌ acyl ACP + CoA

is missing in higher plants, acyl CoA cannot be converted directly to the corresponding acyl ACP. It is for this reason that ^{14}C-palmitic is not elongated, and ^{14}C-stearic acid is not desaturated by a variety of tissue preparations.

When the product of stearyl ACP desaturation by cell-free preparations was examined, it was consistently observed that oleyl ACP did not accumulate, as would be expected, but rather that free oleic acid was the product of the enzyme reaction. Furthermore, it was shown that stearyl ACP was not hydrolyzed rapidly. It was finally observed that a highly reactive acyl ACP thioesterase occurred in higher plant extracts, which hydrolyzed oleyl ACP very rapidly but saturated acyl ACPs only slowly [22].

Since oleyl-CoA is the substrate for the second desaturation step, namely, the formation of linoleyl CoA [25], it was concluded that a switching mechanism must occur that could convert oleyl ACP to oleyl CoA for further metabolic conversions. The switching mechanism would therefore involve two enzyme systems:

acyl ACP thioesterase: oleyl ACP + H_2O → oleic acid + ACP

acyl CoA synthetase: oleic acid + ATP + CoA → oleyl CoA + AMP + PP.

This conclusion gains support from the earlier observation with tissue slices that, whereas stearic acid was totally ineffective as a substrate for desaturation with tissue slices, oleic acid was readily desaturated to linoleic acid (Table 4).

These ideas are summarized in Figure 1, which defines the ACP track, the CoA track, and the switching system. The scheme also suggests that stearic

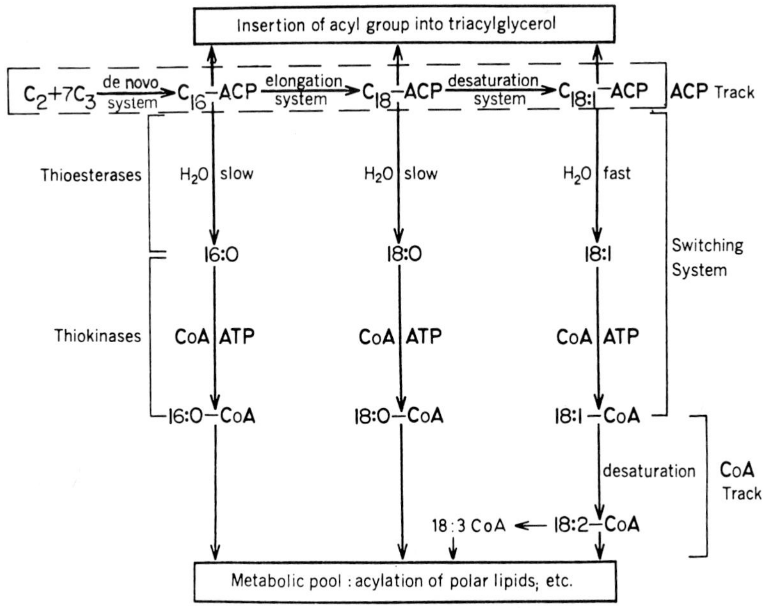

Fig. 1. The interrelation between acyl ACP derivatives and acyl CoA derivatives

acid appears in trace amounts in seed oils (Table 1) because the rate of hydrolysis of stearyl ACP is so slow that stearic acid cannot readily enter the acyl CoA track. In contrast, the rate of desaturation is fast and the rate of hydrolysis of the product, namely, oleyl ACP, is fast. Thus stearic acid (as stearyl ACP) serves as a central intermediate but never accumulates in most seed oils.

III. Localization of Enzymes Involved in Fatty Acid Biosynthesis

In recent years considerable attention has been focused on the precise site for fatty acid synthesis in the plant cell. Beevers and his colleagues have pioneered in the development of techniques suitable for organelle fractionation. Application of these techniques indicates that in the endosperm of developing castor bean seeds a 10,000 *g* particle called a proplastid [16, 27] contains all the enzymes for the generation of ATP by substrate phosphorylation, for the conversion of sucrose to acetyl CoA via a pathway involving UDGP→glucose 1-phosphate →glucose 6-phosphate→glyceraldehyde phosphate→pyruvate→acetyl CoA →malonyl CoA→ACP track→$C_{16}+C_{18}$ fatty acids. Moreover, employing avocado mesocarp tissue, Weaire and Kekwick [26] recently presented evidence that a large particle, which sediments at 2,000 *g*, is the only site for fatty acid biosynthesis. It is conceivable that these organelles, which are considerably enriched with the enzymes associated with fatty acid biosynthesis, are the actual sites for the biogenesis of oil droplets. In addition there may be other sites in the plant cell that participate in the biosynthesis of plasma and organelle membrane lipids rather than storage lipids. Fractionation studies with cotyledons of developing soybean seeds have shown, however, that the fatty acid synthesizing enzymes in these tissues are associated with the cytosolic compartment [19].

E. Conclusions

The data presented here can be summarized in Figure 2, which identifies the various compartments responsible for the smooth flow of carbon from CO_2 to lipids.

In the chloroplast, CO_2 is converted by a series of well-defined reactions to 3-phosphoglyceric acid. This product is then transported out of the chloroplast into the cytosolic compartment of the leaf cell, where a series of reactions eventually convert it to sucrose. Sucrose is transported to the developing seed compartment where it is rapidly broken down to pyruvic acid and thence to acetyl CoA. Acetyl CoA enters the ACP track and is converted to C_{16} and C_{18} ACPS. These substrates are then switched over to acyl CoA derivatives in the appropriate amounts and utilized in the formation of complex lipids. Presumably, there are organelle structures, enriched with the enzymes for fatty acid biosynthesis, that are the sites for the accumulation of massive amounts of fatty acids. Triacylglycerols may be formed rapidly in these organelles as the final product so characteristic of oil-rich seeds.

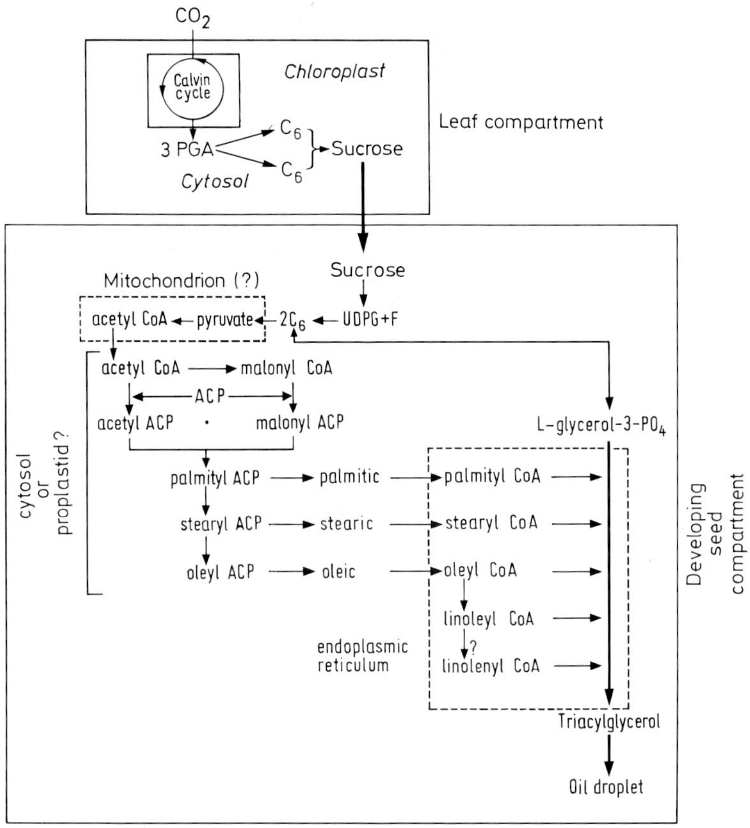

Fig. 2. Schematic summary of the flow of carbon from CO_2 to lipids in a developing seed

References

1. Appelqvist, L.-A.: Biochemical and structural aspects of storage and membrane lipids in developing oil seeds. In: Recent Advances in the Chemistry and Biochemistry of Plant Lipids. Galliard, T., Mercer, E.I. (eds.). London: Academic Press, 1975, pp. 247–286
2. Appleby, R.S., Gurr, M.I., Nichols, B.W.: Studies on seed oil triglycerides. Factors controlling the biosynthesis of fatty acids and acyl lipids in subcellular organelles of maturing *Crambe abyssinica* Seeds. Europ. J. Biochem. **48**, 209–216 (1974)
3. Beevers, Harry: Organelles from castor bean seedlings: Biochemical roles in gluconeogenesis and phospholipid Biosynthesis. In: Recent Advances in the Chemistry and Biochemistry of Plant Lipids. Galliard, T., Mercer, E.I. (eds.). London: Academic Press, 1975, pp. 287–299
4. Canvin, D.T.: Formation of oil in the seed of *Ricinus communis,* L. Canad. J. Biochem. **41**, 1879–1885 (1963)
5. Galliard, T., Stumpf, P.K.: Fat metabolism in higher plants XXX. Enzymatic synthesis of ricinoleic acid by a microsomal preparation from developing *Ricinus communis* seeds. J. Biol. Chem. **241**, 5806–5812 (1966)
6. Gurr, M.I., Blades, J., Appleby, R.S.: Studies on seed oil triglycerides. The composition of *Crambe abyssinica* triglycerides during seed maturation. Europ. J. Biochem. **29**, 362–368 (1972)
7. Gurr, M.I., Blades, J., Appleby, R.S., Smith, C.G., Robinson, M.P., Nichols, B.W.: Studies

on seed oil triglycerides. Triglyceride biosynthesis and storage in whole seeds and oil bodies of *Crambe abyssinica*. Europ. J. Biochem. **43**, 281–290 (1974)
8. Harwood, J.L., Stumpf, P.K.: Fat metabolism in higher plants XL. Synthesis of fatty acids in the initial stages of seed germination. Plant Physiol. **46**, 500–508 (1970)
9. Harwood, J.L., Stumpf, P.K.: Fat metabolism in higher plants LI. Palmitic and stearic synthesis by an avocado supernatant system. Arch. Biochem. Biophys. **148**, 282–290 (1972)
10. Harwood, J.L., Sodja, A., Stumpf, P.K., Spurr, A.R.: On the origin of oil droplets in maturing castor bean seeds, *Ricinus communis*. Lipids **6**, 851–854 (1971)
11. Heber, U.: Metabolite exchange between chloroplasts and cytoplasm. In: Ann. Rev. Plant Physiol. Briggs, W.R., Green, P.B., Jones, R.L. (eds.). Palo Alto: Annual Reviews Inc., 1974, Vol. 25, pp. 393–421
12. Huang, K.P., Stumpf, P.K.: Fat metabolism in higher plants XLIV. Fatty acid synthesis by a soluble fatty acid synthetase from *Solanum tuberosum*. Arch. Biochem. Biophys. **143**, 412–427 (1971)
13. Jaworski, J.G., Stumpf, P.K.: Fat metabolism in higher plants. Properties of a soluble stearyl acyl carrier protein desaturase from maturing *Carthamus tinctorius* seeds. Arch. Biochem. Biophys. **162**, 158–165 (1974)
14. Jaworski, J.G., Goldschmidt, E.E., Stumpf, P.K.: Fat metabolism in higher plants. Properties of the palmityl acyl carrier protein: stearyl acyl carrier protein elongation system in maturing *Carthamus tinctorius* seed extracts. Arch. Biochem. Biophys. **163**, 769–776 (1974)
15. McMahon, V., Stumpf, P.K.: Fat metabolism in higher plants XXVI. Biosynthesis of fatty acids in tissues of developing seeds and germinated seedlings of safflower (*Carthamus tinctorius*). Plant Physiol. **41**, 148–156 (1966)
16. Nakamura, Y., Yamada, M.: Long chain fatty acid synthesis in developing castor bean seeds. I. The operation of the path from acetate to long chain fatty acids in subcellular particulate system. Plant Cell Physiol. (Tokyo) **15**, 37–38 (1974)
17. Norton, G., Harris, J.F.: Compositional changes in developing rapeseed (*Brassica napus,* L.) Planta **123**, 163–174 (1975)
18. Privett, O.S., Dougherty, K.A., Erdahl, W.L., Stolyhwo, A.: Studies on the lipid composition of developing soybeans. J. Am. Oil Chem. Soc. **50**, 516–520 (1973)
19. Porra, R., Stumpf, P.K.: Lipid biosynthesis in developing and germinating soybean cotyledons: The formation of palmitate and stearate by chopped tissue and supernatant preparations. Arch. Biochem. Biophys. **176**, 53–62 (1976)
20. Radwan, S.S., Mangold, H.K.: The lipids of plant tissue cultures. In: Advances in Lipid Research. Vol 14, Paoletti, R., Kritchevsky, D., (eds.). New York: Academic Press, 1976, Vol. 14, pp. 171–211
21. Reubel, A., Rinne, R.W., Canvin, D.T.: Protein, oil, and fatty acid in developing soybean seeds. Crop Sci. **12**, 739–741 (1972)
22. Shine, W.E., Mancha, M., Stumpf, P.K.: Fat metabolism in higher plants. The function of acyl thioesterases in the metabolism of acyl coenzyme A and acyl carrier proteins. Arch. Biochem. Biophys. **172**, 110–116 (1976)
23. Sims, R.P.A., McGregor, W.G., Plessers, A.G., Mes, J.C.: Lipid changes in maturing oil bearing plants II. Changes in fatty acid composition of flax and safflower seed oils. J. Am. Oil Chem. Soc. **38**, 276–279 (1961)
24. Stumpf, P.K.: Biosynthesis of fatty acids in spinach chloroplasts. In: Recent Advances in the Chemistry and Biochemistry of Plant Lipids. Galliard, T., Mercer, E.I. (eds.). London: Academic Press, 1975, pp. 95–113
25. Vijay, I., Stumpf, P.K.: Fat metabolism in higher plants XLVIII. Properties of oleyl CoA desaturase of *Carthamus tinctorius*. J. Biol. Chem. **247**, 360–366 (1972)
26. Weaire, J., Kekwick, R.G.O.: The synthesis of fatty acids in avocado mesocarp and cauliflower bud tissue. Biochem. J. **146**, 425–437 (1975)
27. Yamada, M., Usami, Q., Nakajima, K.: Long chain fatty acid synthesis in developing seeds II. Operation of the path from sucrose to long chain fatty acids in a subcellular particulate system. Plant Cell Physiol. (Tokyo) **15**, 49–58 (1974)

CHAPTER 5

The Cyclopentenyl Fatty Acids

H.K. MANGOLD and F. SPENER

A. Introduction

The seeds of a great number of shrubs and trees belonging to the plant family Flacourtiaceae contain lipids that are characterized by their constituent cyclopentenyl fatty acids [25, 34, 47]. The oils extracted from *Hydnocarpus Kurzii* and *Caloncoba echinata,* called chaulmoogra oil and gorli oil, respectively, as well as those obtained from other Flacourtiaceae species, have been used for centuries in the treatment of leprosy [47]. Table 1 summarizes the systematic names of some members of the family Flacourtiaceae together with the names of the oils obtained from seeds of these plants. Also listed are some of the countries in which these plants are indigenous and those in which they have

Table 1. Some oils producing cyclopentenyl fatty acids [47]

Species	Occurrence	Name of oil
Hydnocarpus anthelminthica	Laos Thailand Vietnam Ceylon [a] Uganda [a] U.S.A. (Hawaii) [a]	Lukrabo oil
H. Kurzii (*Taraktogenos Kurzii*)	Bangladesh Burma India (Assam) Ceylon [a] Brazil (Minas Gerais) [a] U.S.A. (Hawaii) [a]	Chaulmoogra oil
Caloncoba echinata	Guinea Ivory Coast Senegal Sierra Leone Brazil (Sao Paulo) [a] Cuba [a]	Gorli oil
Carpotroche brasiliensis	Brazil (Bahia) (Espirito Santo) (Minas Gerais) (Piaui) (Rio de Janeiro) (Sao Paulo)	Sapocainha oil

[a] Introduced and cultivated

been introduced for cultivation on a large scale. For a review on the older literature on oils containing cyclopentenyl fatty acids the reader is referred to Ref. [47].

B. Chemistry of Cyclopentenyl Fatty Acids

I. Isolation, Characterization, Synthesis

Although the major cyclopentenyl fatty acids, hydnocarpic and chaulmoogric acids, had already been isolated at the beginning of this century [6, 42, 43], and formulae were proposed [6], definite proof of their structures was provided twenty years later [39, 50]. Hydnocarpic and chaulmoogric acids were shown to be 11-(2-cyclopentenyl)undecanoic and 13-(2-cyclopentenyl)tridecanoic acids, respectively. In addition to these, a series of homologous cyclopentenyl fatty acids of lower molecular weights were isolated [14].

Gorlic acid, a cyclopentenyl fatty acid having a double bond in the aliphatic chain, was isolated from gorli oil [2] and, in 1937, its structure was established [11, 38] as 13-(2-cyclopentenyl)-6-tridecenoic acid.

In the seed lipids of Flacourtiaceae [12, 13, 15, 16, 24], hydnocarpic and chaulmoogric acids are, as a rule, the predominant constituent fatty acids, whereas gorlic acid is present at a level of 10 to 25%. Lower homologs of these cyclopentenyl fatty acids occur at levels below 1% each [14].

Physical characteristics of the cyclopentenyl fatty acids and their methyl esters are listed in Table 2.

Chaulmoogric accid can be prepared from hydnocarpic acid by malonic ester synthesis [56]. Conversely, hydnocarpic acid can be obtained from chaulmoogric acid via a Barbier-Wieland degradation [8]. Cyclopentenyl fatty acids having an odd number of carbon atoms, such as homohydnocarpic and homochaulmoogric acids, which have so far not been found in nature, were also prepared [46]. Recently, several even-numbered and odd-numbered cyclopentenyl fatty acids, labeled with ^{14}C in C_1 of the aliphatic chain, were obtained by 'homologous synthesis' [32].

The cyclopentenyl fatty acids exhibit high optical activities because of the presence of an asymmetric carbon atom, C_1 of the cyclopentene ring. The absolute configuration of chaulmoogric acid was established by two independent paths [36], as shown in Figure 1. Firstly, oxidation of (S)-(+)2-cyclopentenyl acetic acid yielded (S)-(−)butane-1,2,4-tricarboxylic acid, i.e., a 3-carboxyadipic acid, whose conformation was known. Secondly, reduction of (R)-(−)2-cyclopentenyl acetic acid to (S)-(−)3-ethylcyclopentene, followed by its oxidation gave (S)-(+)2-ethylglutaric acid, whose configuration was also known.

Several syntheses of racemic cyclopentenyl fatty acids were worked out [21, 37, 40, 41, 58]. Optically active chaulmoogric acid was synthesized, as shown in Figure 2, by subjecting a mixture of (S)-(+)2-cyclopentenyl acetic acid and the monoethyl ester of brassidic acid to Kolbe electrolysis, followed by saponification [36].

Table 2. Physical characteristics of cyclopentenyl fatty acids and their methyl esters

Compound	Chain length: number of double bonds	Molecular formula	Molecular weight	M.P. (°C)	B.P. (°C)	C.S.T. (Nitro-methane) (°C)	C.S.T. (Aceto-nitrile) (°C)	Specific rotation $[\alpha]_D^{25}$
Acids								
Aleprolic acid	6:1cy	$C_6H_8O_2$	112.1					+120.5°
(Homolog not found)	8:1cy	$C_8H_{12}O_2$	140.1					+110.5°
Aleprestic acid	10:1cy	$C_{10}H_{16}O_2$	168.1					+100.5°
Aleprylic acid	12:1cy	$C_{12}H_{20}O_2$	196.2	32.0				+ 90.8°
Alepric acid	14:1cy	$C_{14}H_{24}O_2$	224.2	48.0				+ 77.1°
Hydnocarpic acid	16:1cy	$C_{16}H_{28}O_2$	252.2	58.0–59.0		86.0	52.0	+ 69.3°
Chaulmoogric acid	18:1cy	$C_{18}H_{32}O_2$	280.3	67.0–68.0		97.5	65.0	+ 60.3°
Hormelic acid	20:1cy	$C_{20}H_{36}O_2$	308.3	74.5–75.0				+ 52.0°
Manoaic acid[a]	16:2cy	$C_{16}H_{26}O_2$	250.2					
Gorlic acid[a]	18:2cy	$C_{18}H_{30}O_2$	278.2	8.0–8.5		78.0	42.0	+ 60.7°
Oncobic acid[a]	20:2cy	$C_{20}H_{34}O_2$	306.3					
Methyl esters								
Methyl aleprolate	6:1cy	$C_7H_{10}O_2$	125.1					
(Methylester of homolog not found)	8:1cy	$C_9H_{14}O_2$	153.1					
Methyl aleprestate	10:1cy	$C_{11}H_{18}O_2$	181.1					
Methyl aleprylate	12:1cy	$C_{13}H_{22}O_2$	209.2					
Methyl aleprate	14:1cy	$C_{15}H_{26}O_2$	237.2					
Methyl hydnocarpate	16:1cy	$C_{17}H_{30}O_2$	265.2	7.5–8.0	333	15.0	4.5	
Methyl chaulmoograte	18:1cy	$C_{19}H_{34}O_2$	293.3	19.5	354	37.0	28.0	
Methyl hormelate	20:1cy	$C_{21}H_{38}O_2$	321.3					+ 49.3°
Methyl manoate[a]	16:2cy	$C_{17}H_{28}O_2$	263.2					
Methyl gorlate[a]	18:2cy	$C_{19}H_{32}O_2$	291.2	−2.0	356	14.5	−2.0	+ 57.7°
Methyl oncobate[a]	20:2cy	$C_{21}H_{36}O_2$	319.3					

[a] Contains positional isomers (see Table 4)

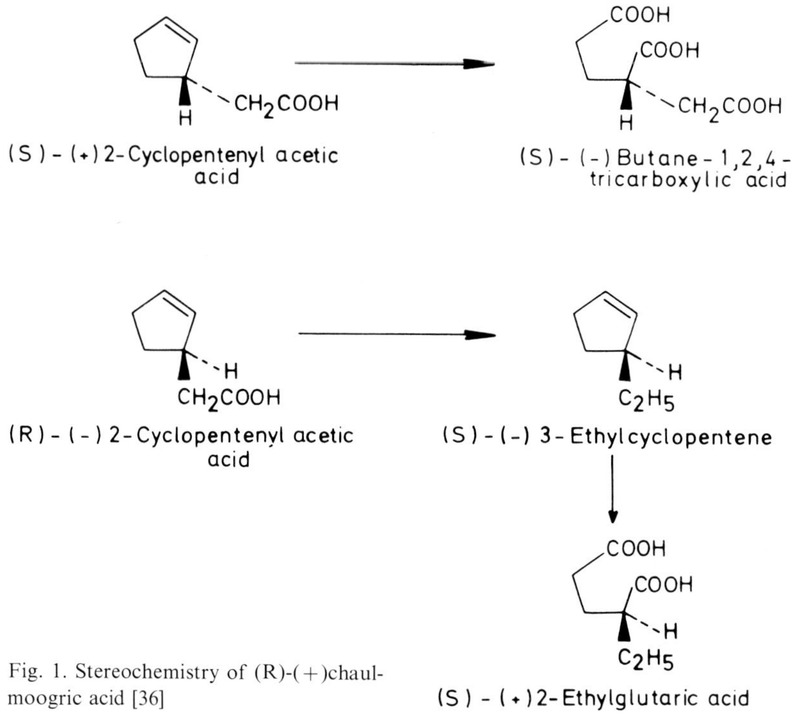

Fig. 1. Stereochemistry of (R)-(+)chaulmoogric acid [36]

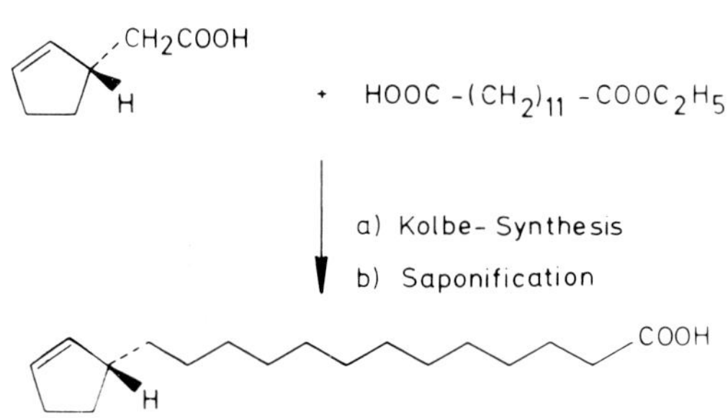

Fig. 2. Synthesis of (R)-(+)chaulmoogric acid [36]

The cyclopentenyl fatty acids and their methyl esters are best characterized by their behavior in various chromatographic systems (vide infra), by their melting points, and by their specific optical rotations (Table 2). Their infrared spectra [5, 20] and nuclear magnetic resonance (NMR) spectra [27] are not very characteristic; their mass spectra [9, 48], however, are of great help in the identification of the cyclopentenyl fatty acids, especially if mass spectrometry is combined with gas chromatography [48]. The mass spectra of methyl esters of the cyclopentenyl fatty acids are characterized by a base peak at m/e 67 (cyclopentene), and m/e 82 (base peak at low voltage), and by a cleavage β to the cyclopentene ring [$^+(CH_2)_nCOOCH_3$].

II. Fatty Acids in Flacourtiaceae

Following the isolation and the elucidation of their structures, interest in the cyclopentenyl fatty acids subsided for many years. In the treatment of leprosy, chaulmoogra oil and similar oils were largely replaced by synthetic drugs that do not lead to the undesired side effects that were observed with these oils. Nevertheless, numerous new derivatives of the cyclopentenyl fatty acids were prepared for possible use in chemotherapy [60].

Interest in the chemistry and biochemistry of the cyclopentenyl fatty acids was regenerated in the sixties, after the structures of the prostaglandins had been elucidated. Like the cyclopentenyl fatty acids, these biologically active fatty acids contain a five-membered ring, though with oxygen functions [45].

1. Methods of Analysis

Studies directed toward the elucidation of the biosynthesis of cyclopentenyl fatty acids were initiated with the development of methods for the analysis and isolation of these unusual fatty acids. It was shown that the methyl esters of saturated, monounsaturated, and diunsaturated straight-chain fatty acids can be separated from the groups of methyl esters of monounsaturated and diunsaturated cyclopentenyl fatty acids both by argentation chromatography [4, 33] and by chromatography of the respective acetoxymercuri-methoxy compounds [4]. In reversed-phase partition chromatography [4, 31], however, the methyl esters of straight-chain fatty acids overlap with the methyl esters of cyclopentenyl fatty acids having the same number of carbon atoms and the same number of double bonds.

It is most practical to separate the methyl esters of saturated and monounsaturated straight-chain fatty acids from all the others by chromatography on silica gel containing silver nitrate, and then to analyze each of the two groups of methyl esters by gas chromatography on a relatively polar stationary phase [53]. The consecutive application of argentation chromatography and gas chromatography [61] if of great value when small amounts of the methyl esters of individual cyclopentenyl fatty acids are to be isolated in order to determine their structures [53] or their radioactivities [54].

In strictly analytical work, capillary gas chromatography may be used to great advantage, because it is possible to resolve a complex mixture of the methyl esters of straight-chain and cyclopentenyl fatty acids in a single run without any prefractionation [48]; an example is shown in Figure 3.

2. Preparative Methods

The classical method for the isolation of the various cyclopentenyl fatty acids, on a preparative scale, involves fractional distillation of the methyl or ethyl esters derived from the total fatty acids of seed oils and subsequent fractional crystallization of the 'free' fatty acids from acetone [12, 13, 15, 16]. This procedure permits oleic acid to be eliminated, but palmitic acid crystallizes with the cyclopentenyl fatty acids [2].

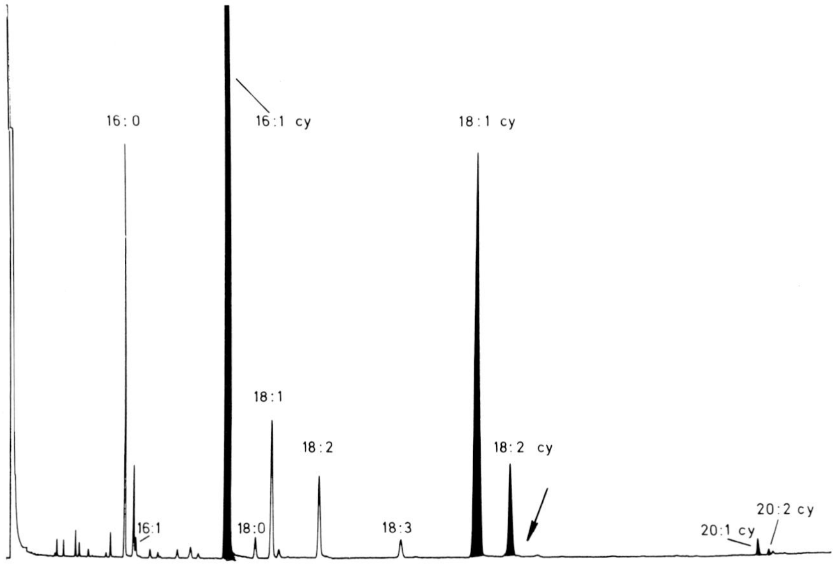

Fig. 3. Gas chromatogram of the methyl esters derived from the lipids of *H. anthelminthica* seeds. (Glass capillary column, 40 m × 0.29 mm, coated with FFAP, 180° C) [48]

Solid-liquid countercurrent distribution with urea [57] can be used to separate straight-chain and cyclopentenyl fatty acids, or their methyl esters, on a preparative scale [20, 35]. Pure hydnocarpic and chaulmoogric acids are best isolated by solid-liquid countercurrent distribution with urea followed by liquid-liquid countercurrent distribution [20], as demonstrated in Figures 4 and 5. Gorlic acid is preferably prepared by fractional distillation of the methyl esters derived from gorli oil, followed by liquid-liquid countercurrent distribution [1] of the 'free' fatty acids; pure chaulmoogric acid is obtained simultaneously [5].

3. Compositional and Structural Studies

New methods of analysis made possible detailed analyses of the lipids and their constituent fatty acids, in various tissues of plants producing cyclopentenyl fatty acids [30, 53, 55, 61]. As an example, Table 3 contains the fatty acid compositions of eleven lipid classes from seeds of *C. echinata*; similar analyses are available of the lipids occurring in seeds of *H. anthelminthica* [53]. Although the cyclopentenyl fatty acids occur mainly in the triacylglycerols, these unusual fatty acids are found also in the 'somatic' lipids [53].

Fatty acid analyses of the triacylglycerols in maturing seeds of *C. echinata* revealed that as early as 14 weeks before maturity, chaulmoogric acid constituted 48% of the total fatty acids. Figure 6 demonstrates that the proportions of chaulmoogric acid increased to 62% at a concomitant loss of unsaturated straight-chain fatty acids, whereas the gorlic acid content did not show any drastic changes. Considering the great increase in oil content with a concomitant increase in the proportion of chaulmoogric acid, the most active synthesis of

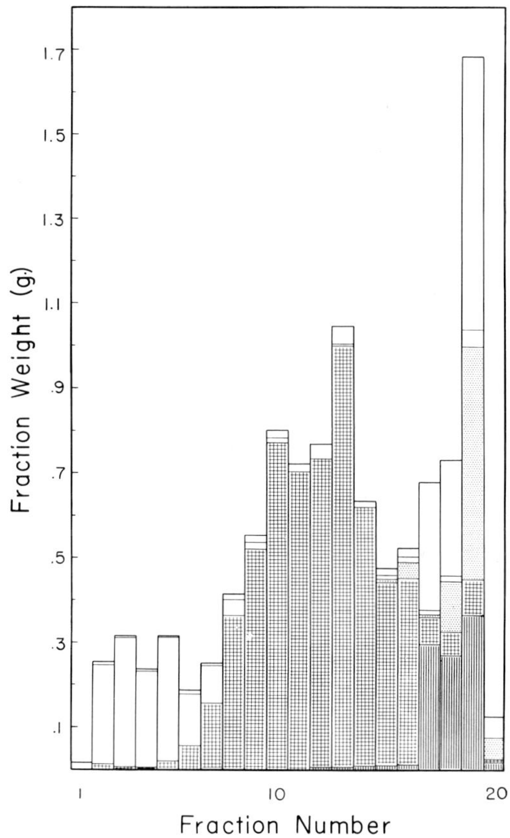

Fig. 4. Fractionation of the methyl esters of straight-chain and cyclopentenyl fatty acids by solid-liquid countercurrent distribution with urea. Sample: 10 g of methyl esters derived from *C. echinata* seed oil (gorli oil). Solvent: methanol-ethyl acetate, 7:3, (vol/vol), 250 ml/fraction; 10 g urea/fraction. White bars, methyl esters of saturated and unsaturated straight-chain fatty acids; cross-hatched bars, methyl chaulmoograte; lined bars, methyl hydnocarpate; dotted bars, methyl gorlate [20]

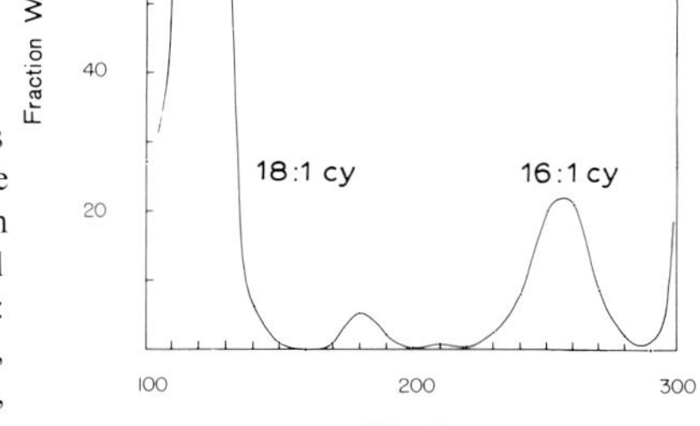

Fig. 5. Isolation of hydnocarpic and chaulmoogric acids by liquid-liquid countercurrent distribution of a mixture obtained by solid-liquid countercurrent distribution with urea. Sample: 3.8 g of a mixture isolated from total fatty acids of *H. wightiana* seed oil (maratti oil). Solvent: heptane–formamide–methanol–acetic acid, 3:1:1:1, (vol/vol/vol/vol); volumes of upper and lower phases, 40 ml each [20]

Table 3. Constituent fatty acids of lipids in *C. echinata* seeds[a, b] [53]

Chain length: No. of double bonds[c]	Triacyl-glycerols	Diacyl-glycerols	Free fatty acids	Ethanol-amine phospho-glycerides	Choline phospho-glycerides	Inositol phospho-glycerides	Glycerol phospho-glycerides	Phos-phatidic acids	Mono-galactosyl diacyl-glycerols	Di-galactosyl diacyl-glycerols	Esterified sterol glycosides
12:0	tr	0.3	1.4	1.5	0.8	1.1	0.6	2.7	0.6	0.1	1.6
12:1cy	tr										
13:0	tr	tr	tr								
14:0	tr	0.2	2.0	1.1	0.7	0.7	1.8	2.0	0.5	0.6	1.2
14:1cy	tr										
15:0	tr	0.1	0.6	0.3	0.4	0.3	0.4	0.3	0.1	0.2	0.4
15:1	tr	tr	0.3	0.2	0.2	tr	0.2	0.2	0.1	0.1	0.1
16:0	6.7	12.1	15.6	19.3	14.6	26.6	36.1	15.4	8.3	10.6	21.8
16:1	1.5	2.2	3.8	1.8	1.7	1.6	2.8	2.1	1.3	1.7	1.8
16:1cy	1.2	1.0	2.2	0.1	1.2	0.7	0.7	1.3	0.2	0.1	0.9
17:0		0.3	0.6	0.5	0.6	0.6	0.5	0.4	0.2	0.5	0.6
17:1		0.3	0.6	0.7	0.6	0.4	1.1	0.5	0.4	0.7	0.4
18:0	0.1	1.2	5.9	4.0	3.2	3.5	5.4	1.3	2.8	7.6	8.2
18:1	1.7	8.0	14.1	11.8	21.2	11.0	12.3	19.8	16.7	12.9	11.1
18:2	1.6	19.0	10.1	39.3	35.0	27.1	16.2	38.6	30.1	22.6	21.1
18:3	0.6	1.0	0.7	3.4	2.4	3.2	2.2	7.0	27.0	28.5	2.3
18:1cy	58.6	35.1	28.1	10.5	9.2	16.3	14.1	5.0	3.6	5.0	18.7
18:2cy	27.2	17.7	14.0	3.2	4.7	5.1	4.5	2.5	5.7	5.7	7.0
20:0		tr			0.5	0.2			0.1	0.2	0.3
20:1	0.1			0.6	1.1	0.5	0.9	0.4	0.7	1.4	0.6
20:1cy	0.2	0.5	tr	1.0	0.7	0.5	0.5	0.4	0.7	0.8	1.2
20:2cy	0.5	1.0		0.7	1.2	0.6	tr	tr	0.9	0.7	0.8

[a] Expressed in wt %; *tr*: trace (<0.1%)
[b] Analyzed as methyl esters at 177°, on a 6 ft × $^1/_8$ in. column filled with 15% DEGS on Anakrom D, 100–120 mesh, carrier gas N_2, 20 ml/min; and at 182° on a 6 ft × $^1/_8$ in. column, filled with 3% OV-1 on Supelcoport, 80–100 mesh, carrier gas N_2, 15 ml/min
[c] The affix "cy" denotes the cyclopentene structure of the acid

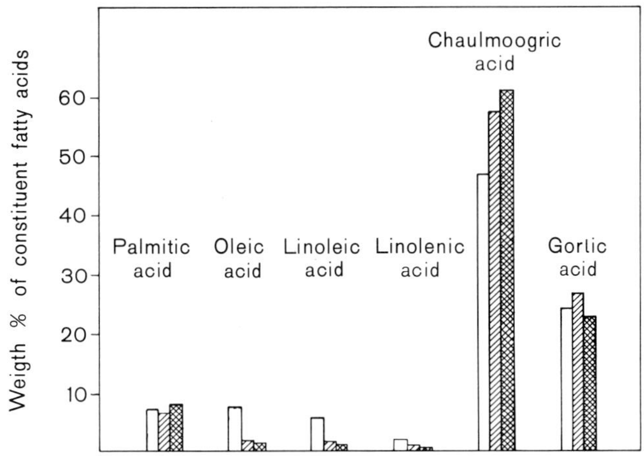

Fig. 6. Major constituent fatty acids of triacylglycerols in seeds of *C. echinata*. White bars, 14 weeks before maturity; hatched bars, 11 weeks before maturity; cross-hatched bars, full maturity [53]. Reprinted with permission from Biochemistry **13**, 2241 (1974). Copyright by the American Chemical Society

cyclopentenyl fatty acids in *C. echinata* seeds occurs during the last 14 weeks before maturity. In *H. anthelminthica* seeds, a similar time span was observed, the major fatty acid, in this case, being hydnocarpic acid.

In contrast to the seed lipids, the total lipids of leaves [55] and cell cultures [52] contain rather low levels of cyclopentenyl fatty acids. As in seeds, the small proportions of cyclopentenyl fatty acids occurring in leaves of *H. anthelminthica* and *C. echinata* are esterified mainly in the triacylglycerols [55].

It can be postulated that the biosynthesis of cyclopentenyl fatty acids involves the cyclization of unsaturated or oxygenated straight-chain fatty acids. In order to check this possibility, the unsaturated straight-chain fatty acids occurring in the lipids of seeds [53], leaves [55], and cell cultures [52] of *H. anthelminthica* and *C. echinata* were isolated and their structures determined by reductive ozonolysis [44]. It was found that the monounsaturated straight-chain fatty acids consisted of mixtures of isomers whereas the di- and triunsaturated straight-chain fatty acids were almost exclusively linoleic and linolenic acids, the polyunsaturated fatty acids most commonly encountered in the plant kingdom.

Unsaturated or oxygenated straight-chain fatty acids, which could conceivably serve as precursors of cyclopentenyl fatty acids, were detected neither in seeds, nor in leaves, nor in cell cultures. Thus, it appears indeed unlikely that the cyclopentenyl fatty acids are formed by cyclization of straight-chain fatty acids [53, 55].

4. New Cyclopentenyl Fatty Acids

In the course of analyses of the constituent fatty acids of various lipid classes in seeds and leaves of Flacourtiaceae, several hitherto unknown cyclopentenyl fatty acids were detected. They were isolated by chromatographic techniques and their structures were determined by chemical degradation according to

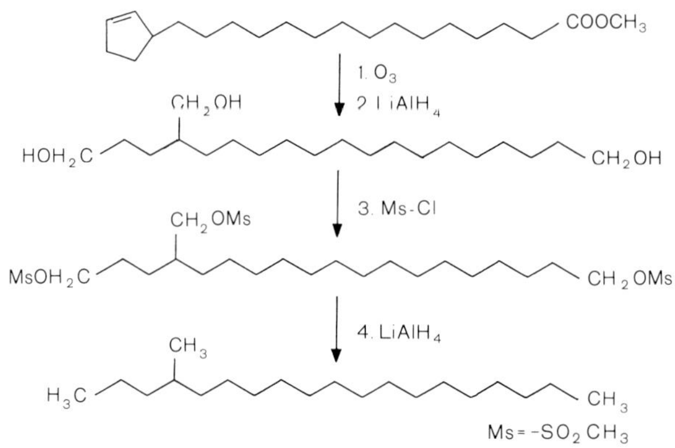

Fig. 7. Degradation scheme of cyclopentenyl fatty acids [53]. Reprinted with permission from Biochemistry **13**, 2241 (1974). Copyright by the American Chemical Society

Table 4. Isomeric polyunsaturated cyclopentenyl fatty acids in triacylglycerols of *C. echinata* seeds [a] [53]

Fatty acid	Positions of double bonds							
	$\Delta 4$	$\Delta 5$	$\Delta 6$	$\Delta 7$	$\Delta 8$	$\Delta 9$	$\Delta 10$	$\Delta 11$
16:2cy	17.3	–	63.6	–	–	19.1	–	–
18:2cy	–	–	85.5	–	–	14.5	–	–
20:2cy	–	–	–	–	81.7	18.3	–	–

[a] Expressed in mole %

the scheme presented in Figure 7. The fragments obtained were identified by gas chromatography and mass spectrometry. The structures of the new cyclopentenyl fatty acids were confirmed by mass spectrometry of their methyl esters and/or the corresponding hydrogenated products.

'Hormelic acid' was isolated from the seed and leaf lipids of several Flacourtiaceae [53, 55]. The structure of this acid was confirmed by synthesis [23, 53]. Reductive ozonolysis proved that gorlic acid is associated with its $\Delta 9$ isomer [53]. In addition, several new cyclopentenyl fatty acids having a double bond in the aliphatic chain were discovered [53]. Table 4 shows the isomer distribution of these diunsaturated cyclopentenyl fatty acids in triacylglycerols of *C. echinata* seeds. 11-(2-Cyclopentenyl)-6-undecenoic acid was named 'manoaic acid' and 15-(2-cyclopentenyl)-8-pentadecenoic acid was named 'oncobic acid'.

C. Biochemistry of Cyclopentenyl Fatty Acids

I. Biosynthesis

In view of the biosynthesis of the prostaglandins, oxidative cyclization of straight-chain unsaturated fatty acids seemed to be a reasonable mechanism

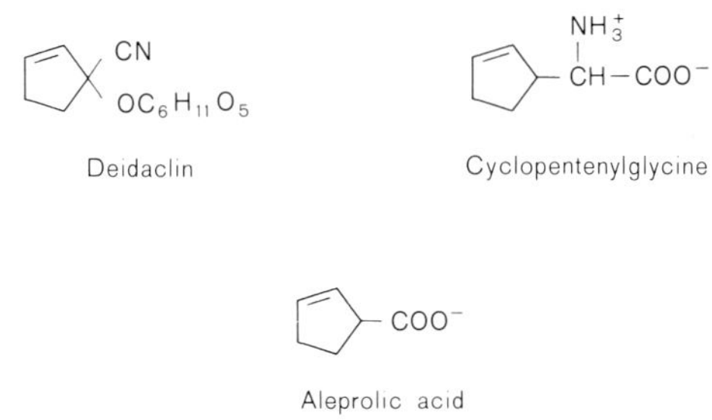

Fig. 8. Potential precursors of cyclopentenyl fatty acids

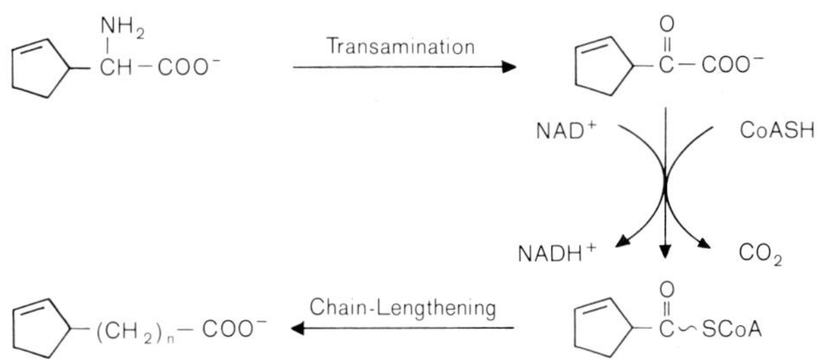

Fig. 9. Scheme for the biosynthesis of cyclopentenyl fatty acids

for the formation of cyclopentenyl fatty acids; in Flacourtiaceae, however, such precursors could not be found [22, 52, 53, 55]. On the other hand, the occurrence of a homologous series of even-numbered cyclopentenyl fatty acids may indicate chain-lengthening as a possible mechanism of biosynthesis.

Various tissues of Flacourtiaceae contain cyclic compounds of low molecular weights, such as deidaclin [10, 59], cyclopentenylglycine [18, 51], and aleprolic acid [14]. The structures of these cyclopentene derivatives are shown in Figure 8. It has been postulated that cyclopentenylglycine may be the precursor of the cyanogenic glycoside deidaclin [10, 26]. The same nonproteinogenic amino acid could possibly serve as precursor of aleprolic acid, which in turn may function as a primer in the biosynthesis of cyclopentenyl fatty acids, a sequence of reactions shown in Figure 9.

1. Aleprolic Acid as Precursor

The role of aleprolic acid as primer in the biosynthesis of cyclopentenyl fatty acids was examined by incubating [1-^{14}C] aleprolic acid [7, 54] with various tissues of Flacourtiaceae; incubations with [1-^{14}C] acetate were carried out for comparison. The data presented in Table 5 clearly show that, in seeds of *H. anthelminthica* and *C. echinata,* [1-^{14}C] aleprolic acid is incorporated almost exclusively into cyclopentenyl fatty acids [17], whereas [1-^{14}C] acetate labels mainly straight-cahin fatty acids. Radioactivity from [1-^{14}C] acetate was found in the aliphatic chain but not in the ring of cyclopentenyl fatty acids [17, 22]. Incorporation of substrates into the cyclopentenyl fatty acids in lipids of leaves was much lower. Most likely, aleprolic acid was degraded to acetate, which was used for the de novo synthesis of straight-chain fatty acids [17].

Table 5. Incorporation of [1-^{14}C] acetate and [1-^{14}C] aleprolic acid into the fatty acids occurring in lipids of Flacourtiaceae [17]

Plant material	Substrate	μmol	Incorporation into total fatty acids (nmol/g fresh tissue)[a]	Radioactivity in cyclopentenyl fatty acids (%)
H. anthelminthica, immature seeds	Acetate	2.0	27.3	17.9
	Aleprolic acid	0.2	1.4	99.6
H. anthelminthica, chopped leaves	Acetate	5.0	49.0	0.4
	Aleprolic acid	1.0	0.9	2.6
C. echinata, mature seeds	Acetate	2.0	15.8	16.4
	Aleprolic acid	0.2	0.9	96.5
C. echinata, chopped leaves	Acetate	7.9	573.0	0.5
	Aleprolic acid	0.6	5.7	2.2

[a] Mean of 2 to 5 determinations each

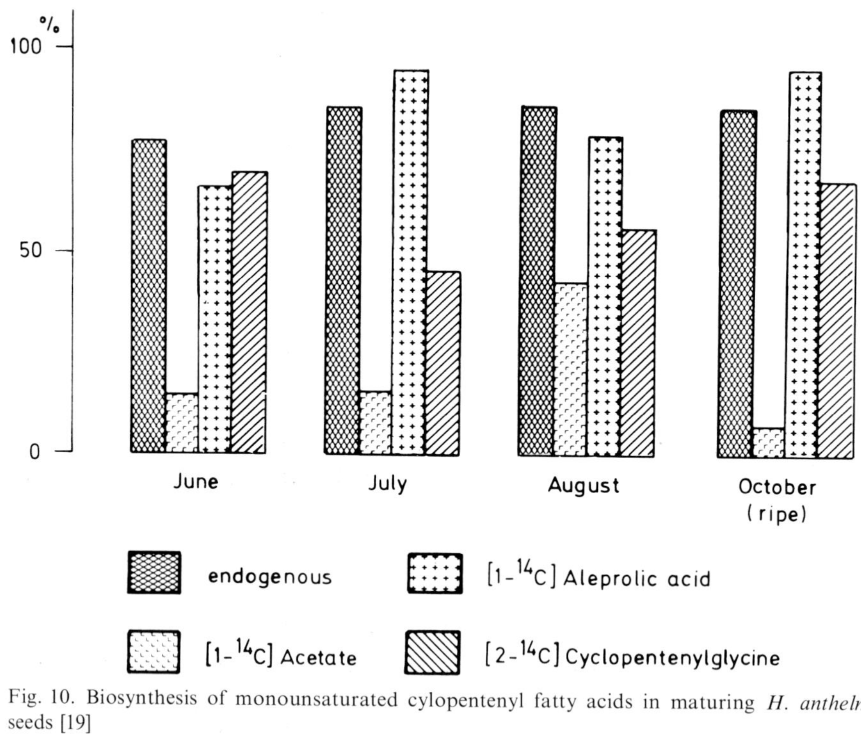

Fig. 10. Biosynthesis of monounsaturated cylopentenyl fatty acids in maturing *H. anthelminthica* seeds [19]

In higher plants, aleprolic acid can serve as a primer only if these plants produce cyclopentenyl fatty acids [17]. It is remarkable that *Chlorella vulgaris*, an alga that does not produce cyclopentenyl fatty acids, can synthesize these unusual fatty acids at a high rate if the cells are provided with aleprolic acid [54].

Table 6. Biosynthesis of cyclopentenyl fatty acids in embryonic tissue of maturing *H. anthelminthica* seeds: Precursor–product relationship between aleprolic acid and cyclopentenylglycine [19]

Substrate	µmol	nmol of labeled substrate incorporated/g fresh tissue [a]
[1-^{14}C] Aleprolic acid	0.2	5.2
[1-^{14}C] Aleprolic acid	0.2	6.2
plus cyclopentenylglycine	1.0	
[2-^{14}C] Cyclopentenylglycine	0.8	15.1
[2-^{14}C] Cyclopentenylglycine	0.8	2.7
plus aleprolic acid	39.0	

[a] Mean of 2 to 4 determinations each

2. *Cyclopentenylglycine as Precursor*

The role of [2-^{14}C] cyclopentenylglycine in the biosynthesis of cyclopentenyl fatty acids was studied in maturing seeds of *H. anthelminthica*; incubations with [1-^{14}C] aleprolic acid and [1-^{14}C] acetate were carried out for comparison [19]. [2-^{14}C] Cyclopentenylglycine was readily incorporated into the cyclopentenyl fatty acids. Figure 10 shows that well over 50% of the radioactivity incorporated into total fatty acids was found in monounsaturated cyclopentenyl fatty acids. When provided with precursors such as aleprolic acid and cyclopentenylglycine, even mature seeds are able to synthesize cyclopentenyl fatty acids. When incubated with [1-^{14}C] acetate, only minute amounts of these unusual fatty acids are labeled. Apparently, in mature seeds such compounds are not produced, whereas the chain-lengthening systems for the biosynthesis of straight-chain fatty acids as well as cyclopentenyl fatty acids are still functioning [19].

Embryonic tissue of *H. anthelminthica* was used for a dilution experiment in order to test the precursor-product relationship between aleprolic acid and cyclopentenylglycine. The data given in Table 6 prove that aleprolic acid is a more direct precursor than cyclopentenylglycine and indicate the existence of a biosynthetic path as outlined in Figure 9.

II. Metabolism

Evidence has been provided [29, 49] that, in higher plants, de novo synthesis of straight-chain fatty acids proceeds to palmitic acid, which is then elongated to higher homologs by different enzyme systems. In *H. anthelminthica* and *C. echinata* seeds, a study on the incorporation of ^{14}C into the carboxyl group of fatty acids, including cyclopentenyl fatty acids, gave similar results, as shown in Table 7. After incubation of the seeds with [1-^{14}C] aleprolic acid and [2-^{14}C] cyclopentenylglycine only minute amounts of activity were found in C_1 of the cyclopentenyl fatty acids. When [1-^{14}C] acetate was used as substrate, 20.4% of the radioactivity incorporated was found in the carboxyl group of hydnocarpic acid, which means that 5 C_2 units were incorporated. High percentages of activity

Table 7. Incorporation of radioactivity from various precursors into C_1 of the constituent fatty acids of lipids in Flacourtiaceae seeds [17]

Fatty acid	Substrate	Percentage of total radioactivity incorporated[a]		
		H. anthel-minthica	*C. echinata*	*Theory*
Total cyclopentenyl fatty acids	[1-^{14}C] Aleprolic acid	0.7	0.2	0
	[2-^{14}C] Cyclopentenylglycine	3.8		0
Hydnocarpic acid	[1-^{14}C] Acetate	20.4		20.0
Chaulmoogric acid	[1-^{14}C] Acetate	60.1	43.4	16.6
Hormelic acid	[1-^{14}C] Acetate	42.9		14.3
Palmitic acid	[1-^{14}C] Acetate	14.0	14.5	12.5
Stearic acid	[1-^{14}C] Acetate	45.5	18.9	11.1
Arachidic acid	[1-^{14}C] Acetate	48.6		10.0

[a] Mean of 2 to 4 determinations each

in the carboxyl group of chaulmoogric and hormelic acids show that endogenous cyclopentenyl fatty acids were the acceptors of labeled C_2 units; the same was found true for the straight-chain fatty acids in this tissue [17].

Regarding the formation of diunsaturated cyclopentenyl fatty acids, it seems likely that a double bond is introduced into the saturated aliphatic chain of monounsaturated cyclopentenyl fatty acids by specific and nonspecific desaturases. Incubation of *H. anthelminthica* seeds with [1-^{14}C] aleprolic acid, under anaerobic conditions, led to monounsaturated cyclopentenyl fatty acids exclusively [19]. When *C. vulgaris* was incubated with [1-^{14}C] aleprolic acid, under aerobic conditions, the ratio of diunsaturated to monounsaturated cyclopentenyl fatty acids was much higher than under anaerobic conditions [54].

D. Conclusions

Classical methods of fractionation and modern separation techniques permit the analysis and isolation of cyclopentenyl fatty acids and lipids containing them. Pure compounds are now available for use in chemical and biomedical investigations. Thus, individual cyclopentenyl fatty acids are being used in the preparation of macrocyclic lactones [3] and in the evaluation of their effects on mycobacteria [28].

According to current knowledge, cyclopentenylglycine is a precursor of the cyclopentenyl fatty acids in Flacourtiaceae. It cannot be excluded that this nonprotein amino acid is a branch in the mainstream of biosynthesis, though amenable to re-utilization [19]. Further work will have to be focused on the biogenesis of the cyclopentene ring itself.

References

1. Ahrens, E.H., Jr., Craig, L.C.: Separation of the higher fatty acids. J. Biol. Chem. **195**, 299–310 (1952)
2. André, E., Jouatte, D.: Contribution à l'étude des huiles chaulmoogriques. Recherches sur l'huile de Gorli [1]. Bull Soc. Chim. France **43**, 347–360 (1928)
3. Baba, S.M., Balakrishnan, V.K., Mathur, H.H., Bhattacharyya, S.C.: Macrocyclic musk compounds, Part **XV**. Synthesis of macrocyclic lactones and related compounds from hydnocarpic and chaulmoogric acids. Indian J. Chem. **4**, 229–234 (1966)
4. Bandi, Z.L., Mangold, H.K.: Chromatography of lipids containing cyclopentenyl fatty acids. Separation Sci. **4**, 83–88 (1969)
5. Bandi, Z.L., Mangold, H.K.: Isolation and characterization of cyclopentenyl fatty acids. II. Preparation of chaulmoogric and gorlic acids by fractional distillation and liquid-liquid countercurrent distribution. Separation Sci., in press
6. Barrowcliff, M., Power, F.D.: The constitution of chaulmoogric and hydnocarpic acids. J. Chem. Soc. **91**, 557–578 (1907)
7. Branner-Jørgensen, S., Berg, A.: Synthesis and reactions of 2-cyclopenten-1-ylmagnesium chloride. Acta Chem. Scand. **20**, 2192–2194 (1966)
8. Buu-Hoi, N.P.: Dégradation des acides cycliques des huiles de Flacourtiaceés. Ann. Chim. [11] **19**, 446–458 (1944)
9. Christie, W.W., Rebello, D., Holman, R.T.: Mass spectrometry of derivatives of cyclopentenyl fatty acids. Lipids **4**, 229–231 (1969)
10. Clapp, R.C., Ettlinger, M.G., Long, L., Jr.: Deidaclin: A natural glucoside of cyclopentenone cyanohydrin. J. Am. Chem. Soc. **92**, 6378–6379 (1970)
11. Cole, H.I., Cardoso, H.T.: Isolation and properties of gorlic acid, an optically active liquid fatty acid. J. Am. Chem. Soc. **60**, 612–614 (1938)
12. Cole, H.I., Cardoso, H.T.: Analysis of chaulmoogra oils. I. *Carpotroche brasiliensis* (Sapucainha) oil. J. Am. Chem. Soc. **60**, 614–617 (1938)
13. Cole, H.I., Cardoso, H.T.: Analysis of chaulmoogra oils. II. *Oncoba echinata* (Gorli) oil. J. Am. Chem. Soc. **60**, 617–619 (1938)
14. Cole, H.I., Cardoso, H.T.: Alepric, aleprylic, aleprestic and aleprolic acids, new homologs of chaulmoogric acid. J. Am. Chem. Soc. **61**, 2349–2351 (1939)
15. Cole, H.I., Cardoso, H.T.: Analysis of chaulmoogra oils. III. *Hydnocarpus wightiana* oil. J. Am. Chem. Soc. **61**, 2351–2353 (1939)
16. Cole, H.I., Cardoso, H.T.: Analysis of chaulmoogra oils. IV. *Hydnocarpus anthelminthica* oil. V. *Taraktogenos kurzii* (chaulmoogra) oil. J. Am. Chem. Soc. **61**, 3442–3445 (1939)
17. Cramer, U., Spener, F.: Biosynthesis of cyclopentenyl fatty acids. (2-Cyclopentenyl) carboxylic acid (aleprolic acid) as a special primer for fatty acid biosynthesis in Flacourtiaceae. Biochim. Biophys. Acta **450**, 261–265 (1976)
18. Cramer, U., Spener, F.: Isolation and identification of (2-cyclopentenyl) glycine from seeds of *Hydnocarpus anthelminthica*. In preparation
19. Cramer, U., Spener, F.: Biosynthesis of cyclopentenyl fatty acids. Cyclopentenylglycine, a non-proteinogenic amino acid as precursor of cyclic fatty acids in Flacourtiaceae. Europ. J. Biochem., in press
20. Cubero, J.M., Bandi, Z.L., Mangold, H.K.: Isolation and characterization of cyclopentenyl fatty acids. I. Preparation of hydnocarpic and chaulmoogric acids by solid-liquid countercurrent distribution and liquid-liquid countercurrent distribution. Separation Sci., in press
21. Diaper, D.G.M., Smith, J.C.: A synthesis of *dl*-hydnocarpic acid. Biochem. J. **42**, 581–583 (1948)
22. van Dorp, D.A.: Cyclization reactions in lipid biochemistry. Lecture, 9th Internat. Congr. Biochem. Stockholm, 1973
23. van Dyke, R.H., Adams, R.: Synthesis of chaulmoogrylacetic acid. V. J. Am. Chem. Soc. **48**, 2393–2395 (1926)
24. Gupta, A.S., Mutha, S.C., Waghrey, A.P.: The component fatty acids of chaulmoogra oil (*Taraktogenos Kurzii,* King). J. Sci. Food Agric. **14**, 457–463 (1963)
25. Hegnauer, R.: Chemotaxonomie der Pflanzen. Vol. 4. Basel: Birkhäuser Verlag, 1966, pp. 155–168

26. Hegnauer, R.: Probleme der Chemotaxonomie, erläutert am Beispiel der cyanogenen Pflanzenstoffe. Pharm. Acta Helv. **46**, 585–601 (1971)
27. Hopkins, C.Y.: Nuclear magnetic resonance in fatty acids and glycerides. In: Progress in the Chemistry of Fats and Other Lipids. Holman, R.T. (ed.). Oxford: Pergamon Press, 1965, Vol. 8, pp. 213–252
28. Jacobsen, P.L., Levy, L.: Antibacterial spectrum of hydnocarpic acid, a component of chaulmoogra oil. Proc. West. Pharmacol. Soc. **1972** (15), 44–46
29. Jaworski, J.G., Goldschmidt, E.E., Stumpf, P.K.: Fat metabolism in higher plants: Properties of the palmityl acyl carrier protein: stearyl acyl carrier protein elongation system in maturing *Carthamus tinctorius* seed extracts. Arch. Biochem. Biophys. **163**, 769–776 (1974)
30. Kaimal, T.N.B., Lakshminarayana, G.: Fatty acid composition of lipids isolated from different parts of *Ceiba pentandra, Sterculia foetida* and *Hydnocarpus wightiana*. Phytochemistry **9**, 2225–2229 (1970)
31. Kaufmann, H.P., Deshpande, M.M.: Die Papier-Chromatographie auf dem Fettgebiet. XXVII: Untersuchung cyclischer Fettsäuren. Fette. Seifen. Anstrichmittel **60**, 645–647 (1958)
32. Mangold, H.K., Becker, H., Cramer, U., Spener, F.: Homologous synthesis: A method for the preparation of mixed substrates and individual compounds. Chem. Phys. Lipids **17**, 176–181 (1976)
33. Mani, V.V.S., Lakshminarayana, G.: The thin-layer chromatographic behaviour of cyclopentanyl and cyclopentenyl fatty acid methyl esters. J. Chromatog. **39**, 182–185 (1969)
34. Markley, K.S.: Alicyclic-substituted fatty acids. In: Fatty Acids. Markley, K.S. (ed.). New York: Interscience Publ. 1960, 2nd ed., Part 1, pp. 213–220
35. Mehta, T.N., Lokras, S.S.: Fractionation of *Hydnocarpus wightiana*—oil fatty acids and their ethyl esters with urea. Indian J. Appl. Chem. **26**, 45–52 (1963)
36. Mislow, K., Steinberg, I.V.: The synthesis and stereochemistry of chaulmoogric acid. J. Am. Chem. Soc. **77**, 3807–3810 (1955)
37. Paasivirta, J.: The alkaline degradation of dehydronorcamphor to isomeric cyclopentenyl acetic acids. Tetrahedron Lett. **1968**, 2867–2870
38. Paget, H.: The liquid acids of Sapucainha oil. J. Chem. Soc. **1937**, 955–960
39. Perkins, G.A.: The oxidation of chaulmoogric acid by permanganate. J. Am. Chem. Soc. **48**, 1714–1725 (1926)
40. Perkins, G.A., Cruz, A.O.: Synthesis of compounds similar to chaulmoogric acid. I. J. Am. Chem. Soc. **49**, 517–522 (1927)
41. Perkins, G.A., Cruz, A.O.: Synthesis of compounds similar to chaulmoogric acid. II. *dl*-Chaulmoogric acid. J. Am. Chem. Soc. **49**, 1070–1077 (1927)
42. Power, F.B., Gornall, F.H.: The constituents of chaulmoogra seeds. J. Chem. Soc. **85**, 838–851 (1904)
43. Power, F.B., Barrowcliff, M.: The constituents of the seeds of *Hydnocarpus wightiana* and of *Hydnocarpus anthelminthica*. Isolation of a homologue of chaulmoogric acid. J. Chem. Soc. **87**, 884–896 (1905)
44. Privett, O.S., Blank, M.L., Romanus, O.: Isolation analysis of tissue fatty acids by ultramicro-ozonolysis in conjunction with thin-layer chromatography and gas-liquid chromatography. J. Lipid Res. **4**, 260–265 (1963)
45. Ramwell, P.W., Shaw, J.E., Clarke, G.B., Grostic, M.F., Kaiser, D.G., Pike, J.E.: Prostaglandins. In: Progress in the Chemistry of Fats and other Lipids. Holman, R.T. (ed.). Oxford: Pergamon Press, 1968, Vol. 9, pp. 231–273
46. Sacks, J., Adams, R.: The synthesis of homochaulmoogric acid, homohydnocarpic acid and chaulmoogrylamines. VI. J. Am. Chem. Soc. **48**, 2395–2399 (1926)
47. Schlossberger, H.: Chaulmoograöl. Geschichte, Herkunft, Zusammensetzung, Pharmakologie, Chemotherapie. In: Handbuch der Pharmakologie, Ergänzungswerk. Heubner, W., Schüller, J. (eds.). Berlin: Springer, 1938, Vol. 5, pp. 1–141
48. Schulte, E., Rehfeldt, A.G., Spener, F.: The direct coupling of glass capillary columns to a mass spectrometer: A method for the complete separation and identification of straight-chain and cyclic fatty acids, in preparation
49. Shine, W.E., Mancha, M., Stumpf, P.K.: Fat metabolism in higher plants: The function of acyl thioesterases in the metabolism of acyl coenzyme A and acyl carrier proteins. Arch. Biochem. Biophys. **172**, 110–116 (1976)

50. Shriner, R.L., Adams, R.: The structure of chaulmoogric and hydnocarpic acids. I. J. Am. Chem. Soc. **47**, 2727–2739 (1925)
51. Spener, F., Dieckhoff, M.: A convenient method for the isolation of individual amino acids from complex mixtures. J. Chromatog. Sci. **11**, 661–662 (1973)
52. Spener, F., Staba, E.J., Mangold, H.K.: Lipids in plant tissue cultures. II. Unusual fatty acids in lipids of *Hydnocarpus anthelminthica*. Chem. Phys. Lipids **12**, 344–350 (1974)
53. Spener, F., Mangold, H.K.: New cyclopentenyl fatty acids in Flacourtiaceae. The straight-chain fatty acids and cyclopentenyl fatty acids in lipids during maturation of the seeds. Biochemistry **13**, 2241–2248 (1974)
54. Spener, F.: Biosynthesis of cyclopentenyl fatty acids. Chain elongation and desaturation. Europ. J. Biochem. **53**, 161–167 (1975)
55. Spener, F., Mangold, H.K.: Straight-chain unsaturated fatty acids and cyclopentenyl fatty acids in leaf lipids of *Caloncoba echinata* and *Hydnocarpus anthelminthica*. Phytochemistry **14**, 1369–1373 (1975)
56. Stanley, W.M., Adams, R.: The synthesis of chaulmoogric acid from hydnocarpic acid. J. Am. Chem. Soc. **51**, 1515–1518 (1929)
57. Sumerwell, W.N.: Solid-liquid countercurrent distribution of fatty acids with urea. J. Am. Chem. Soc. **79**, 3411–3415 (1957)
58. Tanaka, A.: Synthesis of *dl*-aleprylic acid. J. Lipid Res. **10**, 681–682 (1969)
59. Tantisewie, B., Ruijgrok, H.W.L., Hegnauer, R.: Cyanogenic compounds in Parietales and some further species. V. Distribution of hydrocyanic acid in comophytes. Pharm. Weekbl. **104**, 1341–1355 (1969)
60. Wagner-Jauregg, Th.: Aufgaben und Ergebnisse der experimentellen Chemotherapie der Lepra. Angew. Chem. **55**, 195–210 (1942)
61. Zeman, I., Pokorný, J.: Gas chromatographic analysis of cyclopentenyl fatty acids. J. Chromatog. **10**, 15–20 (1963)

CHAPTER 6

Enzymatic Reactions in Galactolipid Biosynthesis

E. HEINZ

A. Introduction

The topic of plant glycolipids has been reviewed several times, and the most recent articles by SASTRY [88], MUDD and GARCIA [69], and VAN HUMMEL [40] document the research in this field up to 1974 in a comprehensive way. Since these results are presumed to be well known and to avoid extensive repetition, emphasis will be placed on developments concerning precursor biosynthesis, cooperation of compartments in assembling the final galactolipids, and the origin of specificities found in the diglyceride portion of these lipids. Galactolipids are composed of galactose, glycerol, and fatty acids; accordingly, the introductory sections will deal with the questions, by which reactions and in which compartments are these components or their activated forms synthesized.

B. Acyl-CoA and Acyl-ACP Thioesters

Details of fatty acid synthesis will not be presented, and only the availability of precursors for fatty acid synthesis in plastids, which contain the bulk of galactolipids in plant cells, will be looked at. According to their synthetic capabilities, plastids may be divided into two groups, both of which contain and probably also synthesize galactolipids. The members of the first group, energetically autonomous chloroplasts, produce ATP and NADPH by photosynthesis, whereas the origin of acetate for acetyl-CoA in mature chloroplasts is an open question [90, 113]. It is known that acetate can freely penetrate chloroplast envelopes in the form of the undissociated acid [33]. Once inside the chloroplast, acetate is activated by acetate thiokinase to acetyl-CoA [55], which then is carboxylated to malonyl-CoA by a prokaryotic type of acetyl-CoA carboxylase system [58]. Immediate products of fatty acid synthesis are acyl-acyl carrier protein (ACP) thioesters. Only very recently, studies on the function of these compounds as acyl donors in lipid synthesis have been started [92].

Established acyl donors for acylation of glycerol phosphate in chloroplasts are acyl-CoA thioesters. Since chloroplasts lack acyl-CoA:ACP acyltransferases [48], which also could catalyze the reverse reaction, acyl-ACP esters are first hydrolyzed by thioesterases and the released fatty acids reactivated to acyl-CoA thioesters by the action of thiokinases [91]. Acyl-ACP thioesterase activity was found in chloroplast stroma and grana [91, 92], whereas long-chain acyl thiokinase [56] was localized in chloroplast envelopes [50]. The observation that long-chain acyl thiokinase activity was found in a 80,000 *g*-supernatant fraction obtained from sonicated *Chlorella* cells [30] may be due to the fragility of the envelope or indicate the existence of soluble and membrane-bound activities. On the other hand, the activation of fatty acids to ACP-thioesters does not occur [48]. The advantage of keeping free fatty acids from re-entering the ACP form and separating ACP and CoA forms by an energy-consuming reaction is not known.

The second group of plastids comprises those forms that depend on imported substrates for energy production, e.g., chromoplasts, nonpigmented plastids in nonphotosynthetic tissues and probably also the very young plastids in meristematic tissues, which are nourished by the rest of the cell. NAD(P)H that cannot pass through the envelope may be derived from permeable substrates by the action of plastidic isoenzymes, such as glyceraldehyde phosphate dehydrogenase and malate dehydrogenase [32]. A limited supply of ATP may be maintained by direct import via the adenylate translocator [35] or may result from the action of phosphoglycerate kinase. In agreement with other results [113] pyruvate kinase [13] and pyruvate dehydrogenase [80] have recently been demonstrated in plastids from nongreen tissues. These enzymes, in photosynthetic tissues established markers for cytoplasmic proteins and mitochondria, would provide the fatty acid synthesizing enzymes in colorless plastids with ATP, NADH and acetyl-CoA. These results indicate further that not only enzymatic activity but even basic enzymatic equipment of organelles may change in relation to the stage of their development. This fact receives increasing attention in investigations of lipid-synthesizing enzymes [17, 31, 54, 101].

C. *sn*-Glycerol 3-Phosphate

The glycerol backbone in glycerolipids is derived from *sn*-glycerol 3-phosphate, whereas the dihydroxyacetone phosphate pathway apparently does not play an important role in plants, as mentioned briefly in one publication [27]. The enzyme most likely responsible for the synthesis of glycerol phosphate from a glycolytic intermediate is glycerol phosphate dehydrogenase, which has been demonstrated in tissues from germinating seeds [112]. This activity could not be detected in leaves either by NADH-dependent photometry [38, 61] or by using an assay based on the formation of tritiated NADH from *sn*-[2-^{3}H] glycerol 3-phosphate [37]. Therefore an alternative pathway from glyceraldehyde phosphate to glycerol phosphate may be constructed, since all the enzymes necessary have been detected in leaf homogenates. This pathway [37] operates in the cytoplasm: glyceraldehyde phosphate is dephosphorylated to glyceraldehyde by a phosphatase such as phosphoglycerate phosphatase, which has been shown to hydrolyze glyceraldehyde phosphate [79]. Glyceraldehyde is then reduced in a NADPH-dependent reaction to glycerol by an aldose reductase [61], and glycerol is finally phosphorylated to glycerol phosphate by glycerol kinase, which may be rate-limiting in this sequence due to its low activity in leaves

[37]. Glycerol phosphate would be available immediately for microsomal acylation systems, whereas the chloroplast system has to be supplied by way of the phosphate translocator [36] whose capacity is high enough to be not rate-limiting in this sequence. This pathway is more complicated than a direct reduction of dihydroxyacetone phosphate, but enzymatic activities exceed 200 pmol/min/mg of protein. This rate has been estimated to be the upper limit, if present at all, for the activity of glycerol phosphate dehydrogenase in leaves [37].

Seed tissues contain small quantities of galactolipids, which after deacylation yield galactosyl ethyleneglycol [77]. The origin of the diol backbone is unknown.

D. UDP-Galactose

Problems similar to those encountered in the detection of routes for glycerol phosphate biosynthesis also exist when looking for enzymes actually providing UDP-galactose for lipid synthesis. UDP-glucose pyrophosphorylase and -epimerase, galactokinase, and UDP-galactose pyrophosphorylase as well as galactose-1-phosphate uridyltransferase are possible candidates. All of them have been demonstrated by radiochemical methods in plant tissues, which in most of the cases had a storage function, whereas in leaves only UDP-glucose pyrophosphorylase and -epimerase have activities high enough to be measured by photometry. Both enzymes were found to be soluble cytoplasmic proteins [6, 61]. Accordingly UDP-galactose is synthesized at highest rates in the cytoplasm and may not be available inside chloroplasts, since the envelope of mature chloroplasts is impermeable to this substance [32]. However, in view of recent results on the distribution of galactolipid-forming galactosyltransferases in chloroplasts, UDP-galactose need not be available in the interior of chloroplasts, since these galactosyltransferases are located in the envelope membranes [53], from which the inner one is the permeation barrier [32]. In view of the permeability of the outer membrane of the envelope, it may be anticipated that UDP-galactose, like sucrose, can reach the inner membrane or at least the outer layer of its bilayer structure. It is interesting to mention that unspecific diffusion of "impermeable" amino acids through the envelope is sufficient to supply chloroplasts with amino acids not synthesized by this organelle [24]. These diffusion rates are of the order of 0.1–1 µmol/mg chlorophyll/h for chloroplast preparations, corresponding roughly to 30–300 pmol/min/mg leaf protein, which are rates measured for many lipid-synthesizing enzymes. Similar to the changing enzymatic equipment of chloroplasts, a change of envelope permeability in various stages of plastid development has been observed [29]. Sucrose, for example, which cannot permeate mature chloroplasts, could easily enter etioplasts. In view of these additional facts and in view of the location of UDP-galactose-using enzymes in the organelle-enclosing membrane, compartmentation and permeability problems might not exist with respect to UDP-galactose.

The conversion of monogalactosyldiglyceride (MGDG) to digalactosyldiglyceride (DGDG) occurs by attaching a second galactopyranose unit in α-galactosidic linkage. Apart from UDP-galactose, galactinol could also serve as galactose donor in this reaction, since this compound is synthesized by chloroplast preparations [42] and functions as α-galactosyl donor in the biosynthesis of oligosaccharides belonging to the raffinose family [62]. However, substitution of UDP-[^{14}C] galactose by [^{14}C]galactinol in a DGDG-forming system did not result in any labeled DGDG [34]. In some cases plant glycosyl diglycerides contain, apart from galactose, appreciable proportions of glucose [46, 60], which may be anticipated to originate from UDP-glucose.

E. Acylation of Glycerol Phosphate

Acylation of glycerol phosphate occurs in several compartments [5, 8, 12, 27, 50, 89]: microsomes, chloroplasts, mitochondria, and fat droplets. In relation to galactolipid synthesis, only microsomal and plastidic systems from leaves will be discussed, since galactolipids are located predominantly in plastids, and microsomal lipid synthesis may be related to galactolipid synthesis in chloroplasts (see below).

The microsomal system from leaves produces phosphatidic and lysophosphatidic acids and the corresponding dephosphorylated products [12, 89], although proportions may vary according to assay time and condition. The system accepts $C_{14:0}$, $C_{16:0}$, $C_{18:0}$ and $C_{18:1}$, which before incorporation have to be converted into CoA-thioesters by membrane-bound thiokinases. Incorporation rates of 1–10 nmol/min/mg protein are among the highest found for lipid-synthesizing enzymes in plant cell homogenates. At present no data are available on the positional specificity of the leaf system, although this reaction may be very important in controlling the positional distribution of fatty acids according to chain length or degree of unsaturation.

When microsomal acyltransferases from storage tissues are assayed by incorporation of radioactivity from acyl groups and not from glycerol phosphate, the incorporation is often independent of glycerol phosphate [92, 106]. Under these conditions acyl groups are found in phospholipids, especially in the *sn*-2-position of phosphatidylcholine (PC). In recently reported experiments with an acyltransferase of this type from avocado microsomes, acyl-CoA and acyl-ACP esters were compared as acyl donors [91]. All acyl groups from ACP-thioesters ($C_{14:0}$, $C_{16:0}$, $C_{18:0}$, $C_{18:1}$, $C_{20:0}$) were primarily incorporated into triacylglycerols with a high specificity for the *sn*-3-position. With stearoyl-CoA a similar specificity in triglyceride synthesis was found. But in contrast to ACP-derivatives, CoA-esters were predominantly incorporated into PC with absolute specificity for the *sn*-2-position as observed before [106]. These results indicate that extraplastidic enzymes use acyl-ACP thioesters, which according to recent concepts would be expected to be confined to the interior of plastids. But stearoyl-ACP desaturase was also found in the stroma as well as in the cytoplasm [43, 49], so that acyl-ACP thioesters can be considered to be physiological substrates in extraplastidic systems.

Chloroplasts contain two acyltransferases [50]. One is a soluble enzyme and catalyzes the acylation of *sn*-glycerol 3-phosphate with palmitoyl- and oleoyl-CoA [5, 50]. The main product from both CoA-esters is lysophosphatidic acid. It is not yet known at which position these acyl groups are esterified.

The second enzyme in chloroplasts is an acyl-CoA: monoacyl glycerol phosphate acyltransferase, which is localized in the envelope and forms phosphatidic acid [50]. Thylakoids were found to have no or only very low acyltransferase activities. It is not known whether the two acyltransferases from chloroplasts can also use acyl-ACP thioesters as acyl donors. Incorporation of acyl groups from acyl-ACP into MGDG by *Euglena* homogenates was increased by the addition of glycerol phosphate, which may indicate the existence of acyl-ACP: glycerol phosphate acyltransferases in this organism [81].

In other publications [56, 92] the incorporation of acyl groups from CoA-thioesters into a variety of lipids has been reported to be catalyzed by spinach grana. With respect to fatty acid specificities in galactolipids, it is interesting that $C_{16:0}$, $C_{18:0}$, and $C_{18:1}$ were incorporated into MGDG. When incorporated into phospholipids by grana, these acids were found exclusively in the *sn*-2-position, similar to the incorporation by microsomes. On the other hand $C_{16:0}$, $C_{18:1}$, and $C_{18:3}$ were also used by chloroplast preparations to acylate monogalactosyl monoglycerides [4, 86]. In that case they were directed mainly into the *sn*-1-position of MGD [4]. The significance of these specificities with respect to the positional distribution of fatty acids in various lipids remains to be investigated. Surprisingly the corresponding acyl-ACP thioesters did not function as acyl donors with grana preparations [92]. Summarizing these results, the whole sequence of phosphatidic acid biosynthesis (and all subsequent reactions in galactolipid assembly) can be considered to operate at and in the envelope membranes. Free fatty acids are provided from the stroma side to be activated by the envelope-bound thiokinase to form a possibly also membrane-bound acyl-CoA. This thioester is combined with glycerol phosphate by the first acyltransferase, both enzyme and the other substrate coming from the stroma side. The resulting lyso-phosphatidic acid is transformed by the use of a second acyl-CoA to phosphatidic acid by the second envelope-bound acyltransferase. Positional or chain-length specificities of the enzymes involved in de novo synthesis are unknown.

F. Diglycerides

Diglycerides are derived from phosphatidic acid by the action of phosphatidic acid phosphatase. Evidence for this enzyme in leaf cells often comes from the distribution of label in products after acylation of glycerol phosphate, which is found to occur in various proportions in mono- and diglycerides. Only a few investigations tried to characterize the corresponding enzyme. In the nonphotosynthetic tissue of castor bean endosperm the enzyme, which was rather specific, as it did not hydrolyze glucose-6-phosphate, was localized primarily on endoplasmic reticulum (ER) membranes [68]. In onion stem explants activity was highest in soluble proteins [67]. In leaf homogenates highest specific activity was found in an unidentified particulate fraction sedimenting at 15,000 *g*, whereas the major part of the total activity was soluble [61]. Whether this soluble activity can be attributed to a specific phosphatidic acid phosphatase is doubtful in view of the activity of a soluble phosphatase from wheat germ, which among various other substrates also hydrolyzed phosphatidic acid [7].

With respect to galactolipid synthesis, the demonstration of phosphatidic acid phosphatase in chloroplast envelopes [50] is most important and is in agreement with the discovery that this membrane contains high levels of diglycerides [51]. These diglycerides amount to 15% of the lipids from spinach envelopes and contain $C_{16:3}$ and $C_{18:3}$ acids [18, 51], which are characteristic for galactolipids. The occurrence of $C_{16:3}$ in leaf diglycerides has been observed in many plants [45] and is important in view of the fact that this fatty acid is restricted to the *sn*-2-position of MGD during further derivatization of diglycerides. If one assumes that the fatty acid composition of diglycerides from whole leaves is representative of those in the envelope, then the diglyceride pool in the envelope could be sufficiently heterogeneous to account for the different diglyceride portions characteristic of various chloroplast lipids. The incorporation of label from acetate into fatty acids from neutral lipids or diglycerides has been investigated only in a few cases [31, 57, 84, 101]. Experiments with chloroplasts showed nearly equal activity in $C_{16:0}$ and $C_{18:1}$ from the beginning of the labeling period, whereas $C_{18:2}$ and $C_{18:3}$ were labeled later and in lower proportions. It would be desirable to analyze the labeling pattern of diglycerides with respect to distribution between molecular species and positions at the glycerol backbone, since such results are important for experiments described below, which are interpreted as indicating desaturation of fatty acids after incorporation of diglycerides into galactolipids.

G. Galactosyltransferases

The final steps in galactolipid biosynthesis are catalyzed by galactosyltransferases with the following characteristics: UDP-galactose is the galactosyl donor for the galactosylation of diglycerides, MGDG, DGDG, and trigalactosyldiglyceride (TGDG) [51, 71]; each reaction may be catalyzed by a separate enzyme [9, 11, 70, 75, 94]; these separate galactosyltransferases are localized in chloroplast envelopes [15, 16, 41, 50, 51, 53, 64]. Widely varying absolute enzymatic activities are presented for these enzymes in leaf homogenates or subcellular fractions, e.g., for envelope preparations, for which activities from 3–45,000 pmol/min/mg are given [41, 51, 64]. K_m-values for UDP-galactose cover a range from 2.2×10^{-7} M to 330×10^{-7} M [11, 75, 94]. In view of the difficulties in assaying these enzymes, it is not suprising that different specificities with respect to acyl groups in substrates (see below) and various proportions of the different galactosyltransferases are ascribed to different subcellular fractions. Balancing

Table 1. Subcellular distribution of enzymatic activities contributing to the biosynthesis of galactolipids

Cytoplasm	Envelope	Chloroplast stroma
Phosphoglycerate phosphatase, 15 [79]	Thiokinase, ≧7 [50]	Enzymes of de novo synthesis of fatty acids, 0.09 [113]
Aldose reductase, 3.5 [61]	Monoacyl glycerophosphate acyltransferase, ≧7 [50]	Thioesterase, 0.17 [91]
Glycerol kinase, 3 [37]	Phosphatidic acid phosphatase	Glycerophosphate acyltransferase, 7 [5]
UDP-Glucose pyrophosphorylase, 30 [61]	1. Galactosyltransferase, 2. Galactosyltransferase, 3. Galactosyltransferase } 45 [51]	
UDP-Glucose 4-epimerase, 3.5 [61]		
Phosphatidic acid phosphatase, 12 [61]		

Only highest activities are given in nmol/min/mg protein of fraction specified, with the exception of "cytoplasmic" proteins that comprise total soluble proteins in leaf homogenates; mg chlorophyll in intact chloroplasts was multiplied by a factor of 6 to obtain mg stroma protein. Values for thiokinase and monoacyl glycerophosphate acyltransferase are included in consideration of rate-limiting steps and distribution of products. The total galactosylation activity in a chloroplast is localized to about 95% in the envelope and to about 5% in the thylakoid system.

of enzymatic activity in subcellular fractions may be disturbed by uneven distribution of lipid-degrading activities in these fractions [39], and the fragility of the envelope may release part of the activity into a soluble form subsequently ascribed to soluble proteins of the cytoplasm [94] or chloroplast stroma [11, 75]. Therefore an important and at present open question is, what percentage of the various galactosyltransferases is not localized in the envelope. It was shown that microsomes do not synthesize galactolipids [52], whereas 40–65% of the total galactolipid-forming enzymes were found in "soluble" proteins [70, 75, 94]. Taking into account the cytoplasmic origin of UDP-galactose, even a cytoplasmic enzyme [94] could galactosylate substrates at the surface of chloroplast envelopes. Different turnover rates of galactose and diglyceride moieties in galactolipids, as suggested by labeling experiments [20, 99], could be due to a reversibility of the galactosylation reactions. However, it has been shown that these reactions are not reversible [11].

Summarizing the results of the preceding Sections, the important role of envelopes in galactolipid synthesis is evident (see Table 1). Soluble substrates and enzymes from the cytoplasma and the stroma meet with envelope-bound substrates and enzymes at or in the envelope to assemble galactolipids. The exclusive formation of galactolipids in the envelope would require the export of all thylakoid-bound galactolipids from the envelope. In young and differentiating plastids, thylakoids could receive the bulk of their galactolipids at the zones where they are budding off from the envelope, whereas in mature chloroplasts existing thylakoids could receive additional galactolipids by lateral diffusion in continuous membrane systems [19]. Details of this picture may become complicated in view of the bilayer structure of membranes, in view of a possible asymmetric distribution of lipids when substrates are delivered from only one side and in view of low rates of lipid transport from one layer to the other in bilayers, as found for example in erythrocyte membranes [82].

H. Fatty Acid Specificities in Galactolipids

The pathway of consecutive galactosylation should result in galactolipids differing only in number of galactosyl residues but not in nature of acyl groups in the diglyceride moiety. Identical diglyceride patterns (Fig. 1A) are found in MGDG, DGDG, and TGDG from the prokaryotic blue-green alga *Tolypothrix* [114]. However, this pattern is unique since other blue-green algae [114] and most eukaryotic organisms show differences in the diglyceride portions of their galactolipids [69]. In most cases, the proportion of palmitic acid increases when MGDG, DGDG, and TGDG from the same tissue are compared [22, 23, 78]. Furthermore, plants forming $C_{16:3}$ (=16:3-plants) direct this acid rather specifically into MGDG and exclude it from other lipids [45]. These facts, known since the first analyses of plant lipids [1], are demonstrated by mass spectra (Fig. 1B, C), showing the diglyceride moieties of MGDG and DGD from 18:3- and 16:3-plants [94]. The high percentage of the 18:3/16:0 combination at m/e 573 in DGDG and its low proportion in MGDG in both groups of plants as well as the reverse situation for the 18:3/16:3 combination in 16:3-plants remain to be explained. Mass spectrometry and gas-liquid chromatography (GLC) of intact lipids have shown that in MGDG and DGDG C_{16}–acids are preferentially combined with C_{18}–acids as evident from the absence or low proportion of C_{16}/C_{16} combinations [3, 10, 85]. In sulfolipids this combination occurs at a higher percentage [104].

The positional distribution of fatty acids in plant glycolipids (Fig. 3), analyzed by lipase hydrolysis [21] or in pure molecular species by mass spectrometry [10], showed significant differences between plants from various taxonomic positions. The importance of positional distribution in relation to labeling of fatty acids in galactolipids will be discussed below. The most obvious specificity found in MGDG is that polyenoic acids with 16 carbon atoms are localized preferentially at C-2 throughout the plant kingdom, pointing to a selection according to chain length [3, 85, 87]. Palmitic acid in DGDG [85, 114] and in sulfolipids [104] shows a distribution that depends on the systematic position of the plant in two ways: (1) the lower this position in taxonomy, the higher the proportion of $C_{16:0}$ at C-2; (2) in 16:3-plants palmitic acid seems to be localized in appreciable proportions at C-2 of DGDG and sulfolipids. $C_{18:4}$ is characteristic for *Boraginaceae* [44] and was found at C-2 in MGDG from *Anchusa* [87]. It should be kept in mind that these generalizations are made on the basis of results obtained from only a limited number of plants.

These facts demonstrate the importance of investigating enzymatic reactions that may control the composition or positional distribution of fatty acids in galactolipids. As mentioned before, specificities of glycerol phosphate acyltrans-

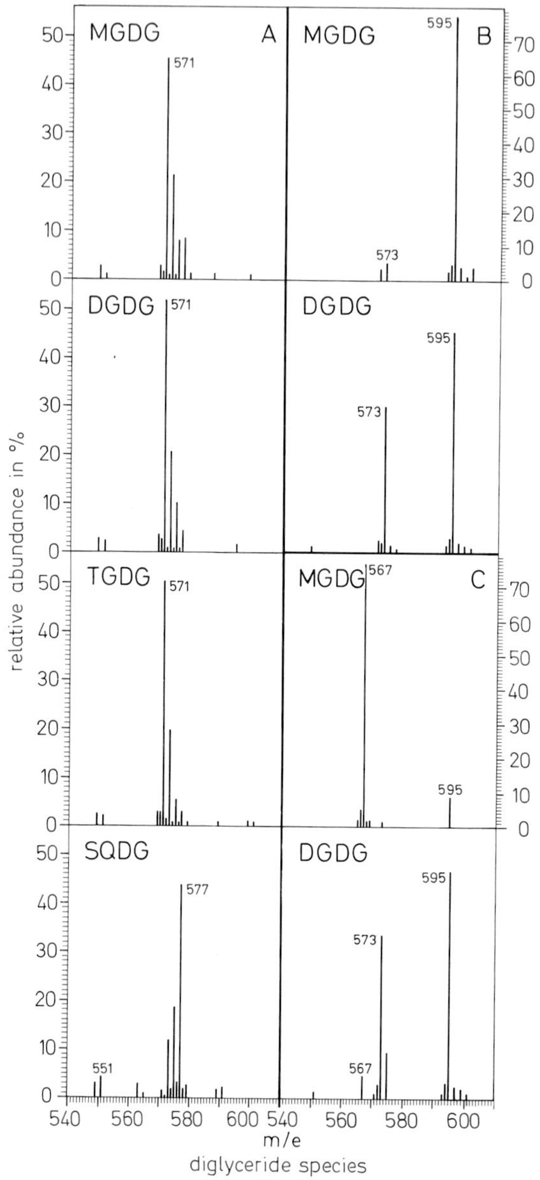

Fig. 1

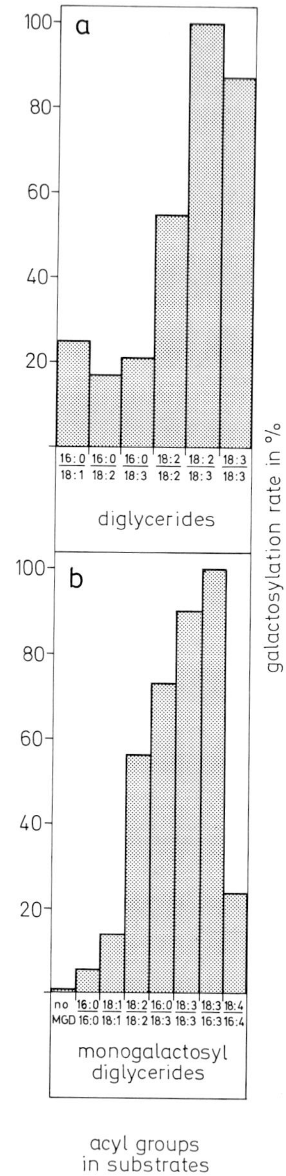

Fig. 2

Fig. 1. Partial mass spectra of acetylated glycolipids from (A), *Tolypothrix*, (B), *Cucurbita*, and (C), *Anthriscus*, showing the diglyceride fragments. Fragments at m/e 551, 567, 571, 573, 577, and 595 are due to 16:0/16:0, 18:3/16:3, 18:4/16:0, 18:3/16:0, 18:1/16:0, and 18:3/18:3 combinations, respectively [114]

Fig. 2. Diglyceride specificity of galactosyltransferases (a), forming MGDG by galactosylation of diglycerides (redrawn from Ref. [70]) or (b), forming DGDG by galactosylation of MGDG (Ref. [94])

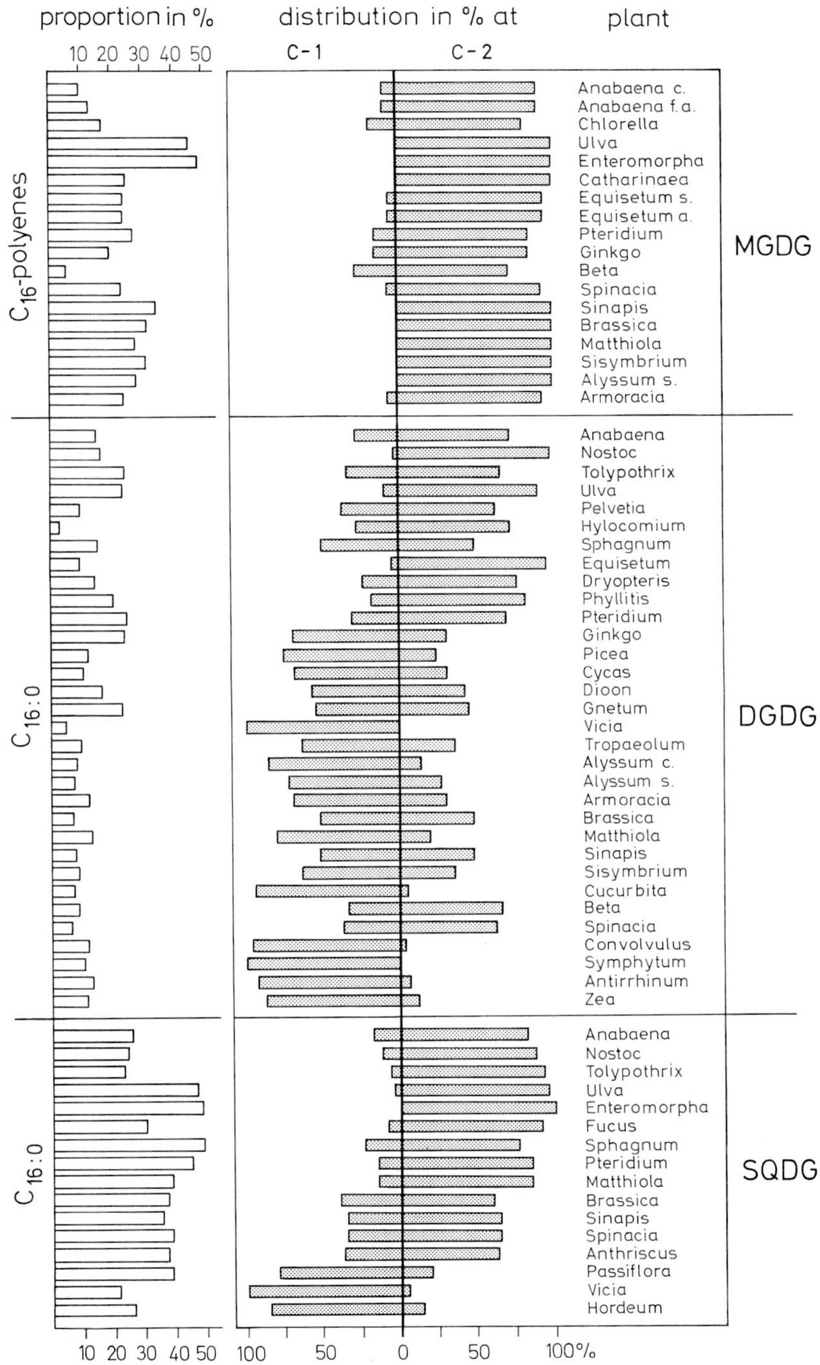

Fig. 3. Proportion (*left side*) and positional distribution of C_{16}—fatty acids in glycolipids. C_{16}—polyenes in MGDG refer to $C_{16:2}$ in *Anabaena c.* and *Chlorella* [87], $C_{16:4}$ in *Ulva* and *Enteromorpha*, and $C_{16:3}$ in the other species. Published [3, 85, 87, 104, 114] and some unpublished results (C. Arnoldt and E. Heinz) are included

ferases have not been investigated. Further possibilities to control fatty acid compositon may be attributed to the various galactosyltransferases, which, due to different specificities, could pick out appropriate substrates and deplete existing pools so that the finally observed patterns would result. The galactosyltransferases forming MGDG and DGDG have been investigated with respect to such specificities. Unfortunately different results were obtained for the enzyme galactosylating diglycerides to MGDG. In one case [18] no specificity for fatty acids in diglycerides was observed, similar to choline phosphotransferase from the same plant [14]; in other publications [9, 70] galactosylation rates increased with increasing numbers of double bonds in diglyceride fatty acids (Fig. 2a). For the enzyme galactosylating MGDG to DGDG the situation is clearer since only one publication exists [94]. Galactosylation rates increased with increasing numbers of double bonds in the fatty acids of MGDG, but beyond six double bonds a decline is again observed (Fig. 2b). The differences in galactosylation rates of 16:0/18:3-, 18:3/18:3-, and 18:3/16:3-MGDG-species cannot be related to the differences in the diglyceride portions observed in natural MGDG and DGDG, and therefore selectivity at this galactosylation step may have no relevance. In relation to experiments discussed below, the low galactosylation rates of the dipalmitoyl glycerol as well as the dipalmitoyl-MGDG-species should be pointed out. It is interesting that other glycosyltransferases involved in the formation of glycosyl diglycerides from various organisms differ greatly in preferring different diglycerides. For some enzymes dipalmitoyl glycerol is a very good substrate (*Streptococcus* [76], *Bifidobacterium* [105], rat brain [107]), whereas for others it is a very inefficient acceptor (a halotolerant bacterium [97], *Micrococcus* [63], *Mycoplasma* [96]).

Other possibilities to explain the preceding differences may be specificities of reacylating enzymes [4, 86], which have not been investigated for DGDG and its difference to MGDG. Finally specific desaturation of fatty acids attached in ester linkage to specific lipids may contribute to these differences, as discussed in the last section.

I. Radioactive Labeling of Galactolipids in vivo

The first experiments of this kind were carried out with *Chlorella* [20]. The results were interpreted as indicating a rapid turnover of galactose moieties as compared to the diglyceride part of galactolipids. This could be due to a reversibility of galactosylation reactions, which was not observed in later experiments [11], or to a regalactosylation after hydrolytic cleavage. On the other hand such labeling could also result from differences in size and turnover of pools providing the various precursors. Since then similar experiments have been carried out with the same or other organisms including leaves of higher plants, offering the same or additional precursors and analyzing the lipids and their fatty acids with increased resolution [108]. The results confirmed the earlier experiments and are in agreement with a consecutive galactosylation sequence in galactolipid biosynthesis [109, 110], whereas differing results were obtained concerning a rapid turnover of galactose moieties [99, 110] in higher plants. On the other hand the half-life for PC has been estimated to be of the order of 20 h [83], and label incorporated from acetate into leaf lipids had disappeared completely after 10 days [100], indicating a continuous breakdown and resynthesis of membrane lipids, which also was evident from a diurnal variation of lipid levels [47].

Detailed analysis of fatty acid labeling became important to the mechanism of fatty acid biosynthesis, since it was discovered that desaturation of fatty acids may occur subsequently to their incorporation into lipids [2, 28, 72–74, 83, 84, 87, 93, 95, 102, 111]. This is concluded from the variation of total and specific activity in those lipid-linked fatty acids ($C_{16:0}$, $C_{16:1}$, $C_{18:0}$, and $C_{18:1}$) that are considered to be intermediates in a desaturation sequence ending in polyenoic acids. Most authors are tempted to suggest that acyl groups bound through oxygen ester linkage in a lipid are direct substrates for a desaturase. So far this mechanism has only been deduced from changing labeling patterns of fatty acids in lipids and awaits demonstration by in vitro reactions with definite exclusion of a rapid shuttling of acyl groups between oxygen ester linkage in lipids and thioester linkage in CoA, ACP, or enzyme protein. Despite these difficulties, an increasing number of experiments with cell-free systems from various organisms, which will not be discussed here, is claimed to be evidence for direct desaturation or modification of lipid-linked acyl chains. On the other hand thioesters of CoA and ACP are undoubtedly substrates for soluble or membrane-bound desaturases in different compartments introducing the first or second double bond into long-chain acyl groups. But since both mechanisms need not be mutually exclusive, desaturation of both thioesters as well as lipid-linked oxygen esters may operate in parallel and contribute in yet unknown proportions to the formation of unsaturated fatty acids. It has been suggested [26] that in contrast to thioester desaturases, which measure their point of attack from the carboxyl group of the acyl chain (Δ-x-desaturase), oxygen ester desaturases could measure from the methyl end (ω-x-desaturase).

Apart from these general agreements there are differences with respect to the lipids that are considered to be the predominant substrates in correlation to desaturation. If it turns out, as suggested by one set of experiments [83, 84, 95, 111], that mainly one lipid or a limited group of lipids not present in all compartments, is involved in desaturation, then a subsequent delivery of desaturation products to other lipids in different compartments will be required. If, as suggested by another group of experiments [2, 72–74, 93, 102], several lipids present in various compartments are substrates in desaturation, acyl transfer between compartments will not be required. Since present experimental results support both models, both mechanisms may operate in parallel, and proportions may depend on tissue and organelle differentiation. The two competing candidates for major substrates in conjunction with desaturation are PC and MGDG.

In many investigations [83, 84, 95, 111] the predominance of labeling of PC at early sampling times and a shift of this label to other lipids during subsequent days or hours suggests that desaturation, at least of $C_{18:1}$ to $C_{18:2}$

[28], occurs predominantly at PC-bound acyl groups, which then are transferred to galactolipids for example, in which no or little desaturation occurs. At present it is not clear which acyl groups are transferred from PC to galactolipids, since galactolipids are characterized by trienoic acids, which hardly become labeled in PC. Therefore it has been suggested [111] that complete diglycerides may be derived from PC for galactosylation. Silver nitrate chromatograms of newly synthesized MGDG-species have been interpreted as showing that diglycerides available for galactosylation are already highly unsaturated [111]. On the other hand acetate labeling of pumpkin leaves showed that total and specific activity of $C_{18:1}$ and $C_{18:2}$ was several times higher in PC than in nonpolar lipids [84], which may be assumed to contain the diglyceride precursor of PC. This may suggest that desaturation does not occur with intermediates of lipid biosynthesis up to and including diglycerides. The unsaturated diglycerides available for galactolipid synthesis could be secondary products derived from PC after desaturation. The location of PC-synthesizing enzymes on ER membranes (whereas PC itself is localized in all but thylakoid membranes) requires that $C_{18:2}$ from PC in ER membranes has to be transferred to galactolipids in thylakoid or envelope membranes with a concomitant conversion to $C_{18:3}$. In fact PC from microsomal fractions contains more radioactivity than that from other fractions after differential centrifugation [95]. The gap to be closed in this model is the fact that it is $C_{18:2}$ that disappears from PC, whereas it is $C_{18:3}$ that shows up in MGDG. This model is supported by the following observations: the conversion of $C_{18:1}$ to $C_{18:2}$ was maximal in microsomes, whereas desaturation of $C_{18:2}$ to $C_{18:3}$ was highest in the plastid fraction from pea seedlings [103]. Polyenoic acid synthesis in chloroplast preparations is increased by addition of other subcellular fractions [31]. ER membranes of endosperm synthesize phospholipids and enzymes that then are exported to membranes of other organelles [25]. Exchange reactions of fatty acids or intact phospholipids between microsomal and other subcellular fractions with the exclusion of plastids have been carried out in vitro and are catalyzed by soluble proteins [66]. The diglyceride portion from PC transported to the chloroplast envelope may be liberated by a reversal of the choline phosphotransferase reaction [59], to be available for galactosylation.

In contrast to this PC-dominated model, other experiments show that besides PC, also MGDG and other lipids may be involved in desaturation [2, 72–74, 93, 102]. In blue-green algae, which lack PC, lipid-linked desaturation ought to occur in other lipids, as for example in phosphatidylglycerol (PG) and MGDG [72]. Therefore it has been suggested that two major lipid-connected pathways of desaturation operate in parallel [2]. One involves PC and is located in microsomes; the other requires MGDG and is confined to thylakoid or envelope membranes.

Support for this model and its extension to higher plants may come from recent experiments with leaves from *Anthriscus* [93], which like all 16:3-plants contains substantial quantities of a characteristic trienoic acid almost exclusively in MGDG. This plant was chosen in view of recently developed schemes on the function of $C_{16:3}$ as a precursor of $C_{18:3}$ [98] and the exclusive involvement of PC in polyenoic acid synthesis. Furthermore, it was hoped that these experiments would elucidate the origin of $C_{16:3}$ and its specific introduction into the *sn*-2-position of MGDG. The labeling sequence from $C_{18:1}$ to $C_{18:3}$ was present in the fatty acids from MGDG at times when $C_{18:3}$ was not labeled in PC, so that there is no need to derive $C_{18:3}$ in MGDG from PC. The labeling patterns of C_{16}-unsaturated acids and their variations with time fit best into a mechanism whereby the complete series of desaturations from $C_{16:0}$ via $C_{16:1}$ and $C_{16:2}$ to $C_{16:3}$ is confined to a pool that can drain off intermediates and products only into MGDG. The exclusion of labeled unsaturated C_{16}-acids from DGDG at times when other acids in DGDG were heavily labeled, suggests, in agreement with other experiments [109, 110], the existence of two pools of MGDG, of which the one with unsaturated C_{16}-acids is not or only to a limited extent available for galactosylation to DGDG. The positional analysis of MGDG showed that labeled C_{18}-acids are excluded from the *sn*-2-position after short-term-labeling, although large quantities of $C_{18:3}$ are present in this position. On the other hand, the complete C_{16}-series was present labeled at the *sn*-2-position. Argentation chromatography showed many species of labeled MGDG and DGDG, in agreement with the demonstration of many diglyceride fragments of minor intensity in the mass spectra [85]. The time-dependent changes in the distribution of label in the various species may be interpreted as showing that the two galactosyltransferases may operate without diglyceride specificity. The question concerning the specific positioning of $C_{16:3}$ is not answered but is related to the separation of a pool of $C_{16:0}$ that is desaturated to $C_{16:3}$ and from which intermediates and end products can only enter efficiently the *sn*-2-position of MGDG.

Acknowledgement. Investigations carried out in the author's laboratory were supported by the *Deutsche Forschungsgemeinschaft.*

References

1. Allen, C.F., Good, P., Davis, H.F., Fowler, S.D.: Plant and chloroplast lipids I. Separation and composition of major spinach lipids. Biochem. Biophys. Res. Commun. **15**, 424–430 (1964)
2. Appleby, R.S., Safford, R., Nichols, B.W.: The involvement of lecithin and monogalactosyl diglyceride in linoleate synthesis by green and blue-green algae. Biochim. Biophys. Acta **248**, 205–211 (1971)
3. Auling, G., Heinz, E., Tulloch, A.P.: Combination and positional distribution of fatty acids in plant galactolipids. Hoppe-Seylers Z. Physiol. Chem. **352**, 905–912 (1971)
4. Bajwa, S.S., Sastry, P.S.: Enzymatic acylation of monogalactosyl monoglyceride by leaf chloroplast acetone powder. Indian J. Biochem. Biophys. **10**, 65–66 (1973)
5. Bertrams, M., Heinz, E.: Experiments on enzymatic acylation of *sn*-glycerol 3-phosphate with enzyme preparations from pea and spinach leaves. Planta **132**, 161–168 (1976)
6. Bird, I.F., Cornelius, M.J., Keys, A.J., Whittingham, C.P.: Intracellular site of sucrose synthesis in leaves. Phytochemistry **13**, 59–64 (1974)
7. Blank, M.L., Snyder, F.: Specificities of alkaline and acid phosphatases in the dephosphorylation of phospholipids. Biochemistry **9**, 5034–5036 (1970)
8. Boehler, B.A., Ernst-Fonberg, M.L.: *sn*-Glycerol-3-phosphate transacylase activity in *Euglena gracilis* organelles. Arch. Biochem. Biophys. **175**, 229–235 (1976)

9. Bowden, B.N., Williams, P.M.: Galactolipid biosynthesis in *Zea Mays* shoots. Phytochemistry **12**, 1059–1064 (1973)
10. Budzikiewicz, H., Rullkötter, J., Heinz, E.: Massenspektroskopische Untersuchungen an Glycosylglyceriden. Z. Naturforsch. **28c**, 499–504 (1973)
11. Chang, S.B., Kulkarni, N.D.: Enzymatic reactions for galactolipid synthesis with a soluble, sub-chloroplast fraction from *Spinacia oleracea*. Phytochemistry **9**, 927–934 (1970)
12. Cheniae, G.M.: Phosphatidic acid and glyceride synthesis by particles from spinach leaves. Plant Physiol. **40**, 235–243 (1965)
13. Dennis, D.T., Green, T.R.: Soluble and particulate glycolysis in developing castor bean endosperm. Biochem. Biophys. Res. Commun. **64**, 970–975 (1975)
14. Devor, K.A., Mudd, J.B.: Biosynthesis of phosphatidylcholine by enzyme preparations from spinach leaves. J. Lipid Res. **12**, 403–411 (1971)
15. Douce, R.: Site of biosynthesis of galactolipids in spinach chloroplasts. Science **183**, 852–853 (1974)
16. Douce, R., Benson, A.A.: Isolation, purification and properties of the envelope of spinach chloroplasts. Portug. Acta Biol. **14**, 45–64 (1974)
17. Duval, J.C., Trémolières, A., Rousseau, B.: Croissance et division du compartiment plastidial et synthèse de la chlorophylle et de l'acide linolénique au cours du développement de la première feuille de pois. Physiol. Vég. **14**, 67–76 (1976)
18. Eccleshall, T.R., Hawke, J.C.: Biosynthesis of monogalactosyl diglyceride by chloroplasts from *Spinacia oleracea* and from some gramineae. Phytochemistry **10**, 3035–3045 (1971)
19. Edidin, M.: Rotational and translational diffusion in membranes. Ann. Rev. Biophys. Bioeng. **3**, 179–201 (1974)
20. Ferrari, R.A., Benson, A.A.: The path of carbon in photosynthesis of the lipids. Arch. Biochem. Biophys. **93**, 185–192 (1961)
21. Fischer, W., Heinz, E., Zeus, M.: The suitability of lipase from *Rhizopus arrhizus delemar* for analysis of fatty acid distribution in dihexosyl diglycerides, phospholipids and plant sulfolipids. Hoppe-Seylers Z. Physiol. Chem. **354**, 1115–1123 (1973)
22. Galliard, T.: Aspects of lipid metabolism in higher plants. I+II. Phytochemistry **7**, 1907–1922 (1968)
23. Gellerman, J.L., Anderson, W.H., Richardson, D.G., Schlenk, H.: Distribution of arachidonic and eicosapentaenoic acids in the lipids of mosses. Biochim. Biophys. Acta **388**, 277–290 (1975)
24. Gimmler, H., Schäfer, G., Kraminer, H., Heber, U.: Amino acid permeability of the chloroplast envelope as measured by light scattering, volumetry and amino acid uptake. Planta **120**, 47–61 (1974)
25. Gonzales, E., Beevers, H.: Role of the endoplasmic reticulum in glyoxysome formation in castor bean endosperm. Plant Physiol. **57**, 406–409 (1976)
26. Gurr, M.I.: The biosynthesis of unsaturated fatty acids. In: Biochemistry of Lipids. Kornberg, H.L., Phillips, D.C. (eds.). London: Butterworths, 1974, Vol. 4, pp. 181–235
27. Gurr, M.I., Blades, J., Appleby, R.S., Smith, C.G., Robinson, M.P., Nichols, B.W.: Studies on seed-oil triglycerides. Triglyceride biosynthesis and storage in whole seeds and oil bodies of *Crambe abyssinica*. Europ. J. Biochem. **43**, 281–290 (1974)
28. Gurr, M.I., Robinson, M.P., James, A.T.: The mechanism of formation of polyunsaturated fatty acids by photosynthetic tissue. The tight coupling of oleate desaturation with phospholipid synthesis in *Chlorella vulgaris*. Europ. J. Biochem. **9**, 70–78 (1969)
29. Hampp, R., Schmidt, H.W.: Changes in envelope permeability during chloroplast development. Planta **129**, 69–73 (1976)
30. Harris, R.V., James, A.T.: Linoleic and α-linolenic acid biosynthesis in plant leaves and a green alga. Biochim. Biophys. Acta **106**, 456–464 (1965)
31. Hawke, J.C., Rumsby, M.G., Leech, R.M.: Lipid biosynthesis by chloroplasts isolated from developing *Zea Mays*. Phytochemistry **13**, 403–413 (1974)
32. Heber, U.: Metabolite exchange between chloroplasts and cytoplasm. Ann. Rev. Plant Physiol. **25**, 393–421 (1974)
33. Heber, U., Kirk, M.R., Gimmler, H., Schäfer, G.: Uptake and reduction of glycerate by isolated chloroplasts. Planta **120**, 31–46 (1974)
34. Heisig, O.: Investigations on the biosynthesis of digalactosyl diglyceride. University of Köln, Thesis, 1975

35. Heldt, H.W.: Adenine nucleotide translocation in spinach chloroplasts. FEBS Lett. **5**, 11–14 (1969)
36. Heldt, H.W., Rapley, L.: Specific transport of inorganic phosphate, 3-phosphoglycerate and dihydroxyacetonephosphate, and of dicarboxylates across the inner membrane of spinach chloroplasts. FEBS Lett. **10**, 143–148 (1970)
37. Hippmann, H., Heinz, E.: Glycerol kinase in leaves. Z. Pflanzenphysiol. **79**, 408–418 (1976)
38. Huang, A.H.: Enzymes of glycerol metabolism in the storage tissues of fatty seedlings. Plant Physiol. **55**, 555–558 (1975)
39. van Hummel, H.C.: Some observations on biosynthesis and ratio regulation of mono- and digalactosyl diglycerides by chloroplasts and other subcellular fractions from spinach leaves. Z. Pflanzenphysiol. **71**, 228–241 (1974)
40. van Hummel, H.C.: Chemistry and biosynthesis of plant galactolipids. In: Progr. Chem. Nat. Org. Prod. Herz, W., Grisebach, H., Kirby, G.W. (eds.). Wien-New York: Springer-Verlag, 1975, Vol. 32, pp. 267–295
41. van Hummel, H.C., Hulsebos, T.J.M., Wintermans, J.F.G.M.: Biosynthesis of galactosyl diglycerides by non-green fractions from chloroplasts. Biochim. Biophys. Acta **380**, 219–226 (1975)
42. Imhoff, V.: Synthesis of galactosides by chloroplasts isolated from pea leaves. Hoppe-Seylers Z. Physiol. Chem. **354**, 1550–1554 (1973)
43. Jacobson, B.S., Jaworski, J.G., Stumpf, P.K.: Fat metabolism in higher plants. Stearyl-acyl carrier protein desaturase from spinach chloroplasts. Plant Physiol. **54**, 484–486 (1974)
44. Jamieson, G.R., Reid, E.H.: The leaf lipids of some members of the *Boraginaceae* family. Phytochemistry **8**, 1489–1494 (1969)
45. Jamieson, G.R., Reid, E.H.: The occurrence of hexadeca-7,10,13-trienoic acid in the leaf lipids of angiosperms. Phytochemistry **10**, 1837–1843 (1971)
46. Jamieson, G.R., Reid, E.H.: The sugar components of the galactosyl diglycerides from green plants. Phytochemistry **15**, 135–136 (1976)
47. Jarvis, M.C., Duncan, H.J.: Diurnal variations in lipids of bracken fronds. Phytochemistry **14**, 77–78 (1975)
48. Jaworski, J.G., Goldschmidt, E.E., Stumpf, P.K.: Properties of the palmityl acyl carrier protein: stearyl acyl carrier protein elongation system in maturing safflower seed extracts. Arch. Biochem. Biophys. **163**, 769–776 (1974)
49. Jaworski, J.G., Stumpf, P.K.: Fat metabolism in higher plants. Properties of a soluble stearyl-acyl carrier protein desaturase from maturing *Carthamus tinctorius*. Arch. Biochem. Biophys. **162**, 158–165 (1974)
50. Joyard, J., Douce, R.: Site of synthesis of phosphatidic acid and diacyl glycerol in spinach chloroplasts. Biochim. Biophys. Acta **486**, 273–285 (1977)
51. Joyard, J., Douce, R.: Mise en évidence et rôle des diacylglycerols de l'enveloppe des chloroplastes d'Épinard. Biochim. Biophys. Acta **424**, 125–131 (1976)
52. Joyard, J., Douce, R.: L'enveloppe des chloroplastes est-elle capable de synthétiser la phosphatidylcholine? C. R. Acad. Sci. **282**, 1515–1519 (1976)
53. Joyard, J., Douce, R.: Préparation et activités enzymatiques de l'enveloppe des chloroplastes d'Épinard. Physiol. Vég. **14**, 31–48 (1976)
54. Kannangara, C.G., Henningsen, K.W., Stumpf, P.K., Appelqvist, L.A., von Wettstein, D.: Lipid biosynthesis by isolated chloroplasts in relation to plastid development. Plant Physiol. **48**, 526–531 (1971)
55. Kannangara, C.G., Jacobson, B.S., Stumpf, P.K.: A comparison of fatty acid-synthesizing enzymes in chloroplasts isolated from mature and immature leaves of spinach. Plant Physiol. **52**, 156–161 (1973)
56. Kannangara, C.G., Stumpf, P.K.: The formation of fatty acyl thioesters during ^{14}C-1-acetate incorporation into long chain fatty acids by isolated spinach chloroplasts. Biochem. Biophys. Res. Commun. **44**, 1544–1551 (1971)
57. Kannangara, C.G., Stumpf, P.K.: Fat metabolism in higher plants. The biosynthesis of polyunsaturated fatty acids by isolated spinach chloroplasts. Arch. Biochem. Biophys. **148**, 414–424 (1972)
58. Kannangara, C.G., Stumpf, P.K.: A procaryotic type of acetyl CoA carboxylase in spinach chloroplasts. Arch. Biochem. Biophys. **152**, 83–91 (1972)
59. Kanoh, H., Ohno, K.: The role of 1,2-diacylglycerol: CDP-choline (ethanolamine)-choline

(ethanolamine) phosphotransferase in forming phospholipid species in rat liver. In: Lipids. Paoletti, R., Porcellati, G., Jacim, G. (eds.). New York: Raven Press, 1976, Vol. 1, pp. 39–48
60. Kondo, Y., Ito, S., Fujino, Y.: Sulfoquinovosyldiglyceride and triglycosyldiglyceride in rice grains. Agr. Biol. Chem. **38**, 2549–2552 (1974)
61. Königs, B., Heinz, E.: Investigation of some enzymatic activities contributing to the biosynthesis of galactolipid precursors in *Vicia faba*. Planta **118**, 159–169 (1974)
62. Lehle, L., Tanner, W.: Synthesis of raffinose-type sugars. In: Methods in Enzymology. Ginsburg, V. (ed.). New York-London: Academic Press, 1972, Vol. 28, pp. 522–530
63. Lennarz, W.J., Talamo, B.: The chemical characterization and enzymatic synthesis of mannolipids in *Micrococcus lysodeikticus*. J. Biol. Chem. **241**, 2707–2719 (1966)
64. Liedvogel, B., Kleinig, H.: Galactolipid synthesis in chromoplast internal membranes of the daffodil. Planta **129**, 19–21 (1976)
65. Macher, B.A., Brown, C.P., McManus, T.T., Mudd, J.B.: Studies on phospholipid-synthesizing enzyme activities during the growth of etiolated cucumber cotyledons. Plant Physiol. **55**, 130–136 (1975)
66. Mazliak, P., Donday, D., Demandre, C., Kader, J.C.: Exchange processes between organelles involved in membrane lipid biosynthesis. In: Recent Advances in the Chemistry and Biochemistry of Plant Lipids. Galliard, T., Mercer, E.I. (eds.). London-New York-San Francisco: Academic Press, 1975, pp. 301–318
67. Morré, D.J., Roland, J.-C., Lembi, C.A.: Comparisons of isolated plasma membranes from plant stems and rat liver. Proc. Indiana Acad. Sci. **79**, 96–106 (1969)
68. Moore, T.S., Lord, J.M., Kagawa, T., Beevers, H.: Enzymes of phospholipid metabolism in the endoplasmic reticulum of castor bean endosperm. Plant Physiol. **52**, 50–53 (1973)
69. Mudd, J.B., Garcia, R.E.: Biosynthesis of glycolipids. In: Recent Advances in the Chemistry and Biochemistry of Plant Lipids. Galliard, T., Mercer, E.I. (eds.). London-New York-San Francisco: Academic Press, 1975, pp. 161–201
70. Mudd, J.B., van Vliet, H.H.D.M., van Deenen, L.L.M.: Biosynthesis of galactolipids by enzyme preparations from spinach leaves. J. Lipid Res. **10**, 623–630 (1969)
71. Neufeld, E.F., Hall, C.W.: Formation of galactolipids by chloroplasts. Biochem. Biophys. Res. Commun. **14**, 503–508 (1964)
72. Nichols, B.W.: Fatty acid metabolism in the chloroplast lipids of green and blue-green algae. Lipids **3**, 354–360 (1968)
73. Nichols, B.W., James, A.T., Breuer, J.: Interrelationships between fatty acid biosynthesis and acyl-lipid synthesis in *Chlorella vulgaris*. Biochem. J. **104**, 486–496 (1967)
74. Nichols, B.W., Moorhouse, R.: The separation, structure and metabolism of monogalactosyl diglyceride species in *Chlorella vulgaris*. Lipids **4**, 311–316 (1969)
75. Ongun, A., Mudd, J.B.: Biosynthesis of galactolipids in plants. J. Biol. Chem. **243**, 1558–1566 (1968)
76. Pieringer, R.A.: The metabolism of glyceride glycolipids I. Biosynthesis of monoglucosyl diglyceride and diglucosyl diglyceride by glucosyltransferase pathways in *Streptococcus faecalis*. J. Biol. Chem. **243**, 4894–4903 (1968)
77. Prokasova, N.V., Todria, C.G., Rozynov, B.V., Vaver, V.A., Bergelson, L.D.: Acylated galactoside of ethyleneglycol—a new type of a natural diol lipid. Dokl. Akad. Nauk USSR **214**, 1448–1451 (1974)
78. Radunz, A.: Localization of tri- and digalactosyl diglyceride in thylakoid membranes with serological methods. Z. Naturforsch. **31c**, 589–593 (1976)
79. Randall, D.D., Tolbert, N.E.: 3-Phosphoglycerate phosphatase in plants I. Isolation and characterization from sugarcane leaves. J. Biol. Chem. **246**, 5510–5517 (1971)
80. Reid, E.E., Lyttle, C.R., Canvin, D.T., Dennis, D.T.: Pyruvate dehydrogenase complex activity in proplastids and mitochondria of developing castor bean endosperm. Biochem. Biophys. Res. Commun. **62**, 42–47 (1975)
81. Renkonen, O., Bloch, K.: Biosynthesis of monogalactosyl diglycerides in photoauxotrophic *Euglena gracilis*. J. Biol. Chem. **244**, 4899–4903 (1969)
82. Renooij, W., van Golde, L.M.G., Zwaal, R.F.A., van Deenen, L.L.M.: Topological asymmetry of phospholipid metabolism in rat erythrocyte membranes. Europ. J. Biochem. **61**, 53–58 (1976)
83. Roughan, P.G.: Turnover of the glycerolipids of pumpkin leaves. The importance of phosphatidylcholine. Biochem. J. **117**, 1–8 (1970)

84. Roughan, P.G.: Phosphatidyl choline: donor of 18-carbon unsaturated fatty acids for glycerolipid biosynthesis. Lipids **10**, 609–614 (1975)
85. Rullkötter, J., Heinz, E., Tulloch, A.P.: Combination and positional distribution of fatty acids in plant digalactosyl diglycerides. Z. Pflanzenphysiol. **76**, 163–175 (1975)
86. Safford, R., Appleby, R.S., Nichols, B.W.: The enzymic conversion of monogalactosyl monoglycerides to monogalactosyl diglycerides by spinach leaf homogenates. Biochim. Biophys. Acta **239**, 509–512 (1971)
87. Safford, R., Nichols, B.W.: Positional distribution of fatty acids in monogalactosyl diglyceride fractions from leaves and algae. Biochim. Biophys. Acta **210**, 46–56 (1970)
88. Sastry, P.S.: Glycosyl glycerides. In: Advances in Lipid Research. Paoletti, R., Kritchevsky, D. (eds.). New York-London: Academic Press, 1974, Vol. 12, pp. 251–310
89. Sastry, P.S., Kates, M.: Biosynthesis of lipids in plants II. Incorporation of glycerophosphate-^{32}P into phosphatides by cell-free preparations from spinach leaves. Can. J. Biochem. **44**, 459–467 (1966)
90. Sherratt, D., Givan, C.V.: The apparent absence of a pathway for synthesis of acetyl coenzyme A in pea chloroplasts: lack of $^{14}CO_2$ incorporation into lipids by the isolated organelle. Planta **113**, 47–52 (1973)
91. Shine, W.E., Mancha, M., Stumpf, P.K.: The function of acyl thioesterases in the metabolism of acyl-coenzymes A and acyl-acyl carrier proteins. Arch. Biochem. Biophys. **172**, 110–116 (1976)
92. Shine, W.E., Mancha, M., Stumpf, P.K.: Differential incorporation of acyl-coenzymes A and acyl-acyl carrier proteins into plant microsomal lipids. Arch. Biochem. Biophys. **173**, 472–479 (1976)
93. Siebertz, H.P., Heinz, E.: Labelling of experiments on the origin of hexa- and octadecatrienoic acids in galactolipids from leaves. Z. Naturforsch. **32c**, in press (1977)
94. Siebertz, M., Heinz, E.: Galactosylation of different monogalactosyldiacylglycerols by cell-free preparations from pea leaves. Hoppe-Seylers Z. Physiol. Chem. **358**, 27–34 (1977)
95. Slack, C.R., Roughan, P.G.: The kinetics of incorporation *in vivo* of [^{14}C] acetate and [^{14}C] carbon dioxide into the fatty acids of glycerolipids in developing leaves. Biochem. J. **152**, 217–228 (1975)
96. Smith, P.F.: Biosynthesis of glucosyl diglycerides by *Mycoplasma laidlawii* Strain B. J. Bacteriol. **99**, 480–486 (1969)
97. Stern, N., Tietz, A.: Glycolipids of a halotolerant, moderately halophilic bacterium II. Biosynthesis of glucuronosyldiglyceride by cell-free particles. Biochim. Biophys. Acta **296**, 136–144 (1973)
98. Stumpf, P.K.: Biosynthesis of fatty acids in spinach chloroplasts. In: Recent Advances in the Chemistry and Biochemistry of Plant Lipids. Galliard, T., Mercer, E.I. (eds.). London-New York-San Francisco: Academic Press, 1975, pp. 95–113
99. Trémolières, A.: Les lipides des tissus photosynthétiques. Ann. Biol. **9**, 113–156 (1970)
100. Trémolières, A.: Mise en évidence, dans la feuille de Trèfle, d'une fraction lipidique à renouvellement rapide, et d'une fraction riche en acide linolénique, à renouvellement très lent. C. R. Acad. Sci. **272**, 2777–2780 (1971)
101. Trémolières, A.: Biosynthèse de l'acide linolénique dans la feuille de pois. Phytochemistry **11**, 3453–3460 (1972)
102. Trémolières, A., Mazliak, P.: Étude de la biosynthèse des galactolipides dans de jeunes feuilles de Pois. C. R. Acad. Sci. **275**, 1883–1886 (1972)
103. Trémolières, A., Mazliak, P.: Biosynthetic pathway of α-linolenic acid in developing pea leaves. *in vivo* and *in vitro* study. Plant Sci. Lett. **2**, 193–201 (1974)
104. Tulloch, A.P., Heinz, E., Fischer, W.: Combination and positional distribution of fatty acids in plant sulfolipids. Hoppe-Seylers Z. Physiol. Chem. **354**, 879–889 (1973)
105. Veerkamp, J.H.: Biochemical changes in *Bifidobacterium bifidum* var. *pennsylvanicus* after cell-wall inhibition VI. Biosynthesis of the galactosyldiglycerides. Biochim. Biophys. Acta **348**, 23–34 (1974)
106. Vijay, I.K., Stumpf, P.K.: Fat metabolism in higher plants. Nature of the substrate and the product of oleoyl coenzyme A desaturase from *Carthamus tinctorius*. J. Biol. Chem. **246**, 2910–2917 (1971)
107. Wenger, D.A., Petitpas, J.W., Pieringer, R.A.: The metabolism of glyceride glycolipids II.

Biosynthesis of monogalactosyl diglyceride from UDP-galactose and diglyceride in brain. Biochemistry **7**, 3700–3707 (1968)

108. Williams, J.P., Watson, G.R., Khan, M., Leung, S., Kuksis, A.: Gas-liquid chromatography of plant galactolipids and their deacylation and methanolysis products. Anal. Biochem. **66**, 110–122 (1975)
109. Williams, J.P., Khan, M., Leung, S.: Biosynthesis of digalactosyl diglyceride in *Vicia faba* leaves. J. Lipid Res. **16**, 61–66 (1975)
110. Williams, J.P., Watson, G.R., Kahn, M.U., Leung, S.: Galactolipid synthesis in *Vicia faba* leaves I. Galactose, glycerol and fatty acid labeling after $^{14}CO_2$ feeding. Plant Physiol. **55**, 1038–1042 (1975)
111. Williams, J.P., Watson, G.R., Leung, S.P.K.: Galactolipid synthesis in *Vicia faba* leaves II. Formation and desaturation of long chain fatty acids in phosphatidylcholine, phosphatidylglycerol, and the galactolipids. Plant Physiol. **57**, 179–184 (1976)
112. Yamada, M.: Studies on fat metabolism in germinating castor beans IV. Glycerol metabolism. Sci. Pap. Coll. Gen. Educ. Univ. Tokyo **10**, 283–323 (1960)
113. Yamada, M., Nakamura, Y.: Fatty acid synthesis by spinach chloroplasts II. The path from PGA to fatty acids. Plant Cell Physiol. (Tokyo) **16**, 151–162 (1975)
114. Zepke, H.D., Heinz, E., Pesch, R.: unpublished results

CHAPTER 7

Light, Function, and Lipids During Plastid Development

M. TEVINI

A. Introduction

The growth and development of higher plants and lower plants are strongly affected and to some extent also regulated by the environmental factor of light. The photomorphogenesis of higher plants consists of several light-triggered single actions that can be divided into morphologic and biochemical photoresponses. The following deals with the biochemical–ultrastructural changes appearing in the chloroplasts of higher plants during the structural and functional development of the photosynthetic apparatus. The main point here is the correlation between lipid changes, functioning, and structural assembly following light-induced thylakoid formation in the etiochloroplast in higher plants. Membrane biogenesis in algae has been extensively studied and reviewed by Ohad [73]. Basic phenomena of structural and functional organization of biomembranes are described and discussed by Prof. Sitte (Chapter 1). Molecular organization of chloroplast thylakoids has been reviewed by Anderson [2] and Kirk [50]. The formation of thylakoids is controlled by genetic material as well as by light. The quality and quantity of the given light act in a regulating capacity during plastid development as discussed in more detail later.

In the leaves of higher plants, chloroplasts normally develop from proplastids when the leaves are illuminated by visible light [70]. In complete darkness the leaves become etiolated and the proplastids develop into etioplasts with the characteristic prolamellar body (Fig. 1). Light treatment of etioplasts induces

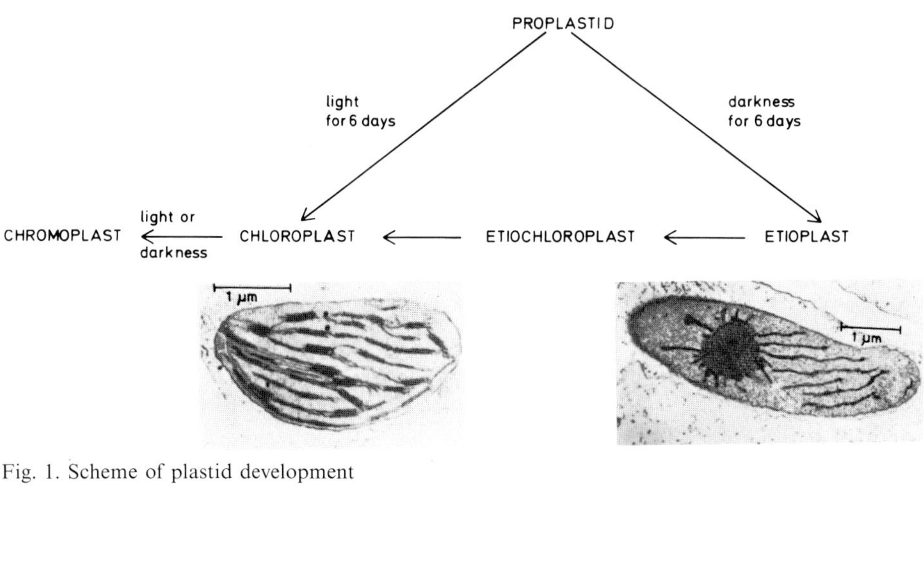

Fig. 1. Scheme of plastid development

the disappearance of the prolamellar body within a few hours and the gradual formation of photosynthetically active thylakoids. After several hours of illumination, the thylakoids overlap and build grana for better photosynthetic efficiency. During physiologic functioning thylakoids finally show aging and ultimately develop into chromoplasts no longer active in photosynthesis (see Chapter 1 and Ref. [91]). The development of functioning chloroplasts from etioplasts can normally be induced by white light. Although we know a great deal about the phytochrome system, which is a very effective regulating system of many photoresponses [68], the plant physiologist wants to know what special light quality is responsible for the regulation of the greening process. The following sections deal in particular with the efficiency of light quality, which is thought to have a regulating capacity during light-driven plastid development.

B. Light and Plastid Structure During Greening

I. General Remarks

As already mentioned, etioplasts of higher plants are derived from proplastids when the plant has been kept in darkness. The etiolated plants show stem elongation and leaf area reduction. They appear pale yellow because they have carotenoids but lack chlorophyll. This means that the proplastid is capable of synthesizing pigments in addition to chlorophylls. As will be shown later, the etioplasts already contain all other lipids in considerable amounts. The etioplasts are surrounded by a double membrane, the envelope. The stroma contains one or more prolamellar bodies, a cluster of plastid ribosomes, and areas with DNA fibrills and in some cases membrane-associated tubules [94]. In nearly all etioplasts one can find a lot of plastoglobuli, which may, together with the prolamellar body, represent a storage pool of several lipids [6, 59, 99].

During the development from proplastid the size of the etioplast is multiplied and the shape is changed. The mechanism of etioplast development from proplastids is ultimately under genetic control. However, we should bear in mind that prolamellar body rearrangements are also possible by means of special red-light treatment [11, 103]. The fate of the prolamellar body is the most interesting part of the ultrastructural changes during development and has already been described by several authors [118, 119].

II. The Prolamellar Body

The prolamellar bodies develop from intrusions of the inner envelope, which first migrate into the stroma and aggregate to form perforated sheets [30, 85], which are then incorporated into tubuli-like prolamellar bodies by a contraction procedure [116]. The structure of the prolamellar body may be paracrystalline or disordered. The regular crystalline order of prolamellar bodies seems to increase with the age of the etioplast, but is also species specific. In *Hordeum*

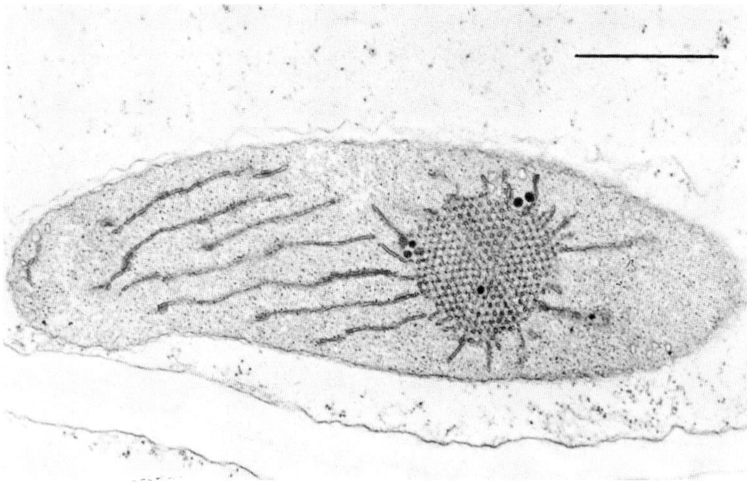

Fig. 2. Electron micrograph of a barley etioplast. Bar: 1 μm

etioplasts, 16% to 20% of the prolamellar bodies are paracrystalline in form, whereas in *Avena* etioplasts, 99% are [11]. The structure of the tubules has been described extensively by several authors [30, 115, 118]. Most of the crystalline networks contain many plastoglobuli ([96]; Fig. 2).

III. Thylakoid Formation

During illumination by visible light the paracrystalline structure is lost and a gradual reorganization of primary thylakoids occurs (Fig. 3a, b). A model of primary thylakoid development was presented some years ago by Wellburn and Wellburn [120]. Illumination with low light intensities (150 lux) at 4°C makes the membrane less liquid, and growth of tubes out of the prolamellar body and the fusion of these tubules into perforated thylakoids can be seen (Fig. 3e). The holes in the perforated lamellae are later closed by the insertion of special polypeptides. Two proteins not found in the prolamellar bodies are found in the membranes of developing maize plastids: one of 65,000 daltons and the other of 46,000 daltons, of which the latter is thought to be the component to fill the holes [15]. Some other high molecular weight proteins are lost during development [15]. As will be shown, the amount of lipids is diminished during the first hours of illumination, but later, lipids are synthesized rapidly. This rapid synthesis parallels thylakoid elongation and grana formation. From the protein and lipid data, we conclude that during the first two hours of illumination the prolamellar body partly breaks down and is gradually rebuilt. The prolamellar bodies and the plastoglobuli contained therein are, respectively, protein and lipid pools for the primary thylakoids. The light-stimulated decrease of membrane substance of the prolamellar body results in an increase of primary thylakoids. This process is reversible in darkness [37]. The development of etiochloroplasts proceeds with the overlapping of single thylakoids, a procedure

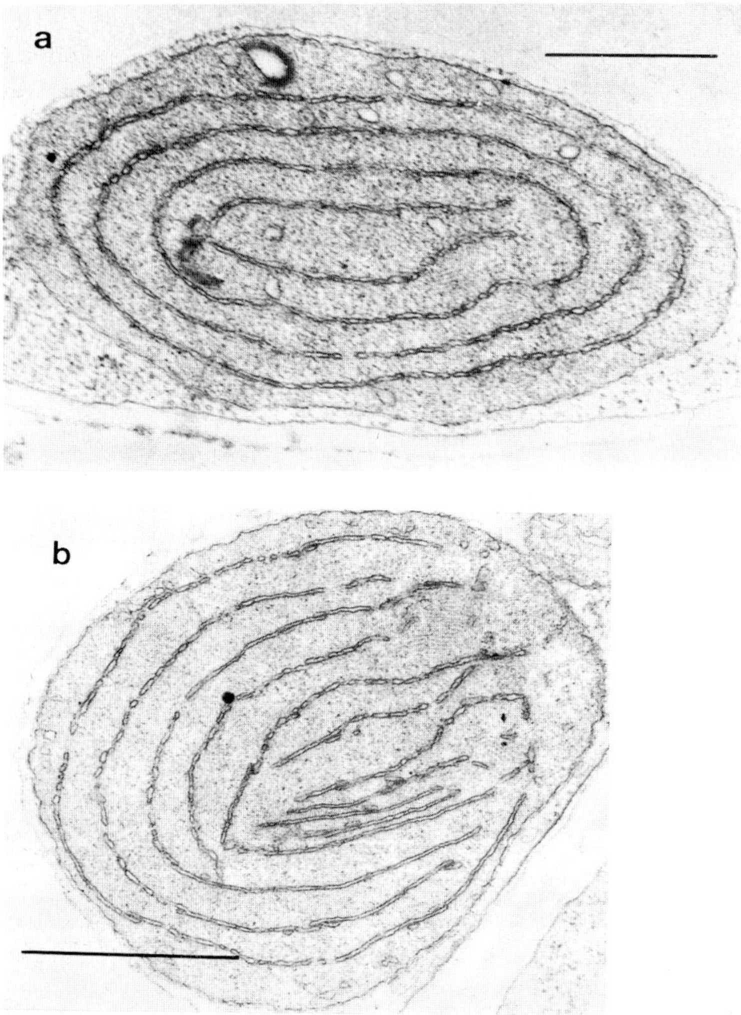

Fig. 3a–e. Electron micrograph of developing barley etioplasts illuminated with blue light (a), after 2 h; (b), after 4 h; (c), after 12 h; (d), after 24 h; (e), etiochloroplast after 2 h with a view on perforated primary thylakoids. Bars: 1 μm

called stacking (Fig. 3b, c). Fully greened chloroplasts contain grana with up to ten thylakoid layers (Fig. 3d). Membrane stacking is correlated with the appearance of photosystem II (PS II), which is characteristic for several plants [14], but this is not the case with barley, where PS II activity precedes stacking. Experiments with intermittent light show clearly that stacking is not necessary for PS II activity [20, 41]. On the other hand the presence of chlorophyll protein complex 2 seems to be correlated with membrane stacking [83], with the exception of barley mutants lacking chlorophyll b, which lack chlorophyll protein complex 2, but exhibit membrane stacking [83, 104]. The molecular aspects of stacking are well discussed by Anderson [2], and the analysis of the structural proteins connecting two thylakoids was recently presented by Staehlin [97].

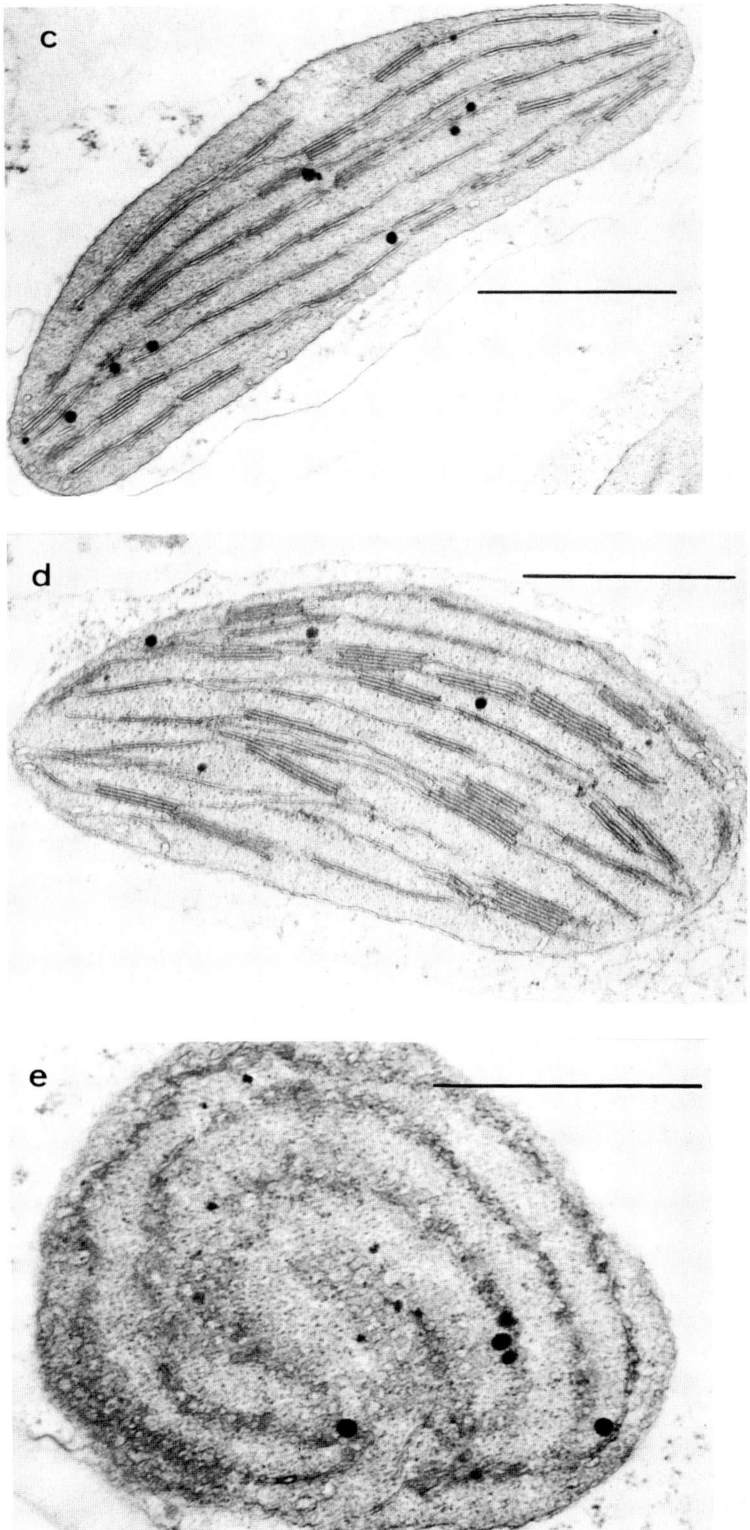

Fig. 3c–d

IV. Light Regulation

Prolamellar body formation from proplastids seems to be initiated and intensified by active phytochrome P_{fr} [47]. The mechanism of light regulation of prolamellar body transformation is not known. Early papers described close correlations between 'tube transformation' and protochlorophyllide changes [13, 52, 113] or Shibata-shift [38]. However, Wellburn and Wellburn [119] observed phytochrome-mediated prolamellar body transformation. They showed that tube transformation is also stimulated by red-light pulses of 5 min duration and inhibited by the following far-red pulses in isolated etiochloroplasts. Tube transformation proceeds for only 2 h in vitro, whereas in intact tissues, it continues without interruption. This means that the regulation of primary thylakoid formation completed during the first two hours is an etioplast-directed procedure in which the subsequent regulation is increasingly influenced by the nucleus or the cytoplasm. The same authors found no significant effect of blue light on thylakoid formation whereas Vlasova et al. [115] showed that grana formation in barley seedlings illuminated with blue light proceeds more quickly than in those exposed to red light. These data can be supported by our own findings concerning ultrastructure and lipid changes during development (see Section C.V.).

Prolamellar body rearrangement is also effected by various light and dark treatments. Red-light pulses of 2 min followed by 3 or 6 h of darkness produced an increase of crystalline arrangements. This is correlated with the capacity of protochlorophyllide synthesis during the dark period, which is dependent on the age of the plant [11, 37]. Recrystallization starts after 90 min darkness when protochlorophyllide synthesis has finished. Thus the new synthesis is a basis for recrystallization. White- and blue-light pulses are more effective on recrystallization than red light. Light intensity also induces changes during chloroplast development [117]. It is known that the chloroplasts in plants grown under weak light are larger, containing more chlorophyll and a greater chlorophyll a/b ratio than chloroplasts in plants grown under strong light [28, 106]. Therefore the photosynthetic unit has to be altered, but no comparable ultrastructural data are available.

Recrystallization of prolamellar bodies is also possible during illumination, provided that ethionin, an inhibitor of precursor synthesis during chlorophyll formation, is added [123]. An interpretation is not possible at present.

C. Light and Lipids

I. General Remarks

Light causes and controls not only ultrastructural changes but also biochemical events during plastid development, such as biosynthesis, metabolism, and catabolism of structural components. Mohr's group has shown that red and far-red light regulate many such events, for instance the early stages of chlorophyll formation and prolamellar body reorganization [47, 48, 65]. Other groups have shown that there must be blue light-inducible photoreceptors. The best known example is the role of blue light in phototropistic movements in higher plants, in respect to chloroplast formation and lipid changes, about which only very little information is available in the literature. First, Bergfeld [10] demonstrated that blue light enhances protein synthesis in *Dryopteris* chloroplasts; later, the Russian group of Voskresenskaya observed more intensive grana formation with the stimulation of chlorophyll synthesis by blue light than by red light [115].

II. Lipids in Plastids and Plastid Fractions

Etioplast and chloroplast development starts from proplastids that have been isolated from young tissues of *Zea mays* [34, 58, 77]. The results show that undifferentiated plastids similar to chloroplasts already contain the major lipids MGDG and DGDG, followed by decreasing amounts of sulpholipid, PG, PC, PI, and DPG [77] (for abbreviations used in this Chapter, see explanations given in subject index). PE has been additionally found in young leaves, partly due to the presence of PE in mitochondria and cell membranes [35]. In light-grown young plants of *Zea* and of *Hordeum,* the synthesis of all glyco- and phospholipids is enhanced in comparison to those grown in darkness (Table 1). All phospholipids including PE seem to be involved in chloroplast development. Mature chloroplasts of *Zea* contain about 50% of all lipids as glycolipids and about 6% of all lipids as phospholipids [56]. Spinach chloroplasts and the

Table 1. Galacto- and phospholipid contents of barley leaves during growth under white light and in darkness. *I–IV:* tissue age after sowing—*I:* 80 h (0.5-cm pieces); *II:* 92 h (2-cm pieces); *III:* 114 h (5-cm pieces); *IV:* 162 h (12-cm pieces). Values in μg/piece

Lipids		I	II	III	IV
MGDG	green	39.2	44.1	138.0	365.0
	etiolated	5.4	23.6	102.0	268.0
DGDG	green	26.4	29.5	101.0	198.0
	etiolated	2.6	16.1	84.0	185.0
PC	green	2.4	22.7	59.0	86.0
	etiolated	1.4	18.6	49.0	74.3
PE	green	1.0	8.1	21.0	66.0
	etiolated	0.9	3.5	16.0	33.6
PG	green	0.6	4.7	17.0	50.0
	etiolated	tr	0.8	6.0	20.1
PI	green	0.8	3.8	5.0	38.0
	etiolated	0.2	1.9	10.0	30.2
PS	green	tr	3.4	0.2	33.0
	etiolated	tr	tr	5.4	23.9

lamellar system of spinach chloroplasts also contain PG, PC, PI, and little PE, even after many washings [108]. Isolated greening etiochloroplasts of barley have high amounts of PC, PG, and PE during the first hours of illumination (Table 2).

Because of better separation techniques, it has been possible in the last few years to separate envelope fractions and lamellae fractions by density gradient centrifugation [22, 62, 63, 78, 95]. The envelopes of spinach chloroplasts have been characterized as sites of biosynthesis of galactolipids (see also Chapter 6) [21, 44]. The envelope proteins were analyzed by several groups [19, 46, 67, 75, 95]. The permeability of the plastid envelope during greening has been studied recently [32]. The results show that increased transport of metabolites occurs during greening, which also may be of interest for transport and biosynthesis of lipids. Mackender and Leech [63, 64] first reported on the lipid composition of envelopes, and more recently Bahl et al. [4] isolated envelope and prolamellar body fractions of *Triticum*. Comparing these two fractions (Table 2), one can see large amounts of DGDG, PC, and PG in the envelope fraction, whereas in the prolamellar body fraction, the PC-level is much lower when expressed as percent of the total glycerolipids. The plastoglobuli of etio- and chloroplasts contain lipophilic quinones [59] and many galacto- and phospholipids (Tevini, unpublished results).

Table 2. Lipid composition of etioplast and chloroplast fractions (isolated plastids, envelopes, prolamellar bodies, lamellae) (% of total glycerolipids)

	MGDG	DGDG	SL	PC	PG	PE	PI	Ref.
Isolated etioplasts (*Hordeum vulgare*)	42.3	25.3	10.3	9.0	7.2	5.9	tr	[102]
Envelope, etioplasts (*Triticum sativum*)	22.0	44.0	10.0	14.0	9.3	0	–	[4]
Prolamellar-bodies (*T. sativum*)	36.0	41.0	10.0	3.7	8.6	0	–	[4]
Envelope, chloroplasts (*Vicia faba*)	29.1	32.4	–	29.6	8.9	0	–	[64]
Envelope, chloroplasts (*Spinacia oleracea*)	32.8	33.6	0.8	9.3	2.9	1.4	1.1	[22]
Lamellae, chloroplasts (*S. oleracea*)	55.0	31.6	5.2	2.6	4.5	0.9	0.2	[a]
Lamellae, chloroplasts (*S. oleracea*)	51.0	26.0	7.1	3.2	9.1	0	1.4	[22]
Isolated etiochloroplasts, 2 h blue light (*H. vulgare*)	31.3	21.1	6.1	19.7	6.8	14.9	tr	[a]
8 h blue light (*H. vulgare*)	27.1	16.9	16.9	18.1	7.5	13.5	–	[a]
16 h blue light (*H. vulgare*)	36.4	25.4	16.3	7.8	11.0	3.2	tr	[a]
Isolated chloroplasts, 24 h blue light (*H. vulgare*)	43.4	23.7	14.5	5.9	10.4	1.9	tr	[102]
Grana fraction 24 h white light (*T. sativum*)	43.0	40.0	8.0	1.5	7.0	–	–	[4]

[a], Tevini, not yet published elsewhere

III. Formation of Glycerolipids During Development

General information about the biosynthesis of glyco- and phospholipids has recently been well summarized and will not be repeated here [27, 31, 43, 49, 66, 69]. During plastid development in etiolated leaves in white light, the following tendencies can be noted:

1. During the lag phase of chlorophyll synthesis, there is a significant decrease in the galactolipids [88, 100, 102]
2. After 2 h of white light, galactolipid synthesis starts anew, in parallel to the log phase of chlorophyll synthesis and grana formation [14, 64, 88, 100]
3. During the first hours of illumination phospholipids show a rapid increase that is due to PC and PG synthesis. With high light intensities (7,000–10,000 lux), one can observe a steady increase whereas with low light intensities (up to 1,000 lux), the phospholipid increase proceeds more slowly. In isolated plastids from greening pea leaves, a much greater amount of label was found in PC than in the major chloroplast lipids MGDG, DGDG, and PG during the first hours of illumination [74].

Illumination of *Hordeum* leaves with blue and red light in general has the same effect on lipid metabolism (Fig. 4) that white light does, but we assume that these light qualities, at the intensities used (red light 0.5 Wm^{-2}, blue light 0.8 Wm^{-2}), retard the developmental process in the same manner as intermittent light, which is also used for 'slow-motion' studies of lipids [5, 107]. On the other hand, it is interesting to note that longwave ultraviolet (UV) light (interference filter with $\lambda_{max}=366$ nm, 0.5 Wm^{-2}) enhances the galactolipid synthesis during the first hours of illumination (Fig. 4). This may be a reflection of partly inhibited prolamellar body breakdown and low rate of chlorophyll synthesis, so that all newly formed galactolipids can be measured. Protochlorophyll absorption measurements show no 'Shibata-shift', meaning that no phytylation takes place under these low-light-intensity conditions. In contrast to all other light qualities, UV causes the inhibition of phospholipid synthesis during the first hours of illumination, and later on even the breakdown of phospholipids can be seen by the formation of lysoglycerides.

To exclude the possibility that the changes described above may occur in organelles other than plastids, we tried to prove the results with etioplasts and etiochloroplasts isolated from tissues illuminated with red and blue light, by using sucrose density gradient centrifugation. The analysis of UV-illuminated etiochloroplasts has been started but not yet finished. The plastid fractions were collected from the sucrose layer (1.15–1.17 g/ml) after centrifugation of plastid suspensions with 1,600 *g* for 20 min. The fractions were checked for other organelles by cytochrome c-reductase and catalase-activity measurements, which were in both cases extremely low in the etioplast fraction (less than 2%). The isolated plastids were counted in a Coulter counter. Returning to our question concerning which individual lipids are found in the etioplast before, during, and after development of active thylakoids and what correlation with function can be drawn from the data, Table 3 shows the baseline quantities of individual lipids present in etioplasts and the quantities present after 24 h

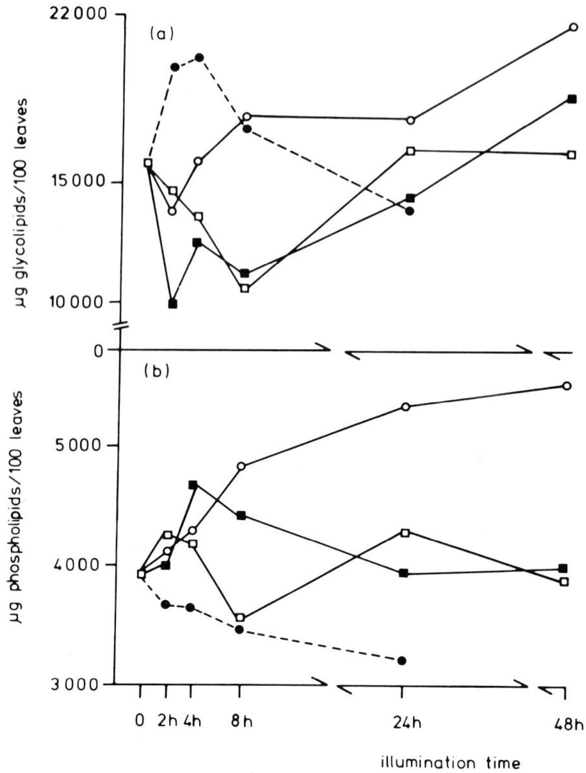

Fig. 4. (a) Total glycolipid (MGDG, DGDG, SL) content of etiolated and illuminated barley leaves; (b) total phospholipid (PC, PE, PG, PI, PS) content of etiolated and illuminated barley leaves. (a) and (b) after Ref. [102]; values given as µg/100 leaf pieces (=4-cm pieces); light qualities are □, blue light; ○, white light; ■, red light; ●, UV light

Table 3. Lipid concentrations (µmol/10^9 plastids) in etioplasts and blue/red-light chloroplasts. Last two columns: photosynthetic activity in blue- and red-light chloroplasts (µmol · Chl · h). (FECY = Ferricyanide)

	MGDG	DGDG	MGDG/DGDG	SL	Total glycolipids
Etioplasts	1.32	0.79	1.67	0.32	2.43
Chloroplasts					
24 h red	1.53	0.91	1.68	0.48	2.92
24 h blue	2.41	1.31	1.83	0.80	4.52

	PC	PE	PG	Total phospholipids	Chl. a+b	DCPIP reduction	FECY-reduction
Etioplasts	0.28	0.22	0.19	0.69	0	0	0
Chloroplasts							
24 h red	0.27	0.20	0.49	0.96	1.05	85	80
24 h blue	0.33	0.11	0.57	1.01	1.25	105	144

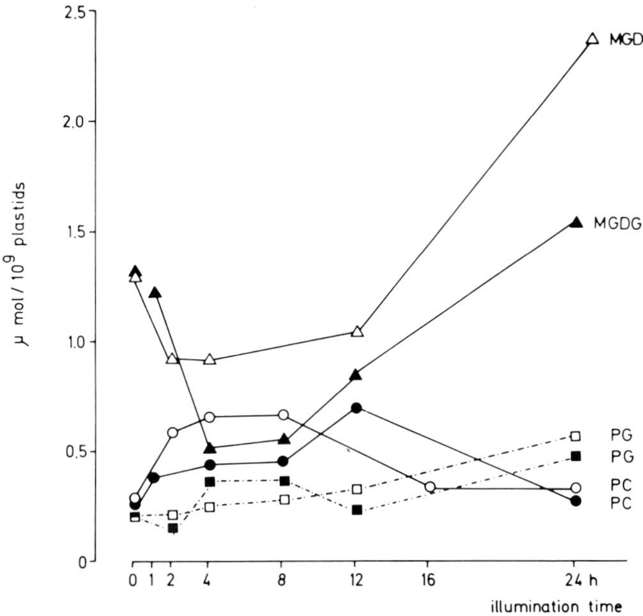

Fig. 5. Molar MGDG, PG, and PC concentrations in developing "red-light" plastids (open symbols) and "blue-light" plastids (black symbols); values given as μg/10^9 plastids

of illumination in red and blue light. Initially 10^9 etioplasts already contain about 2 μmol of galactolipids and about 0.7 μmol of phospholipids. This means that the synthesis of these compounds is initially light independent. After 24 h the lipid concentrations are much higher than before, with the exception of PE. The same results were obtained with white-light illumination in wheat etioplasts [5]. Related to the same plastid number, blue-light illuminated plastids contain higher amounts of glycolipids than red-light illuminated chloroplasts, even when they receive fewer quanta in blue than in red light. The photosynthetic activity measured as DCPIP-reduction is about 25% higher, with ferricyanide about 80% higher than in red-light illuminated plastids.

What happens with lipids in between the two baselines under blue and red light? Table 2 shows the relative percentage changes of total glycerolipids found in 'blue-light' etioplast fraction. Molar changes of lipid composition were shown in more detail in Figure 5 with the selected examples MGDG, PC, and PG because they are the most interesting as far as their metabolic function is concerned. Isolated etiochloroplasts contain during the first stages of greening much less MGDG than do the etioplasts that are present. This may reflect the hydrolysis of galactolipids during the disorganization of prolamellar bodies. The breakdown of prolamellar bodies with regard to lipids is completed in blue light after 2 h, in red light, after 4 h illumination, as can be seen from the minimum values. During the early greening periods, the synthesis of PC is again enhanced much more in blue- than in red-light illuminated plastids. At the point when grana stacking first begins—that is, in blue light after 4 h to 8 h—a rapid increase of MGDG synthesis is seen, whereas the phospholipid content begins to fall. PE changes are similar to PC changes during the first hours of illumination, whereas in later stages the PE content of chloroplasts is extremely low. We assume that PE and PC are used during the formation of structural galactolipids, but we cannot say whether the phospholipids come from outside (see Mazliak, Chapter III) or inside the etioplast. Experiments with flashing light also show that the total phospholipid content, when related to 10^9 plastids [4, 103], is stimulated (80%) much more by very short pulses than is the galactolipid content (20%). We can therefore conclude that the time-course of plastid development under flashing light is similar to that under continuous red or blue light of low intensity. The phospholipids PC and PG may have a fatty acid carrier function for MGDG and DGDG especially during prolamellar body breakdown [82, 122]. For the discussion of this carrier function based on recent findings, see the following Chapter.

IV. Formation of Fatty Acid

In etiolated barley leaves the glycerolipids considered in this chapter contain large amounts of unsaturated fatty acids; α-linolenic acid ($C_{18:3}$) is predominantly associated with the galactolipids and linolenic acid ($C_{18:2}$), with most of the phospholipids. Palmitic acid ($C_{16:0}$) is distributed equally in galactolipids and phospholipids [40]. Light treatment of etiolated leaves stimulates acetate incorporation into fatty acids [29, 74, 92] and production of α-linolenic acid. It is not known whether greening etioplasts isolated from blue- or red-light illuminated tissues have any differences in their fatty acid compositions. The knowledge is of interest insofar as one can correlate fatty acid differences with membrane fluidity during thylakoid formation. All light qualities examined cause a decrease in α-linolenic acid portions of the individual glycerolipids MGDG, DGDG, PG, and SL within the first hours of illumination, probably to the advantage of palmitic acid. In Table 4, only MGDG, PG, and PC are presented as examples. The loss of α-linolenic acid during the early greening stages may be due to lipid hydrolysis and to the fatty acid transfer from phospholipids to galactolipids [122]. After 8 h light the percentage amounts of α-linolenic acid increase, correlated with increasing rates of photosynthetic activity. Only trace amounts of trans-3-hexadecenoic acid, which is predominantly connected with phosphatidylglycerol, were found in barley etioplasts. On the other hand, some authors find $\Delta 3$ trans $C_{16:1}$ in etiolated tissues, but it seems possible that this acid might be $\Delta 9$ cis $C_{16:1}$, palmitoleic acid. This acid could not be detected either in chlorotic mutants of Vicia faba or in white variegated tissues of *Lolium perenne* (P. Bolton and J.L. Haarwood, personal communication). This acid first appears in blue-light etioplasts after a lag phase of 2 h light, in red-light etioplasts, somewhat later. In both cases the highest synthesis rates are measured between 8 and 12 h illumination, which can be correlated with the rapid grana stacking during this time. The idea that trans-3-hexadecenoic acid may be involved in grana stacking has already been discussed by Tuquet et al. [107].

On the other hand, α-linolenic acid is a favorite for such assembly functions.

Table 4. Percentage of fatty acid composition of MGDG, PG, and PC in etioplasts and etiochloroplasts isolated from dark-grown seedlings and seedlings illuminated with blue and red light for 2, 4, 12, and 24 h

			$<C_{14}$	14:0	16:0	16:1	18:0	18:1	18:2	18:3
MGDG	darkness		tr	2.0	15.2	0.6	1.9	0.5	14.6	67.6
	2 h	blue light	–	2.7	26.4	–	–	13.0	11.0	46.9
		red light	tr	4.9	16.9	–	tr	7.2	11.7	59.3
	4 h	blue light	–	–	31.5	–	4.0	4.0	12.3	48.2
		red light	–	6.7	23.8	–	0.7	3.7	11.9	53.2
	12 h	blue light	–	2.3	7.8	–	0.6	0.8	11.6	76.8
		red light	15.6	9.8	4.9	–	–	4.1	13.3	52.4
	24 h	blue light	3.7	4.9	4.1	–	–	3.7	8.1	79.3
		red light	2.0	3.7	7.4	–	tr	–	11.9	71.2
PG	darkness		–	8.7	49.7	tr[a]	3.9	tr	7.0	30.6
	2 h	blue light	–	8.9	47.0	tr[a]	4.7	tr	15.0	24.4
		red light	8.3	10.0	50.2	tr[a]	3.5	tr	9.6	18.5
	4 h	blue light	tr	0.7	60.7	tr[a]	8.4	4.2	8.7	15.2
		red light	–	5.3	51.8	2.1[a]	4.7	2.1	16.0	20.1
	12 h	blue light	1.3	4.7	31.4	12.7[a]	1.0	5.6	12.4	31.0
		red light	8.2	7.3	24.0	12.2[a]	–	0.6	22.5	25.1
	24 h	blue light	–	0.9	32.6	14.7[a]	3.3	3.2	8.1	37.8
		red light	2.1	3.8	29.2	13.9[a]	3.3	6.9	6.8	33.8
PC	darkness		–	3.3	35.1	–	3.1	4.7	40.3	13.4
	2 h	blue light	–	1.8	35.5	0.2[b]	3.5	6.7	39.7	12.9
		red light	–	0.9	33.7		3.2	3.9	41.0	17.3
	4 h	blue light	–	2.5	41.4	–	4.1	1.0	37.2	13.6
		red light	–	2.7	37.2	–	1.7	4.2	35.1	19.1
	12 h	blue light	tr	3.2	35.2	–	3.7	3.7	41.4	12.8
		red light	tr	0.7	33.8	2.3[b]	1.9	3.3	30.8	26.9
	24 h	blue light	–	3.9	36.8	–	2.9	2.8	41.0	12.4
		red light	–	3.2	44.9	–	2.4	5.1	29.2	15.2

[a] 3-trans-hexadecenoic acid

[b] $C_{16:x}$

The maximum synthesis of linolenic acid parallels that of chlorophyll. The function of α-linolenic acid molecules is seen in their participation in giving the right order to the chlorophyll molecules [23, 105].

In general we can say that blue and red light have qualitatively the same effect on fatty acid metabolism during greening following continuous illumination. One point worth noting is the rapid increase of palmitic acid from 15% to 60% in the sulfolipid in blue light (Fig. 6). This may reflect a special blue light effect that cannot be interpreted with certainty at the present time. We should consider the question of whether PC could be a fatty acid carrier for MGDG. As is evident from Table 4, no drastic changes are seen in PC. Therefore

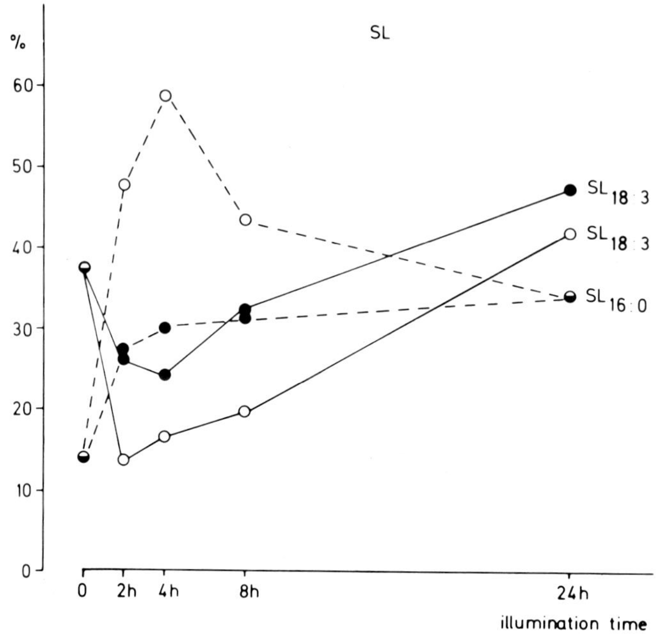

Fig. 6. Relative percentage amounts of α-linolenic acid and palmitic acid in SL of greening "blue- and red"-light plastids

we cannot deduce any carrier function of PC from the percentage fatty acid composition or we would have to assume a rapid turnover of oleic and linoleic acid in these lipids.

V. Light Regulation of Glycerolipid Metabolism

For studying regulation phenomena we used, instead of continuous illumination, 5-min pulses of red, far-red, blue, and UV light, given separately or following each other. After illumination the barley plants were kept in the dark for 24 h. As is known from classical phytochrome theory, repeated irradiations at short intervals with red light induce galactolipid synthesis in *Sinapis alba* that can be reversed by administering far-red light immediately after the red light [110]. We can support this for barley plants using only a pair of 5-min red and 5-min far-red pulses. Furthermore, it can easily be shown that in barley leaves, synthesis of the phospholipids PG, PC, PE, and PI is clearly phytochrome regulated, whereas the other glycerolipids SL, PS, and some fatty acids have another regulating system (Table 5, Fig. 7). The regulation of phospholipids by phytochrome does not necessarily have to be viewed in connection with plastids, because evidence has recently been published for some phytochrome regulation in other cell organelles [7, 86]. This system could be a combination of a UV-absorbing form with the phytochrome system, as we showed in experiments in which UV was combined with red and far-red light. When

Table 5. Lipid content of etiolated and illuminated barley plants

	MGDG	DGDG	SL	PC	PG	PE	PI	PS	UPL
CD	805	716	116	617	71	251	116	80	98
UV	873	812	147	648	78	255	147	95	95
Red	1,243	1,078	139	733	159	306	176	83	111
Far-red	911	866	165	552	110	267	110	120	92

UV: 5′UV light; red: 5′ red light; far-red = 5′ far-red light (μg/g fresh wt)
UPL = unidentified phospholipid

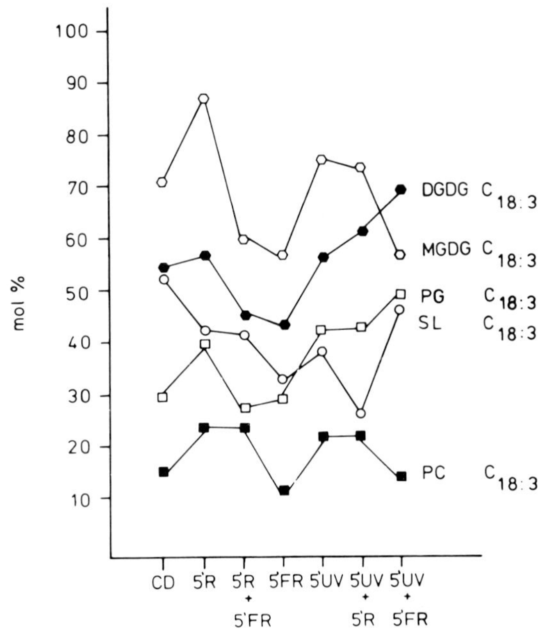

Fig. 7. Molar percentage amounts of α-linolenic acid in MGDG, DGDG, PG, PC, and SL of greening barley leaves illuminated with different light qualities. *CD:* continuous darkness; *R:* red; *FR:* far red

we give 5 min of UV following 5 min of far-red light, the etiolated system produces the highest percentage amounts of α-linolenic acid in the digalactolipid (70% of total acids) and the phosphatidylglycerol (about 50%) (Fig. 7). No other combinations were as effective for this special fatty acid production. Therefore we must discuss a UV-absorbing system regulating special pathways in lipid biosynthesis during etiochloroplast development.

D. Light and Function

I. General Remarks

As we were able to show in the preceding sections, the etioplast is already well established with lipid but does not function photosynthetically. Etioplasts also contain precursors of chlorophyll or even pools (Liljenberg, Chapter 14) of diverse forms of protochlorophyllides. On illumination the protochlorophyllide is photoconverted within a short time, and parallel to the structural disorganization of the prolamellar body, the new chlorophyll protein complexes with simpler molecular organization are built [2, 16, 17, 52]. These complexes later develop to secondary photosystems with full photosystem I and photosystem II activity [2]. The fully active particles can be separated by several disintegration methods [60, 71, 101, 111, 112]. The photosynthetic systems normally function together [24]. The sequence of events during plastid development leading to photosystem-function will be reviewed in the next section.

II. Chlorophylls

Biosynthesis of chlorophyll was summarized recently by Schneider [84] and the phytylation step is discussed by Liljenberg (Chapter 14). In this Section we shall deal only with protochlorophyll changes during greening. Etiolated leaves and etioplasts respectively contain mainly protochlorophyllide together with small amounts of protochlorophyll, which is thought to be the phytyl- or geranylester of protochlorophyllide ([61], see Chapter 14). The protochlorophyllides exist in etioplasts in several absorption forms depending on the age of the leaves. In four-day-old bean leaves and in dark-cultured *Euglena,* mainly PC635 which upon illumination can be converted directly to PC673, is present [18, 53]. The pigment level, however, is low, and the plastids have no prolamellar

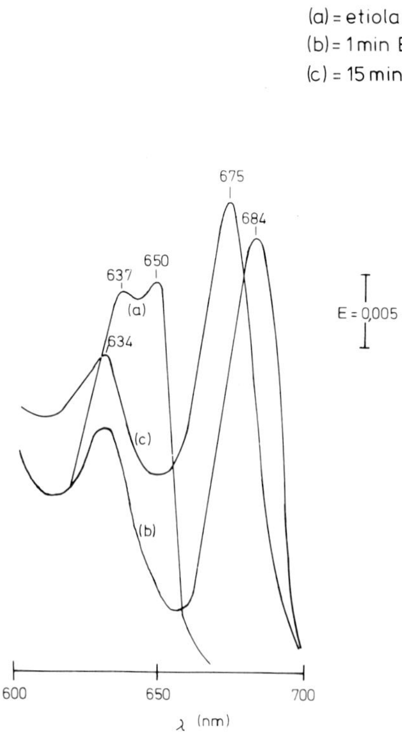

Fig. 8. (Proto)chlorophyll(ide) absorption spectra of etiolated and illuminated barley leaves

Table 6. Total chlorophyll amounts in barley leaves illuminated with white (800 erg/cm^{2}xsec), red (500), blue (800), and UV light (500)

	WL	RL	BL	UV
2 h	45.2	45.8	57.8	47.8
4 h	142.4	185.0	244.0	83.2
8 h	469.2	386.0	482.6	150.9
24 h	1,293.0	909.7	1,103.0	525.0
48 h	1,741.0	1,350.0	1,370.0	–

Values given as μg/100 leaves

body [54]. In older tissues the pigment synthesis is very active and protochlorophyllide PC650 predominates, which is photoconverted to chlorophyllide C685 [12, 51, 93]. In a slower stage this C685 is shifted back to C672, a shift named after the investigator Shibata [54, 87]. The Shibata-shift is completed after 30 min and is probably correlated to phytylation [39, 90]. One of the intermediates, C678, is interesting insofar as it can be converted in the dark to C684 and reverted to C678 by red light [8]. PC634 is not capable of shifting and therefore remains unchanged on illumination, as shown in Figure 8, with blue light of the intensity used in our lipid experiments. These quick absorbance changes have nothing to do with the main additional synthesis of chlorophyll, which cannot be measured until after a lag phase of 2 h. Within the first two hours, first the reaction center chlorophylls are made and afterward the bulk chlorophyll. Chlorophyll a synthesis operates with great efficiency so that the production of chlorophyll b is slower. Consequently, the chlorophyll a/b ratios are up to 20 at the commencement of greening, but decrease very quickly and exhibit normal values of 2.7–3.0 in barley leaves after 20 h of white light illumination [36, 98]. Chlorophyll b synthesis is apparently correlated with active thylakoid formation and is also light-dependent [98].

We found that blue light enhanced chlorophyll synthesis in all the stages examined during the first 24 h, as compared with red light (Table 6). After 48 h the chlorophyll levels are nearly equivalent. Normal chlorophyll a/b ratios are reached earlier with blue light than with red light, which supports the findings of Vroskresenskaya [115].

In the UV region of 320–370 nm, chlorophyll is also synthesized but in amounts that are only 30% of the blue light values. This is due to the low intensities used in our experiments (interference filter λ_{max} 366 nm) and to a possible damaging effect of UV light. Similar to that with UV light, chlorophyll synthesis with far-red light is also very slow [20]. Chlorophyll b is not measurable until after 5 days, although the photosynthetic activity is measurable earlier.

III. Photosynthetic Activity

The functioning of photosystem activity is based on the components of the electron transport chain. The electron transport carriers are placed in pigment system I or II, which is the water-splitting system. The components cyt f, cyt b_6, or plastocyanin are located in or near to PS I and are already present in etioplasts [121]. The main component for PS I, chlorophyll P700, first appears after 6 h of white light illumination. After 8 h the ratio of P700 to normal chlorophylls in bean leaves is 1:600; after 24 h it is reduced to 1:350 [1]. It is likely that during the first few hours chlorophylls other than P700 take part in photosynthetic activity. Cytochrome b 559_{HP}, a potent electron carrier in green chloroplasts and located near photosystem II, is absent in barley etioplasts. It has been first detected after 4 h of continuous illumination, whereas PS II activity with water as electron donor can be induced earlier. From the results obtained one can suggest that this cytochrome cannot be directly involved in the water oxidation process [79]. During rapid grana formation, cyt b 559_{HP} content shows a manifold increase [26]. PS II activity can already be detected within 2 h [36, 81]. For proper O_2 measurements it is necessary to use actinic light (2.5×10^5 erg/cm^2) [51]. The relatively high O_2 rates between 1 and 3 h of light support the fact that during the first greening stages the 'photosynthetic unit' needs much less chlorophyll. On the other hand cyt b 559_{HP} is present only in very small amounts, proving that there is no direct correlation with high photosynthetic activity. Later on, the newly formed chlorophyll is used more for filling light harvesting units. Measurements of the Hill-reaction with DCPIP showed a parallel with O_2 evolution first appearing after one hour of white light illumination [36], but DCMU does not completely inhibit the Hill-reaction at this time.

PS I activity is measurable in general earlier than O_2 evolution [25, 72]. The results of phosphorylation measurements are in fact very different in barley leaves. Phung et al. [80] measured cyclic phosphorylation, which might be of no account in vivo, after 4 h, noncyclic phosphorylation with NADP as electron acceptor after 10 h light. Plesnicar and Bendall [76] measured these two pathways earlier.

The functional role of lipids during the development of photosynthetic activity is not known. Fully green chloroplasts of spinach treated with galactolipase require both galactolipids for full photosynthetic activity [3, 55, 57, 108]. Phospholipase D experiments clearly demonstrate that phospholipids are also necessary for full action of photosynthesis [108]. The effect of phospholipase D-hydrolysis products on photochemical activities indicate that PC and to some extent the other phospholipids play a role in photophosphorylation by maintaining the surface charges of the thylakoid membrane [109]. Fatty acids, particularly α-linolenic acid, inhibit photosystem II electron transport at appropriate concentrations [89].

IV. Light Regulation

The photosynthetic activity of tissues illuminated with different light qualities has not been examined in detail. Some years ago it was shown that far-red illumination causes the same developmental sequence of photosynthetic activity as white light, but much more slowly. Cyclic phosphorylation appears after

Table 7. DCPIP-reduction rates of etiochloroplasts isolated from barley seedlings illuminated with red and blue light

Light treatment	4 h	8 h	12 h	24 h
Red light[a]	187	124	268	90
Blue light[a]	230	151	177	131
Red light[b]	13	15.5	26	42
Blue light[b]	8.5	9	62	114

[a] μmol DCPIP/μmol chl·h
[b] μmol DCPIP/mg protein·h

12 h illumination (not DCMU-insensitive), the noncyclic together with O_2 evolution after 24 h (DCMU-sensitive) [72]. More recently it was reported that plastids isolated from plants grown under blue light show twofold greater DCPIP- and ferricyanide-reduction than 'red' plastids [33]. This was supported by Vroskesenskaya [114] and by our own results (Table 7). It has been shown that the first measurable reduction rates are higher in red light than in blue light with protein as a basis; after 8 h light, however, the reduction rates under blue light are much higher. This would mean that blue light makes more collection pigments and more photosynthetic units than red light, which was demonstrated by chlorophyll stimulation under blue light. On the other hand, in yellow-green mutants of *Chlorella fusca,* a reduction of photosynthetic capacity in blue light of high intensities as compared with red light, was observed [9]. The inhibiting influence is due to inhibition of the synthesis of chlorophyll a/b–protein complexes and to conformational changes of the thylakoid proteins. Treatment with glutardialdehyde stops the conformational changes and the cited blue light inhibition disappears.

E. Final Remarks

In summarizing all the results, we have shown that the continuous illumination of etiolated tissues with blue and red light causes a breakdown of the prolamellar bodies as well as white light does. Correlated with this fact is the breakdown of several lipids that are constituents of these prolamellar systems and of plastoglobuli. The total amount of unsaturated fatty acids is also reduced in etioplasts during these early stages of greening. It seems possible from these data that not only membrane fluidity but also hydrolysis of proteins and lipids takes place during the first hours of illumination. When primary thylakoid formation commences, galactolipids are synthesized in greater amounts, perhaps by the stimulation of envelope synthesizing capacity. The transfer of phospholipids through the envelope may also be a problem during this period, as a result of which the synthesizing rates are lower. Photosynthetic activity is also correlated with the presence of primary thylakoids and later, for full efficiency, with the formation of grana. At present we do not know the real components responsible for the correct ordering of the molecular structure in primary thylakoids and grana. Some components such as MGDG, α-linolenic acid, or trans-3-hexadecenoic acid are thought to be regulators that maintain the right position for the insertion of chlorophyll or grana assembly. Short-term illumination causes the regulation of synthesis of several lipids and fatty acids. From reversion experiments, especially with UV and far-red, we know that the manipulation of fatty acids is possible. These combination experiments, besides the use of mutants [9], herbicides [42, 45] etc., may therefore be a useful tool for studying lipids and function in developing plastids. In particular, we have to look for a special UV light-driven regulation system that works together with the phytochrome system during plastid development.

Acknowledgments. This work was supported by the *Deutsche Forschungsgemeinschaft*. I am very grateful to Mrs. H. Matzkeit and to Mrs. U. Widdecke for their valuable assistance as well as to my colleagues, Mr. K. Herm, Mr. H.D. Leonhardt, and Dr. H. Uhrig for supplying experimental data.

References

1. Alberte, R.S., Thornber, J.P., Naylor, A.W.: Time of appearance of photosystem I and II in chloroplasts of greening Jack bean leaves. J. Exp. Botany **23**, 1060–1069 (1972)
2. Anderson, J.M.: The molecular organization of chloroplast thylakoids. Biochim. Biophys. Acta **416**, 191–235 (1975)
3. Anderson, M.M., McCarty, R.E., Zimmer, E.A.: The role of galactolipids in spinach chloroplast lamellar membranes. Plant Physiol. **53**, 699–704 (1974)
4. Bahl, J., Francke, B., Moneger, R.: Lipid composition of envelopes, prolamellar bodies and other plastid membranes in etiolated, green and greening wheat leaves. Planta **129**, 193–201 (1976)
5. Bahl, J., Lechevallier, D., Monéger, R.: Etude comparée de l'évolution à l'obscurité et à la lumière des lipides plastidiaux de feuilles étioliées de Blé. Physiol. Vég. **12**, 229–249 (1974)
6. Bailey, J.L., Whyborn, A.G.: The osmiophilic globules of chloroplasts: II. Globules of the spinach beet chloroplast. Biochim. Biophys. Acta **78**, 163–174 (1963)
7. Bajracharya, D., Falk, H., Schopfer, P.: Phytochrome-mediated development of mitochondria in the cotyledons of mustard (*Sinapis alba L.*) seedlings. Planta **131**, 253–261 (1976)
8. Bauer, S., Siegelmann, H.W.: Photoconversion of chlorophyllide 684 to chlorophyllide 678. FEBS Lett. **20**, 352–354 (1972)
9. Bauer, K., Wild, A.: Die Wirkung von Blaulicht auf den photosynthetischen Elektronentransport bei gelbgrünen Mutanten von Chlorella fusca. Z. Pflanzenphysiol. (in press)
10. Bergfeld, R.: Die Wirkung von hellroter und blauer Strahlung auf die Chloroplastenausbildung. Z. Naturforsch. **18b**, 328–331 (1963)
11. Berry, D.R., Smith, H.: Red-light stimulation of prolamellar body recrystallization and thylakoid formation in barley etioplasts. J. Cell Sci. **8**, 185–200 (1971)
12. Boardman, N.K.: Protochlorophylls. In: The chlorophylls. Vernon, L.P., Seeley, R.S. (eds.). New York: Academic Press, 1966, pp. 437–479
13. Boardman, N.K., Anderson, J.M.: Studies on the greening of dark-grown bean plants. I. Formation of chloroplasts from proplastids. Australian J. Biol. Sci. **17**, 86–92 (1964)
14. Boardman, N.K., Anderson, J.M., Hiller, R.G., Kahn, A., Roughan, P.G., Treffry, T.E., Thorne, S.W.: Biosynthesis of the photosynthetic apparatus during chloroplast development in higher plants. In: Proc. 2nd Int. Congr. Photosynthesis Research. Forti, G., Avron, M., Melandri, A. (eds.). The Hague: Dr. W. Junk, N.V., 1972, Vol. 3, pp. 2265–2287

15. Bogorad, L.: Eukaryotic Intracellular relationships. In: Membrane Biogenesis. Tzagoloff, A. (ed.). New York and London: Plenum Press, 1976, pp. 201–245
16. Bovey, F., Ogawa, T., Shibata, K.: Photoconvertible and non photoconvertible forms of protochlorophyll (ide) in etiolated bean leaves. Plant Cell Physiol. (Tokyo) **15**, 1133–1137 (1974)
17. Bradbeer, J.W.: Plastid development in primary leaves of *Phaseolus vulgaris*. The effect of short blue, red, far-red, and white light treatments on dark grown plants. J. Exp. Botany **22**, 382–390 (1971)
18. Butler, W.L., Briggs, W.R.: The relation between structure and pigments during the first stages of proplastid greening. Biochim. Biophys. Acta. **112**, 45–53 (1966)
19. Cobb, A.H., Wellburn, A.R.: Changes in plastid envelope polypeptides during chloroplast development. Planta **121**, 273–282 (1974)
20. De Greef, J., Butler, W.L., Roth, T.E.: Greening of etiolated bean leaves in far-red light. Plant Physiol. **47**, 457–464 (1971)
21. Douce, R.: Site of biosynthesis of galactolipids in spinach chloroplast. Science **183**, 852–853 (1974)
22. Douce, R., Holtz, R.B., Benson, A.A.: Isolation and properties of the envelope of spinach chloroplasts. J. Biol. Chem. **248**, 7215–7222 (1973)
23. Duval, J.C., Trémolières, A., Rousseau, B.: Croissance et division du compartiment plastidial et synthèse de la chlorophylle et de l'acide linolénique au cours du développement de la première feuille de Pois. Physiol. Vég. **14**, 67–76 (1976)
24. Duysens, L.M.U., Amesz, J., Kamp, B.M.: The photochemical systems in photosynthesis. Nature (London) **190**, 510–511 (1961)
25. Egneus, H., Reffel, S., Seuden, G.: The appearance and development of photosynthetic activity in etiolated barley leaves and isolated etiochloroplasts. Physiol. Plantarum **27**, 48–55 (1972)
26. Farineau, J.: Simultaneous appearance of C-550 and cytochrome $b_{559\ HP}$ (high potential form) in barley leaves greened under intermittent light. Physiol. Plantarum **36**, 393–398 (1976)
27. Galliard, T.: Phospholipid metabolism in photosynthetic plants. In: Form and Function of Phospholipids B.B.A. Library. Ansell, G.B., Hawthorne, J.N., Dawson, R.M.C. (eds.). Amsterdam-London-New York: Elsevier, 1973, Vol. 3, pp. 253–288
28. Grahl, H., Wild, A.: Die Variabilität der Größe der Photosyntheseeinheit bei Licht- und Schattenpflanzen. Z. Pflanzenphysiol. **67**, 443–453 (1972)
29. Gray, I.K., Rumsby, M.G., Hawke, J.C.: The variations in linolenic acid and galactolipid levels in Gramineae species with age of tissue and light environment. Phytochemistry **6**, 107–113 (1967)
30. Gunning, B.E.S., Jagoe, M.P.: The prolamellar body. In: Biochemistry of Chloroplasts. Goodwin, T.W. (ed.). New York and London: Academic Press, 1967, Vol. 2, pp. 655–676
31. Gurr, M.I., James, A.T.: Lipid Biochemistry: An Introduction, 2nd ed. London: Chapman and Hall, 1975
32. Hampp, R., Schmidt, H.W.: Changes in envelope permeability during chloroplast development. Planta **129**, 69–73 (1976)
33. Harnischfeger, G., Treharne, K., Feierabend, G.: Studies on the primary photosynthetic processes of plastids from wheat grown under light of different spectral quality. Plant Sci. Lett. **3**, 61–66 (1974)
34. Hawke, J.C., Leech, R.M., Rumsby, M.G.: Lipid biosynthesis by chloroplasts isolated from developing *Zea mays*. Phytochemistry **13**, 403–413 (1974a)
35. Hawke, J.C., Rumsby, M.G., Leech, M.L.: Lipid biosynthesis in green leaves of developing maize. Plant Physiol. **53**, 555–561 (1974b)
36. Henningsen, K.W., Boardman, N.K.: Development of photochemical activity and the appearance of the high potential form of cytochrome b 559 in greening barley. Plant Physiol. **51**, 1117–1126 (1973)
37. Henningsen, K.W., Boynton, J.E.: Macromolecular physiology of plastids VII. The effect of a brief illumination on plastids of dark-grown barley leaves. J. Cell Sci. **5**, 757–793 (1969)
38. Henningsen, K.W., Boynton, J.E.: Macromolecular physiology of plastids. VIII. Pigment and membrane formation in plastids of barley greening under low light intensity. J. Cell Biol. **44**, 290–304 (1970)

39. Henningsen, K.W., Thorne, S.W., Boardman, N.K.: Properties of protochlorophyl(ide). Holochromes from etiolated and greening leaves. Plant Physiol. **53**, 419–425 (1974)
40. Herm, K., Tevini, M.: Die Wirkung von Rotlicht auf die Synthese der Glyko- und Phospholipide bei der Keimung von *Hordeum vulgare*. Z. Pflanzenphysiol. **73**, 349–359 (1974)
41. Hiller, R.G., Pilger, D., Genge, S.: Photosystem II activity and pigment-protein complexes in flashed bean leaves. Plant Sci. Lett. **1**, 81–88 (1973)
42. Hilton, J.L., John, J.B.St., Christiansen, M.N., Norris, K.H.: Interaction of lipoidal materials and a pyridazinone inhibitor of chloroplast development. Plant. Physiol. **48**, 171–177 (1971)
43. Hitchcock, C., Nichols, B.W.: Plant Lipid Biochemistry. London and New York: Academic Press, 1971
44. van Hummel, H.C.: Some observations on biosynthesis and ratio regulation of mono- and digalactosyl diglycerides by chloroplasts and other subcellular fractions from spinach leaves. Z. Pflanzenphysiol. **71**, 228–241 (1974)
45. John, J.B.St.: Manipulation of galactolipid fatty acid composition with substituted pyridazinones. Plant Physiol. **57**, 38–40 (1976)
46. Joy, K.W., Ellis, R.J.: Protein synthesis in chloroplasts. IV. Polypeptides of the chloroplast envelope. Biochim. Biophys. Acta **378**, 143–151 (1975)
47. Kasemir, H., Bergfeld, R., Mohr, H.: Phytochrome-mediated control of prolamellar body reorganization and plastid size in mustard cotyledons. Photochem. Photobiol. **21**, 111–120 (1975)
48. Kasemir, H., Masoner, M.: Control of chlorophyll synthesis by phytochrome. II. The effect of phytochrome on amino-levulinate dehydratase in mustard seedlings. Planta **126**, 119–126 (1975)
49. Kates, M., Marshall, M.O.: Biosynthesis of phosphoglycerides. In: Recent Advances in the Chemistry and Biochemistry of Plant Lipids. Galliard, T., Mercer, E.J. (eds.). London-New York-San Francisco: Academic Press, 1975
50. Kirk, J.T.O.: Biochemical aspects of chloroplast development. Ann. Rev. Plant Physiol. **21**, 11–42 (1970)
51. Kirk, J.T.O., Goodchild, D.J.: Relationship of photosynthetic effectiveness of different kinds of light to chlorophyll content and chloroplast structure in greening wheat and ivy leaves. Australian J. Biol. Sci. **25**, 215–241 (1972)
52. Klein, S., Bryan, G., Bogorad, L.: Early stages in the development of plastid fine structure in red and far-red light. J. Cell Biol. **22**, 433–442 (1964)
53. Klein, S., Schiff, J.A., Holowinsky, A.W.: Events surrounding the early development of *Euglena* chloroplasts. 2. Normal development of fine structure and the consequences of preillumination. Plant Physiol. **45**, 339–347 (1971)
54. Klein, S., Schiff, J.A.: The correlated appearance of prolamellar bodies, protochlorophyll(ide) species and the Shibata-shift during development of bean etioplasts in the dark. Plant. Physiol. **49**, 619–626 (1972)
55. Kochubei, S.M., Shadchina, T.M., Kononenko, A.A., Lukashev, E.P., Timofeev, K.N., Matorin, D.N.: Effect of galactolipase on photoreactions in light chloroplast fragments. Photosynthetica **9**, 255–260 (1975)
56. Koenig, F.: Konzentration einiger Lipide in den Chloroplasten von *Zea mays* und *Antirrhinum majus*. Z. Naturforsch. **26b**, 1180–1187 (1971)
57. Krupa, Z., Baszynski, T.: Requirement of galactolipids for photosystem I activity in lyophilized spinach chloroplasts. Biochim. Biophys. Acta **408**, 26–34 (1975)
58. Leech, R.M., Rumsby, M.G., Thomson, W.W.: Plastid differentiation, acyl lipid and fatty acid changes in developing green maize leaves. Plant Physiol. **52**, 240–245 (1973)
59. Lichtenthaler, H.K.: Die Plastoglobuli von Spinat, ihre Größe und Zusammensetzung während der Chloroplastendegeneration. Protoplasma **68**, 315–326 (1969)
60. Lichtenthaler, H.K., Tevini, M.: Die Verteilung der Pigmente, Plastidenchinone und Plastoglobuli auf verschiedene aus beschallten Spinatchloroplasten gewonnene Partikelfraktionen. Z. Pflanzenphysiol. **62**, 33–50 (1970)
61. Liljenberg, C.: Characterization and properties of a protochlorophyllide-ester in leaves of dark grown barley with geranyl-geraniol as esterifying alcohol. Physiol Plant. **32**, 208–213 (1974)
62. Mackender, R.O., Leech, R.M.: Isolation of chloroplast envelope membranes. Nature (London) **228**, 1347–1349 (1970)

63. Mackender, R.O., Leech, R.M.: The isolation and characterization of plastid envelope membranes. In: Proc. 2nd Internat. Congr. Photosynthesis Research. Forti, G., Avron, M., Melandri, A. (eds.). The Hague: Dr. W. Junk, N.V., 1972, Vol. 2, pp. 1431–1440
64. Mackender, R.O., Leech, R.M.: The galactolipid, phospholipid, and fatty acid composition of the chloroplast envelope membranes of *Vicia faba L.* Plant Physiol. **53**, 496–502 (1974)
65. Masoner, M., Kasemir, H.: Control of chlorophyll synthesis by phytochrome. I. The effect of phytochrome on the formation of 5-aminolevulinate in mustard seedlings. Planta **126**, 111–117 (1975)
66. Mazliak, P.: Lipidmetabolism in plants. Ann. Rev. Plant Physiol. **24**, 287–310 (1973)
67. Mendiola-Morgenthaler, L.R., Morgenthaler, J.-J.: Proteins of the envelope and thylakoid membranes of spinach chloroplasts. FEBS Lett. **49**, 152–155 (1974)
68. Mohr, H.: Lectures on Photomorphogenesis. Berlin-Heidelberg-New York: Springer, 1972
69. Mudd, J.B., Garcia, R.E.: Biosynthesis of glycolipids. In: Recent Advances in the Chemistry and Biochemistry of Plant Lipids. Galliard, T., Mercer, E.J. (eds.). London-New York-San Francisco: Academic Press, 1975
70. Mühlethaler, K., Frey-Wyssling, A.: Entwicklung und Struktur der Proplastiden. J. Biophys. Biochem. Cytol. **6**, 507–512 (1959)
71. Niewen, N.C.: Electrophoretic characterization of membrane of proteins during chloroplast development of barley. Europ. J. Biochem. **50**, 611–624 (1975)
72. Oelze-Karow, H., Butler, W.L.: The development of photophosphorylation and photosynthesis in greening leaves. Plant Physiol. **48**, 621–625 (1971)
73. Ohad, J.: Biogenesis of chloroplast membranes. In: Membrane Biogenesis. Tzagoloff, A. (ed.). New York and London: Plenum Press, 1976, pp. 279–350
74. Panther, R.A., Boardman, N.K.: Lipid biosynthesis by isolated plastids from greening pea, *Pisum sativum*. J. Lipid Res. **14**, 664–671 (1973)
75. Pineau, B., Douce, R.: Analyse Electrophoretique des Protéines de l'Enveloppe des Chloroplastes d'Epinard. FEBS Lett. **47**, 255–259 (1974)
76. Plesnicar, M., Bendall, D.S.: The development of photosynthetical activities during greening of etiolated barley. In: Proc. 2nd Internat. Congr. Photosynthesis Research. Forti, G., Avron, M., Melandri, A. (eds.). The Hague: Dr. W. Junk, N.V., 1972, Vol. 3, pp. 2367–2374
77. Poincelot, R.: Isolation and lipid composition of spinach chloroplast envelope membranes. Arch. Biochem. Biophys. **159**, 134–142 (1973)
78. Poincelot, R.P., Day, P.R.: An improved method for the isolation of spinach chloroplast envelope membranes. Plant Physiol. **54**, 780–783 (1974)
79. Phung Nhu Hung, S.: Absence of correlation between the high-potential cytochrome b-559 and photosystem II activity. Z. Pflanzenphysiol. **72**, 389–394 (1974)
80. Phung, S., Hoarau, A., Moyse, A.: Etude de l'évolution en chloroplastes des plantes étole d'orge: II Photophosphorylation et photoréduction du NADP, formatin de ferredoxin, en enclairment contenu et par l'action déclairs. Z. Pflanzenphysiol. **62**, 245–258 (1970)
81. Robertson, D., Laetsch, W.M.: Structure and function of developing barley plastids. Plant Physiol. **54**, 269–290 (1971)
82. Roughan, P.G.: Phosphatidyl choline: Donor of 18-carbon unsaturated fatty acids for glycerolipid biosynthesis. Lipids **10**, 609–614 (1975)
83. Sagromsky, H., Döbel, P.: Zur Bedeutung von Chlorophyll b für die Chloroplastendifferenzierung. Biochem. Physiol. Pflanzen **166**, 371–376 (1974)
84. Schneider, H.A.W.: Chlorophylle: Aspekte der Biosynthese und ihre Regulation. Ber. Deut. Botan. Ges. **88**, 83–124 (1975)
85. Schnepf, E.: Über Zusammenhänge zwischen Heitz-Leyon-Kristallen und Thylakoiden. Planta **61**, 371–373 (1964)
86. Schopfer, P., Bajacharya, D., Falk, H., Thien, W.: Phytochromgesteuerte Entwicklung von Zellorganellen (Plastiden, Microbodies, Mitochondrien). Ber. Deut. Botan. Ges. **88**, 245–268 (1975)
87. Shibata, K.: Spectroscopic studies on chlorophyll formation in intact leaves. J. Biochem. **44**, 147–173 (1957)
88. Selldén, G., Selstam, E.: Changes in chloroplast lipids during the development of photosynthetic activity in barley etio-chloroplasts. Physiol. Plantarum **37**, 35–41 (1976)
89. Siegenthaler, P.A.: Inhibition of photosystem II electron transport in chloroplasts by fatty acids and restoration of its activity by Mn^{2+}. FEBS Lett. **39**, 337–340 (1974)

90. Sironval, C., Michel-Wolwertz, M.R., Madsen, A.: On the nature and possible function of the 673- and 684 mμ forms in vivo of chlorophyll. Biochim. Biophys Acta **94**, 334–354 (1965)
91. Sitte, P.: Plastiden-Metamorphose und Chromoplasten bei *Chrysosplenium*. Z. Pflanzenphysiol. **73**, 243–265 (1974)
92. Slack, C.R., Roughan, P.G.: The kinetics of incorporation in vivo of [^{14}C] acetate and [^{14}C] carbon dioxide into the fatty acids of glycerolipids in developing leaves. Biochem. J. **152**, 217–228 (1975)
93. Smith, J.H.C., French, C.S.: The major and accessory pigments in photosynthesis. Ann. Rev. Plant Physiol. **14**, 181–224 (1963)
94. Sprey, B.: Membranassoziierte Tubuli während der Chloroplastengenese von *Hordeum vulgare*. Protoplasma **84**, 194–203 (1975)
95. Sprey, B., Laetsch, W.M.: Chloroplast envelopes of *Spinacia oleracea L.* I. Polypeptides of chloroplast envelopes and lamellae. Z. Pflanzenphysiol. **75**, 38–52 (1975)
96. Sprey, B., Lichtenthaler, H.K.: Zur Frage der Beziehungen zwischen Plastoglobuli und Thylakoidgenese in Gerstenkeimlingen. Z. Naturforsch. **21 b**, 697–699 (1966)
97. Staehlin, L.A.: Chloroplast membrane structure. Intramembranous particles of different sizes make contact in stacked membrane regions. Biochem. Biophys. Acta **408**, 1–11 (1975)
98. Tevini, M.: Die Phospho- und Glycolipidänderungen während des Ergrünens etiolierter *Hordeum*-Keimlinge. Z. Pflanzenphysiol. **65**, 266–272 (1971 a)
99. Tevini, M.: Der Einfluß von Phosphat- und Nitratmangel auf die Synthese der Phospho- und Glykolipide bei Impatiens balsamina. Ber. Deut. Botan. Ges. **84**, 595–606 (1971 b)
100. Tevini, M.: The formation of lipids following illumination of etiolated seedlings. In: Proc. 2nd Internal. Congr. Photosynthesis Research. Forti, G., Avron, M., Melandri, A. (eds.). The Hague: Dr. W. Junk, N.V., 1972, Vol. 3, pp. 2471–2478
101. Tevini, M., Lichtenthaler, H.K.: Untersuchungen über die Pigment- und Lipochinonausstattung der zwei photosynthetischen Pigmentsysteme. Z. Pflanzenphysiol. **62**, 17–32 (1970)
102. Tevini, M., Herm, K., Leonhardt, H.D.: Lipids and function of etiochloroplasts after ultra-violet, blue and red light illumination. Biochem. Soc. Transactions, Vol. **5**, 95–98 (1977)
103. Treffry, T.: Chloroplast development in etiolated peas: Reformation of prolamellar bodies in red light without accumulation of protochlorophyllide. J. Exp. Botany **24**, 185–195 (1973)
104. Thornber, J.P., Highkin, H.R.: Composition of the photosynthetic apparatus of normal barley leaves and a mutant lacking chlorophyll b. Europ. J. Biochem. **41**, 109–116 (1974)
105. Trémolières, A., Jacques, R., Mazliak, P.: Régulation par la lumière de l'accumulation de l'acide linolénique dans la jeune feuille de Pois. Spectre d'action, influence de l'intensité de l'éclairement et rôle du phytochrome. Physiol. Vég. **11**, 239–251 (1973)
106. Tsel-Niker, Y.L.: Effect of light intensity on the number and dimensions of chloroplasts in woody plants. Soviet. Plant Physiol. **22**, 214–219 (1975)
107. Tuquet, C., Guillot-Salomon, T., Farineau, J., Signol, M.: Biogenèse des membranes plastidiales dans les feuilles étiolées d'Orge soumises à des éclairs répétés: développement d'accolements de thylacoides, synthèse de phosphatidglycérol et appération de cytochrome b 559 (forme haut potentiel). Physiol. Vég. **14**, 11–30 (1976)
108. Uhrig, H., Tevini, M.: Effekte der Phospholipase D auf den Elektronentransport und die Lipidzusammensetzung isolierter Spinatchloroplasten. Planta **128**, 173–178 (1976 a)
109. Uhrig, H., Tevini, M.: The effects of phospholipase D hydrolysis products on photochemical activities of spinach chloroplasts. In: 'Lipids and Lipid Polymers in Higher Plants' Symposium abstracts, Botanical Institute, University of Karlsruhe, 1976, pp. 44–45
110. Unser, G., Mohr, H.: Phytochrome-mediated increase of galactolipids in mustard seedlings. Naturwissenschaften **27**, 358–359 (1970)
111. Vernon, L.P., Shaw, E.R., Ke, B.: A photochemically active particle derived from chloroplasts by action of the detergent Triton X-100. J. Biol. Chem. **241**, 4101–4109 (1966)
112. Vernon, L.P., Klein, S., White, F.G., Shaw, E.R., Mayne, B.C.: Properties of a small photosystem II particle obtained from spinach. In: Proc. 2nd Internat. Congr. Photosynthesis Research. Forti, G., Avron, M., Melandri, A. (eds.). The Hague: Dr. W. Junk, N.V., 1972, Vol. 1, pp. 801–802
113. Virgin, H.I., von Wettstein, D., Kahn, A.: The physiology of chlorophyll formation in relation to structural changes in chloroplasts. Photochem. Photobiol. **2**, 83–91 (1963)
114. Voskresenskaya, N.P., Drosdova, I.S., Krendeleva, T.E.: The effect of light quality on the

organization of photosynthetic electron transport chain of pea seedlings (*Pisum sativum L.*). In: VII. Internat. Congr. on Photobiology, Abstract P 86. Rome (1976)

115. Vlasova, M.P., Drosdova, I.S., Voskresenskaya, N.P.: Modification of chloroplast fine structure in barley plants greening under blue and red light. (russ.) Fiziol. Rast. **18**, 5–11 (1971)
116. Weier, T.E., Brown, D.L.: Formation of prolamellar body recrystallisation and thylakoid formation in barley etioplasts. J. Cell Sci. **8**, 185–200 (1970)
117. Weier, T.E., Sjoland, R.D., Brown, D.L.: Changes inducted by low light intensities on the prolamellar body of 8-day dark-grown seedlings. Am. J. Botany **57**, 276–284 (1970)
118. Wellburn, A.R., Hampp, R.: Movement of labelled metabolites from mitochondrial to plastids during development. Planta **131**, 17–20 (1976)
119. Wellburn, F.A.M., Wellburn, A.R.: Developmental changes occurring in isolated intact etioplasts. J. Cell Sci. **9**, 271–287 (1971)
120. Wellburn, A.R., Wellburn, F.A.M.: Developmental changes of etioplasts in isolated suspensions and in situ. Ann. Botany **37**, 11–19 (1973)
121. Whatley, F.R., Gregory, P., Haslett, B.G., Bradbeer, J.W.: Development of electron transport intermediates in greening chloroplasts. In: Proc. 2nd Internat. Congr. Photosynthesis Research. Forti, G., Avron, M., Melandri, A. (eds.). The Hague: Dr. W. Junk, N.V., 1972, Vol. 3, pp. 2375–2382
122. Williams, J.P., Watson, G.R., Leung, S.P.K.: Galactolipid synthesis in *Vicia faba* leaves II. Formation and desaturation of long chain fatty acids in phosphatidylcholine, phosphatidylglycerol, and the galactolipids. Plant Physiol. **57**, 179–184 (1976)
123. Wrischer, M.: Protein crystalloids in the stroma of bean plastids. Protoplasma **77**, 141–150 (1973)

CHAPTER 8

The Biosynthesis of Plant Sterols

L.J. GOAD

A. Introduction

The great variety of plant sterols and related compounds have for many years been a source of interest to natural products chemists because of the challenges they have offered in structural determination and development of analytical techniques. Interest in the biosynthesis of animal sterols spans the past thirty years, but plant sterol biosynthesis has been extensively investigated only since about 1965. Apart from the additional mechanisms required for the introduction of the extra carbon atoms into the side of typical plant sterols, there was probably the expectation that the pathway in plants would be much the same as the animal route leading from acetyl-CoA through mevalonic acid, squalene, and lanosterol to cholesterol. It soon became apparent, however, that such is not the case, and the studies of phytosterol biosynthesis have been rewarded by the recognition of several intriguing mechanistic differences not only between animals and plants but also between particular classes of plants. Moreover some of these differences appear to correlate with other evidence showing phylogenetic relationships of various classes of lower and higher plants.

Several reviews on plant sterols have been published, and these should be consulted for information on earlier work, since in this Chapter a comprehensive converage of the literature will not be attempted. Reviews on plant sterol distribution include those of Bean [18], Bergmann [23], Goad [58], Heftman [74], and Stoll and Jucker [153]. Phytosterol biosynthesis is covered by the reviews of Goad [59], Goad and Goodwin [65], Goad et al. [68], Grundwald [72], Jacobsohn [84], and Weete [166]. The possible physiological roles of plant sterols are considered in the reviews of Bean [18], Grundwald [72], Heftman [75], and Nes [110].

B. Structure and Distribution

Cholesterol (*1a*), well known as the major sterol of animals, has a C_8 side chain and a Δ^5 double bond. The sterols produced by most, but not all, plant taxa also have a Δ^5 bond, but there is usually an additional C_1 or C_2 unit at C-24 of the side chain that is regarded as the distinguishing feature of phytosterols. Campesterol (*2a*), sitosterol (*3a*), and stigmasterol (*7a*) are generally cited as the typical plant sterols, and they certainly are the most frequently reported sterols in tracheophytes. However, there are several types of side-chain and ring systems known, and the possible permutations can give a considerable number of potential sterols. In a previous review about 90 known plant sterols were listed [65]. The number of naturally occurring sterols has now increased to somewhere in the region of 150, and many more can now be predicted with the recognition of several new patterns of side-chain structure. The recent interest in marine invertebrate natural products has contributed to the discovery of a significant number of these new sterols, but it seems likely that many, if not all, of these compounds are of dietary origin in marine animals and that they are probably produced by algae or perhaps in some cases by fungi [61].

The structures of all the known phytosterols will not be presented here, and the reader is referred to previous reviews for further information [18, 61, 65]. The structures of some of the known sterol side chains and tetracyclic ring systems most pertinent to this discussion of phytosterol biosynthesis are shown in Figures 1 and 2, respectively. The majority of sterols have been given a trivial name, usually derived from the name of the original source, which is often totally uninformative concerning the exact structure of the compound. However, the trivial name is generally short and more convenient to use than that obtained by application in the recommended IUPAC-IUB nomenclature. Trivial names therefore will be used throughout this Chapter, the systematic name being given only at the first mention.

Introduction of an alkyl group at C-24 of the side chain renders this carbon chiral and thus two epimers are possible. Side chains *2* and *3* have the 24α- or 24*R*-configuration while side chains *4* and *5* are the 24β- or 24*S*-epimers. In side chains *6* and *7*, although 24α-, the change in relative priorities of C-23, C-25, and C-28 caused by introduction of the Δ^{22} bond [26] means that they are now described as 24*S*, not 24*R*. Likewise the 24β-side chains *8* and *9* become 24*R*-. The *RS*-convention is recommend by the IUPAC-IUB 1971 Definitive Rules for Steroid Nomenclature and while this permits unambiguous assignments of configuration it is not always so convenient for conveying biosynthetic relationships between sterols. For this reason the 24α/24β-system [49] will be used in this text following the suggestions of other authors [152, 157].

Turning to the ring system, compounds with methyl groups attached to C-4 and C-14, e.g., *d* to *h*, have in the past been referred to as tetracyclic triterpenes. However, with the realisation that such compounds are precursors of sterols, they are now frequently referred to as 4,4-dimethyl or 4-monomethyl sterols. Also in some contexts sterols with ring systems *a, b, c,* and *i* are now often classed as 4-desmethylsterols.

It is now recognised that cholesterol (*1a,* cholest-5-en-3β-ol) is widespread in the plant kingdom, but it usually constitutes only a few percent of the sterol mixture. In the Rhodophyceae, however, it is most often the predominant sterol and it is often accompanied by desmosterol (*10a,* cholesta-5,24-dien-3β-ol) and 22-dehydrocholesterol (*11a,* cholesta-5,22-dien-3β-ol) [61, 116].

Sterols with a 24β-methyl or ethyl group are generally considered to be characteristic of algae belonging to the Chlorophyta, Chrysophyta, and Euglenophyta [116]. In fungi 24β-methyl sterols predominate, with ergosterol (*8c,* ergosta-5,7,22-trien-3β-ol) as the major sterol component in most Ascomycetes

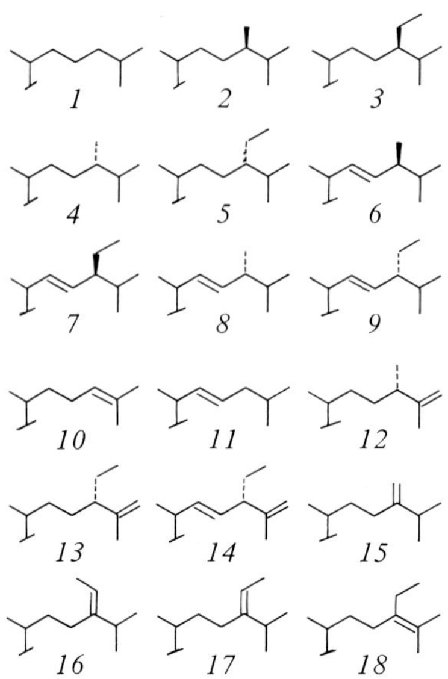

Fig. 1. Structures of plant sterol side chains

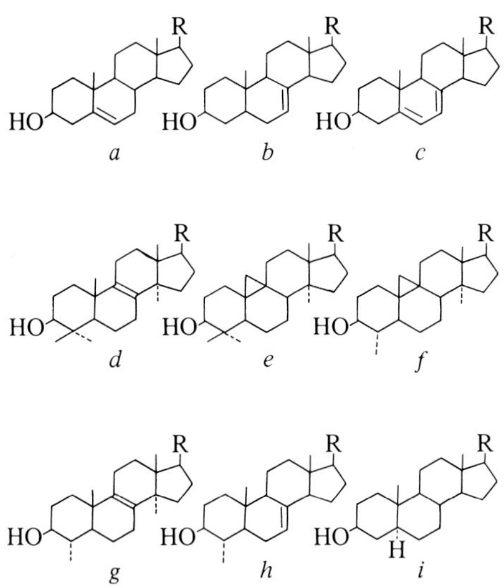

Fig. 2. Ring structures of plant sterols. *R* represents one of the side chains shown in Figure 1

and Basidiomycetes [166, 167]. Diatoms and slime moulds appear, on the present available evidence, to be unique among the lower organisms as the only ones producing C-24 alkylated sterols with the 24α-configuration. Thus slime moulds produce compestane (*2i*) and stigmastane (*3i*, *7i*) compounds [44, 96], while (24*S*)-24-methylcholesta-5,22-dien-3β-ol (*6a*) is present in some diatoms [114, 142].

Sterols with a 25-methylene group, codisterol (*12a*, ergosta-5,25-dien-3β-ol) and clerosterol (*13a*, (24*S*)-24-ethylcholesta-5,25-dien-3β-ol), are present [141] as the major components of the alga *Codium fragilis* (Siphonales), and *13a, 14a,* and related sterols have also been reported in higher plants of the Cucurbitaceae, Verbenaceae, and Compositeae families [65].

Several 24-methylene sterols (side chain *15*) are widespread in small amounts in fungi, algae, and higher plants [61, 65, 116, 166]. The 24-ethylidene compound fucosterol (*16a,* stigmasta-5,E-24(28)-dien-3β-ol) is the major component of the brown seaweeds, the Phaeophyta [61, 116], and the isomer, isofucosterol (*17a,* stigmasta-5,Z-24(28)-dien-3β-ol), is abundant in marine green alga of the Ulvales [61, 116]. Isofucosterol (*17a*), and other sterols with this side chain (*17*), are found in small amounts in the sterol mixtures from many higher plants.

Examples of sterols with a Δ^{22} bond are ergosterol (*8c*), found in fungi and some algae, brassicasterol (*8a,* ergosta-5,22-dien-3β-ol), a component of algal and some higher plant sterol mixtures, and stigmasterol (*7a,* stigmasta-5,22-dien-3β-ol), found in many higher plants.

Until recently it was generally believed that the sterols of higher plants probably all had the 24α-configuration as exemplified by campesterol (*2a,* (24*R*)-24-methylcholest-5-en-3β-ol) and sitosterol (*3a,* stigmast-5-en-3β-ol), the two most commonly occurring ones. However, the 24α- and 24β-isomeric sterols are difficult to differentiate since they are inseperable by chromatographic techniques, and identifications based upon melting point and optical rotation measurements require pure samples not easily obtained from the complex sterol mixtures occurring in most plants. Also such measurements can give only the stereochemistry of the major component if mixtures of C-24 epimers are present. Thus in most examinations of higher plant sterols, the C-24 configurations have been assumed to be 24α- in the absence of adequate analytical techniques to determine otherwise. The recent application of 220 MHz nuclear magnetic resonance (NMR) spectroscopy to the analysis of plant sterols has shown that the 24α- and 24β-epimers give diagnostically different chemical shifts for the side-chain methyl groups that permit unambiguous determination of the C-24 configuration using a few mg of sterol [109, 113, 140, 143, 157]. Moreover this NMR method permits approximate quantitation of mixtures of C-24 epimers. Using this technique it has been found that the 24-ethyl sterols in several vascular plants are indeed the 24α-epimers sitosterol (*3a*) and stigmasterol (*7a*) [109, 113]. By contrast the 24-methyl sterols are mixtures of the 24α- and 24β-epimers, campesterol (*2a*) and dihydrobrassicasterol (*4a,* ergost-5-en-3β-ol). From the limited evidence so far available, it seems that the proportion of 24α-sterol (campesterol, *2a*) in the mixture increases as the evolutionary hierarchy is ascended. Thus it was demonstrated [112] that the primitive tracheophyte *Lycopodium compolanatum* contains the 24β-sterol, ergosterol (*8c*), with only a minor amount of the 24α-epimer (*6c,* (24*S*)-24-methylcholesta-5,7,22-trien-3β-ol). Similarly several mosses and liverworts contain predominantly the 24β-epimers dihydrobrassicasterol (*4a*) and brassicasterol (*8a*) as their C_{28} sterols, while the 24α-compounds sitosterol (*3a*) and stigmasterol (*7a*) constitute their C_{29}-sterols [165]. In the gymnosperms and angiosperms so far examined, however, the 24α-epimer campesterol (*2a*) is present at about twice the concentration of the 24β-compound (*4a*) [109, 113].

Considering the sterol ring systems found in plants, the Δ^5 4-desmethyl sterols are the most commonly encountered in algae and higher plants. Δ^7-Sterols are found in some algae and members of Cucubitaceae and, together with $\Delta^{5,7}$-sterols, they are characteristic of most fungi [110]. The various 4,4-dimethyl and 4-monomethyl sterols are usually minor sterol components in plant sterol mixtures, as would be expected of possible precursor compounds. However, relatively high concentrations of some of these compounds do occur in some plants. For example, the latex of *Euphorbia* species is rich in such compounds as cycloartenol (*10e*, 4,4,14α-trimethyl-9β,19-cyclo-5α-cholest-24-en-3β-ol), 24-methylenecycloartanol (*15e*, 4,4,14α-trimethyl-9β,19-cyclo-5α-ergost-24(28)-en-3β-ol), and obtusifoliol (*15g*, 4α,14α-dimethyl-5α-ergosta-8,24(28)-dien-3β-ol) [120]. Cycloartenol (*10e*) and cyclolaudenol (*12e*, 4,4,14α-trimethyl-9β,19-cyclo-5α-ergost-25-en-3β-ol) occur in fern rhizomes [24], and some seed oils are relatively rich in cycloartenol (*10e*), 24-methylenecycloartanol (*15e*), obtusifoliol (*15g*), 24-methylenelophenol (*15h*, 4α-methyl-5α-ergosta-7,24(28)-dien-3β-ol), and 24-ethylidenelophenol (*17h*, 4α-methyl-5α-stigmasta-7,Z-24(28)-dien-3β-ol) [83].

Although the preceding brief account of sterol structure and distribution is very incomplete, it does suffice to pose some of the main problems which it is hoped that a comprehensive understanding of phytosterol biosynthesis will eventually explain. For example, what are the biosynthetic relationships, if any, between the various 4,4-dimethyl and 4-monomethyl sterols and the principal 4-desmethyl sterols? Are cycloartenol (*10e*) and the related 9β,19-cyclopropane compounds, and indeed many of the other phytosterols, simply the end products of aberrant biosynthetic pathways or are any of them essential sterol precursors? Why do particular sterols accumulate in the various plant classes? What are the biosynthetic relationships between the various sterol side chains? Are any of the unsaturated side-chain sterols precursors of the saturated side-chain compounds or vice versa? What alkylation mechanisms are operative to ensure that 24β-sterols predominate in algae and fungi while the C_{29} sterols of higher plants are usually the 24α-compounds? Also by what routes do mixtures of the 24α- and 24β-methyl epimers arise in higher plants and does this have any evolutionary significance?

The following account will show that the answers to some of these questions have been obtained. It is also hoped to indicate that we may now be within reach of a sufficient understanding of the basic phytosterol biosynthetic mechanisms to permit some rationalisation of the elaboration and occurrence of the, at times apparently bewildering, array of sterols produced by the plant kingdom.

C. Squalene Formation and Cyclisation

The initial stages of sterol production in both animals and plants proceed along the well-established isoprenoid biosynthetic route from acetyl-CoA through hydroxymethylglutaryl-CoA to mevalonic acid and thence via phosphorylated intermediates to farnesyl pyrophosphate and squalene. This pathway

Fig. 3. Conversion of farnesyl pyrophosphate into squalene. The carbon marked is derived from C-6 of mevalonic acid

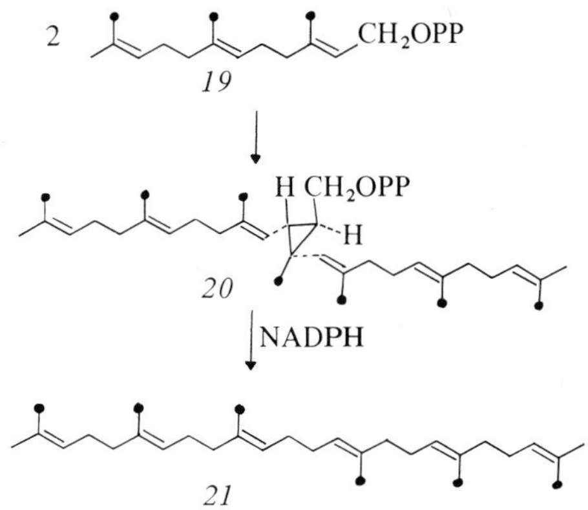

is described in several reviews [34, 59, 60, 108, 129–131], and its operation in plants is discussed in detail in Chapter 10. The stereochemical aspects of the reaction mechanisms have been elucidated using animal preparations [123], but all the available evidence indicates that the synthesis of squalene in plants follows the same stereochemical course [65]. Recently it was shown [154] that in squalene production by pea seedlings (*Pisum sativum*) the C-6 of mevalonic acid retains its individuality throughout the pathway to give the cis-terminal methyl groups of squalene.

Work with animal and yeast preparations showed that the coupling (Fig. 3) of the farnesyl pyrophosphate (*19*) molecules first gives presqualene pyrophosphate (*20*), which has a cyclopropane ring [42, 47, 124, 137]. Subsequent reduction of the presqualene pyrophosphate (*20*), using NADPH and particulate enzyme, gives the C_{30} hydrocarbon squalene (*21*) [6, 126, 144). Several possible mechanisms for these reactions have been proposed, and they are presented in the reviews of Rees and Goodwin [129, 131].

The participation of presqualene pyrophosphate in squalene synthesis in photosynthetic plant tissues has been revealed by the use of a cell-free system prepared from a bramble tissue culture [78]. In the presence of NADH, mevalonic acid was incorporated into squalene, but in the absence of NADH, squalene was hardly labelled and radioactivity accumulated in a compound identified, after treatment with alkaline phosphatase, as presqualene alcohol. Moreover the conversion of presqualene pyrophosphate into squalene has been achieved using a cell-free system from the pea, *Pisum sativum* [20]. The squalene synthetase system of etiolated *Phaseolus vulgaris* leaves is located in the endoplasmic reticulum [73]. Likewise it is microsomal in yeast [146].

Cyclisation of the C_{30} hydrocarbon chain of squalene requires introduction of atmospheric oxygen, in the presence of NADPH, and a microsomal enzyme system, to give first squalene-2,3-oxide (*22*) [155, 169]. Proton-initiated cyclisation then produces either the cyclopentanophenanthrene ring system of sterols [36, 38, 163] or the various 3β-hydroxy pentacyclic ring systems of the numerous plant triterpenes [136, 147].

The intermediacy of squalene-2,3-oxide was first established in animals, but several investigations now support its role in plant sterol production. The incorporation of [1-^{14}C] acetate into squalene-2,3-oxide was demonstrated with a tissue culture of tobacco [21], and radioactivity accumulates in this compound in plants incubated with radioactive precursors in the presence of squalene-2,3-oxide cyclase inhibitors [35, 41, 46, 76, 135].

The nature of the sterol precursor produced by squalene-2,3-oxide cyclisation shows a most striking dichotomy between photosynthetic and non-photosynthetic organisms. Fungi, like animals, produce lanosterol (*10d*), which is then modified by a sequence of reactions involving C-4 and C-14 demethylation, C-24 methylation, and double bond rearrangements to produce ergosterol (*8c*) [10, 53, 166]. By contrast photosynthetic plants produce, not lanosterol, but the 9β,19-cyclopropane compound cycloartenol (*10e*), which undergoes similar modifications, though including a subsequent opening of the cyclopropane ring, to yield the characteristic phytosterols [65, 72].

The occurrence of lanosterol (*10d*) and acidic compounds with the lanostane skeleton in fungi has been reviewed [166, 170], but there have been no reports of cycloartenol (*10e*) or any other 9β,19-cyclopropane steroids in yeast or other fungi [119]. Cell-free systems have been obtained from yeast, *Phycomyces blakesleeanus,* and *Chepalosporium caerulens,* which catalyse the cyclisation of squalene-2,3-oxide to lanosterol [13, 15, 87, 104, 145]. The yeast cyclase, like the rat liver and pea seedling cyclases, utilises the *S* epimer of squalene-2,3-oxide [13], and the two hydride migrations involved in the cyclisation are consecutive 1,2-shifts [15] as postulated by the Ruzicka isoprene rules [48; see Fig. 4].

Some years ago, perhaps in the belief that once a biosynthetic pathway was evolved it generally remains uniform throughout nature, it was assumed that lanosterol (*10d*) would also participate in sterol elaboration in chlorophyll-containing plants. However, there were very few reports of lanosterol (*10d*) in green plants, and careful reinvestigations have resulted so far in the authentication of lanosterol in only a few higher plants, notably in the latex of the Euphorbiaceae [69, 70, 120–122]. The failure to isolate a particular compound cannot always be regarded as de facto evidence that it does not participate in a biosynthetic sequence since after production it may be so rapidly metabolised that it does not accumulate in sufficient amounts for detection by the available analytical techniques. This problem can often be resolved by recourse to suitably designed radioactive labelling experiments. Several attempts have now been made to demonstrate the incorporation of labelled acetate or mevalonate into lanosterol (*10d*) by algae and higher plant tissues but with negative results [1, 22, 43, 46, 62–64, 80, 90, 101, 132]. However cycloartenol (*10e*) has consistently been identified among the labelled products which is in accord with it proposed role as a phytosterol precursor [22, 62, 164].

Several lines of evidence support the intermediary role of cycloartenol (*10e*) and its formation directly by cyclisation of squalene-2,3-oxide rather than by modifications of preformed lanosterol (*10d*). Thus cell-free preparations have been obtained from the bean (*Phaseolus vulgaris*), tobacco (*Nicotiana tobacum*) and a chrysophyte alga (*Ochromonas malhamensis*) which contain a microsomal squalene-2,3-oxidecycloartenol cyclase active under anaerobic conditions [19, 76, 133, 134]. In the absence of oxygen further metabolism by C-4 or C-14 demethylation cannot occur and therefore the products of cyclisation must

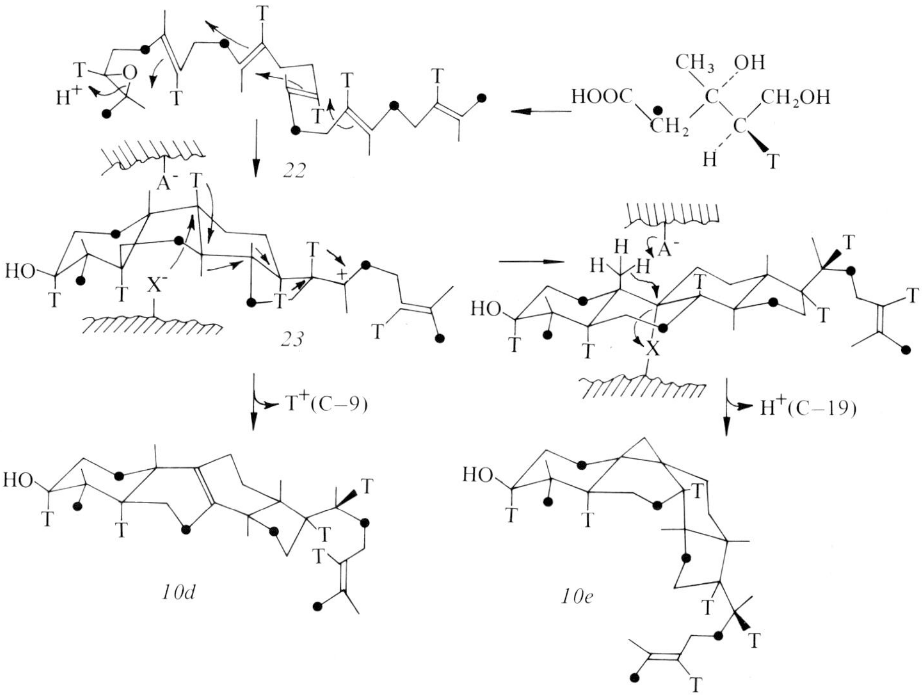

Fig. 4. The cyclisation of squalene-2,3-oxide to give either lanosterol (*10d*) or cycloartenol (*10e*). The labelling pattern of these compounds from [2-^{14}C(4*R*)4-$^{3}H_1$] mevalonic acid is indicated

accumulate. However, lanosterol (*10d*) was not found in any of the preceding cases nor in an anaerobic in vivo incubation of squalene-2,3-oxide with *N. tobaccum* tissue culture [46].

Proof that cycloartenol (*10e*) is not produced by rearrangement of lanosterol (*10d*) has been obtained using [2-^{14}C(4*R*)4-$^{3}H_1$] mevalonic acid. This substrate incubated with animal tissue gives lanosterol (*10d*) with a ^{3}H:^{14}C atomic ratio of 5:6 as indicated in Figure 4, but with plant tissues the cycloartenol (*10e*) formed has a ^{3}H:^{14}C atomic ratio of 6:6, showing that it is produced directly from the cyclised cation (*23*) by a series of methyl and hydride migrations that culminate in the closure of the 9β,19-cyclopropane ring [9, 132].

The folding of squalene-2-3-oxide in a chair-boat-chair-boat conformation, presumably at the active site of the cyclase enzyme, and the ring closures following proton addition to the epoxide comply with the postulates of Eschenmoser et al. [48]. The methyl and hydrogen migrations also follow the required antiplaner cationic 1,2-rearrangement pattern upto the establishment of the positive charge at C-9. At this point closure of the cyclopropane ring requires loss of a proton from the C-19 methyl group, which, however, is cis to the departing C-9 hydride. To overcome this anomaly it is suggested that participation of an electron-donating X^- group on the enzyme, or a carrier protein, intervenes to transiently neutralise the C-9 charge [9]. Subsequent withdrawl of X^- then permits closure of the cyclopropane ring in a trans manner, the C-19 proton perhaps being abstracted by a basic radical of the enzyme or carrier protein (Fig. 4 [77]).

Cornforth [37], in a discussion of squalene-2,3-oxide cyclisation, has considered the possibility of enzyme X^- group participation and concluded that the initial C-20 cation (*23*) might also be neutralised, for example, by a water molecule at an anionic enzyme site. This would produce (20*R*)-protost-24-en-3β,20-diol (*24*), which by reorientation on the enzyme could allow the ^-OH group to approach a catonic site, and its removal would then promote the 1,2-trans rearrangements to give the stabilised 3β-hydroxytriterpene. Labelled (20*R*)-protost-24-en-3β,20-diol (*24*) has been synthesised [82], but its conversion to lanosterol (*10d*) could not be demonstrated using a rat liver homogenate. However, this observation does not necessarily discount X^- group participation at C-20. Possibly the positive charge need only be transiently neutralised to achieve the desired result, and an X^- group other than OH^- may be involved. Thus modification of the scheme in Figure 4 may be required to allow for another electron-donating group of the enzyme interacting with C-20.

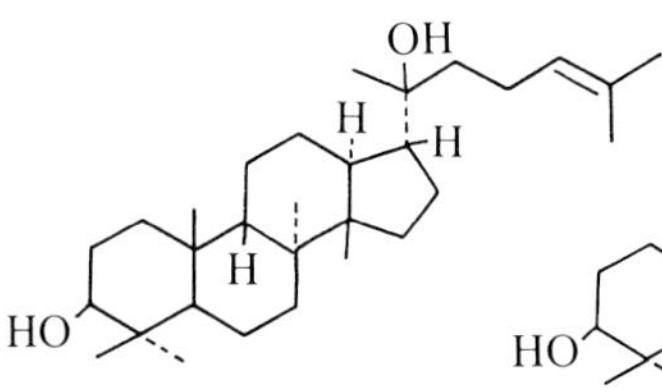

(*24*)

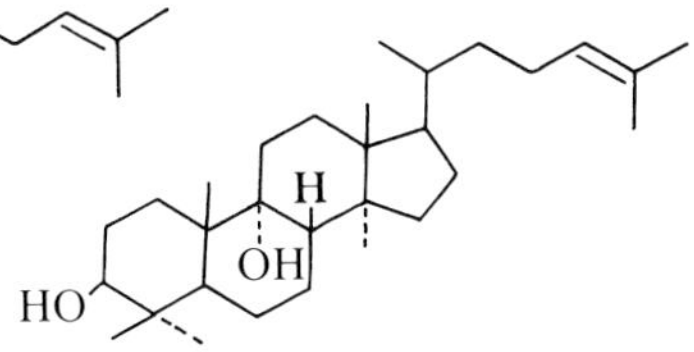

(*25*)

Neutralisation of the C-9 cation by OH^- in a manner analogous to that discussed above has been considered in relation to cycloartenol (*10e*) production. 5α-Lanost-24-en-3β,9α-diol (*25*) labelled with tritium has been synthesised and incubated with the alga *Ochromonas malhamensis* [115]. It was incorporated efficiently into poriferasterol (*9a*, (24*R*)-24-ethylcholesta-5,22-dien-3β-ol), the major sterol, thus showing that this organism has the capacity to abstract water from *25* and so introduce unsaturation in the manner depicted in Figure 4 for X^- group removal. However, it is not yet clear whether the first product is cycloartenol (*10e*) or lanosterol (*10d*), the two likely products, assuming that OH^- removal occurs before other modifications.

Consideration of the enzyme-triterpene intermediate (Fig. 4) suggests how lanosterol (*10d*) may perhaps arise in *Euphorbia* latex (and indeed fungi). A small conformational change in the enzyme active site, perhaps induced, e.g., by a slightly different amino acid sequence or disulphide bond formation, could result in the group A^- being more favourably placed for removal of the 8β-hydrogen than a C-19 proton. The production of both lanosterol and cycloartenol from [1-^{14}C]-acetate by *E. helioscopia* latex [120] would then require the presence of two cyclases rather than the cycloartenol cyclase and cyclopropane opening enzyme systems suggested by the observation that labelled cycloartenol is converted to lanosterol by *E. lathyris* latex [121, 122]. Small changes in the placement of A^- relative to the triterpene rings could also favour loss of either the 11β- or 5β-hydrogens, thus leading to production of parkeol or cucurbitacins, respectively [65].

D. Sterol Side-Chain Formation

The mechanisms involved in phytosterol side-chain elaboration have been a focus of attention for some time. As the work unfolded it became apparent that the original mechanism proposed by Nes [30, 31], although serving as a basic blueprint, needed modifications to accommodate new experimental observations. In particular these studies have again served to illustrate how biosynthetic routes can be subtly modified in different organisms, although the end product may well be the same.

The extra C-28 and C-29 carbon atoms of phytosterols arise by transmethylation reactions requiring S-adenosylmethionine as the methyl donor [94]. Most fungi synthesise 24β-methyl sterols by a sequence commencing with the transfer of the electrophilic S-methyl group of S-adenosylmethionine to C-24 of a Δ^{24} sterol precursor. The cation produced is stabilised by loss of a C-28 proton to give a 24-methylene compound, which is then reduced stereospecifically to a 24β-methyl sterol (Fig. 5). The operation of this pathway is amply supported by several lines of evidence. Only two deuteriums from [CD_3]-methionine are incorporated into the ergosterol (*8c*) produced by various fungi [68, 94, 102, 105]. A number of 24-methylene sterols occur in fungi, and labelled 24-methylene sterols are converted into ergosterol (*8c*) by yeast [4, 11, 13, 16, 51, 52, 166]. A preparation with S-adenosylmethionine-Δ^{24} sterol methyl transferase activity has been obtained from yeast, which yields a 24-methylene product [106, 156], and an NADPH-dependent microsomal 24-methylene reductase has also been demonstrated in yeast [85]. Finally the required hydride migration from C-24 to C-25 has been verified [3, 105].

The stereospecific reduction of the $\Delta^{24(28)}$ bond to give a 24β-methyl group can occur either before or after introduction of the Δ^{22} bond, as indicated in Figure 5. The spectrum of sterols found in fungi and also metabolic studies show that both routes are operative, but the balance of evidence favours the production of a $\Delta^{22,24(28)}$ diene intermediate in yeast [50, 51, 85].

The production of 24-methylene dihydrolanosterol (*15d*, 4,4,14α-trimethyl-5α-ergosta-8,24(28)-dien-3β-ol) in some fungi [71, 102] reveals that the transmethylation reaction occurs in these fungi at the C_{30}-sterol level. This contrasts

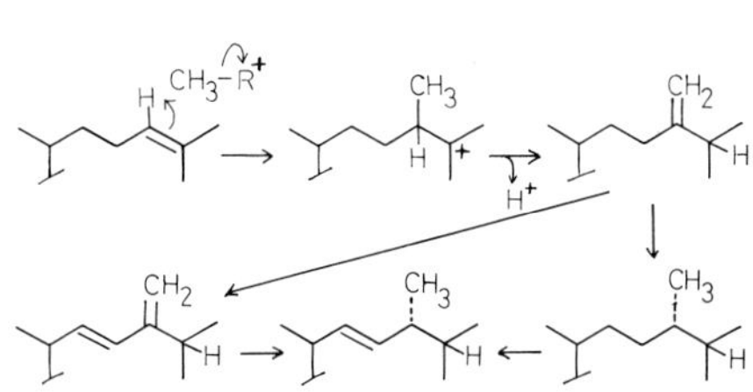

Fig. 5. The alkylation mechanism involved in 24β-methyl sterol synthesis in fungi

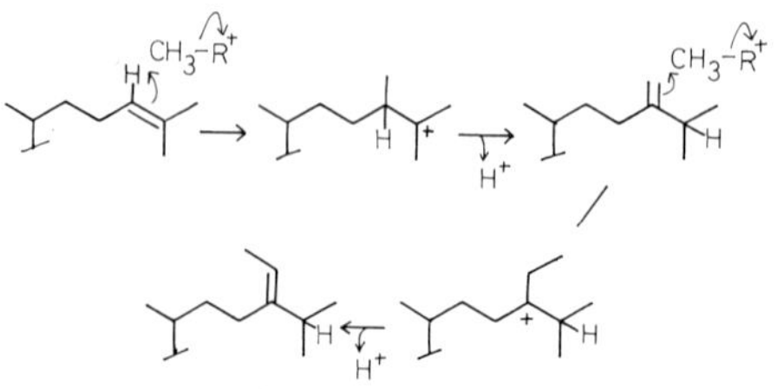

Fig. 6. Mechanism for the elaboration of a 24-ethylidene side chain

sharply with yeast in which the alkylation reaction is delayed to a later point in the sequence following the C-14 and C-4 demethylations [14, 53].

Most of the Ascomycetes, Basidiomycetes, Deuteromycetes, and Phycomycetes examined produce 24β-methyl sterols, showing that while the first transmethylation reaction is active, the second transmethylation step leading to the C_{29} sterols found in most other classes of plant is inoperative. There are, however, at least two exceptions. In the uredospores of rust fungi C_{29} sterols (*3b, 17b*) predominate [99, 167], and a number of sterols including fucosterol (*16a*) have been identified in *Achlya bisexualis* [125] and other Phycomycetes [68]. The production of C_{29} sterols by some orders of Phycomycete is in accord with other evidence suggesting their phylogenetic affinity to some algae [17, 89].

In the brown algae (Phaeophyta) the 24-ethylidene sterol, fucosterol (*16a*), is produced as a result of the operation of the mechanism outlined in Figure 6. A 24-methylene sterol produced by the first transmethylation serves as the substrate for the second transmethylation to give a 24-ethyl bearing cation which is stabilised by a stereospecific loss of a C-28 proton to yield the *E*-configuration of the 24-ethylidene product. As in the case of fungi, the first transmethylation step results in migration of hydrogen from C-24 of the Δ^{24} sterol precursor to C-25, where it remains after the second transmethylation [64]. The Phaeophyta are apparently unable to reduce the 24-ethylidene moiety to a 24-ethyl group presumably due to the absence of an active $\Delta^{24(28)}$-reductase.

The chrysophyte alga *Ochromonas malhamensis* produces poriferasterol (*9a*) into which a maximum of four deuteriums are incorporated when the alga is cultured in the presence of [CD_3]-methionine [150]. This is consistent with the production of a 24-ethylidene intermediate, which is reduced stereospecifically to give a 24β-ethyl sterol as shown in Figure 7. Support for the operation

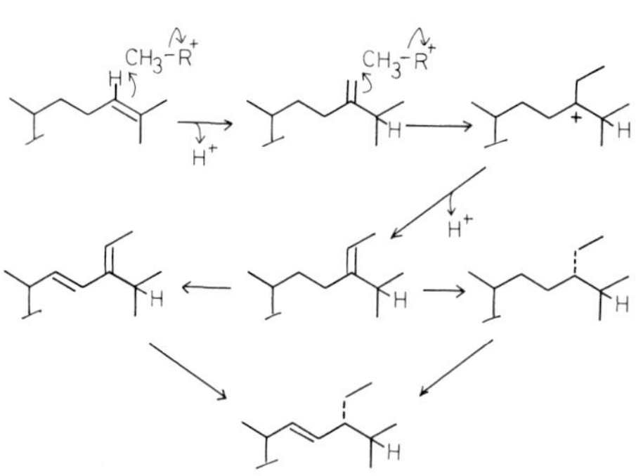

Fig. 7. Route for the production of 24β-ethyl sterols in *O. malhamensis*

of this route is provided by the validation of the C-24 hydrogen migration [151] and by the incorporation of labelled 24-ethylidene sterols into poriferasterol by *O. malhamensis* [91, 97]. The fact that a 24-ethylidene sterol with the *Z*-configuration (isofucosterol *17a*) was better incorporated into poriferasterol than the isomer with the *E*-configuration (fucosterol, *16a*) allowed the tentative suggestion that the 24*Z*-ethylidene compound may be the true precursor. Its more efficient utilisation may be due to the greater specificity of the $\Delta^{24(28)}$-reductase for the Z-isomer. Alternatively, by analogy with the mechanism in yeast, a $\Delta^{22,24(28)}$ intermediate may be produced from the 24*Z*-ethylidene sterol, but with the 24*E*-isomer the C-29 group may sterically hinder hydrogen abstraction from C-22 and C-23 [111]. The small conversion of the 24*E*-isomer would then be explained by reduction to the 24β-ethyl-saturated side chain, as found in clionasterol (*5a*, (24*S*)-24-ethylcholesta-5,22-dien-3β-ol), an *Ochromonas* constitutent, followed by introduction of the Δ^{22} bond. A decision between these routes might be possible when the naturally occurring 24-ethylidene sterol has been isolated from *Ochromonas* and its configuration elucidated. *O. malhamensis* produces a small quantity of the 24β-methyl compound brassicasterol (*8a*) via a 24-methylene intermediate (Fig. 7) [68]. The same pathways (Fig. 7) for 24β-methyl and 24β-ethyl sterol production occur in other *Ochromonas* species; the euglenophyte, *Euglena gracilis* [68] and the Xanthophyte, *Monodus subterraneus* [103].

Evidence that 24-ethylidene sterols are not obligatory precursors of 24-ethyl sterols was provided by the demonstration that *Chlorella* species incorporate five deuterium atoms from [CD_3]-methionine into their 24β-ethyl sterols [2, 158, 159]. Moreover the incorporation of three deuteriums into the 24-methyl sterols revealed that *Chlorella* could not be reducing a 24-methylene intermediate, as occurs in fungi and in the chrysophyte algae. These observations were extended to other chlorococcalean algae (*Scenedesmus* and *Trebouxia*), and the recognition of a strong hydrogen-isotope effect favouring 24β-methyl sterol production at the expense of 24β-ethyl sterols suggested mechanisms whereby 25-methylene intermediates are produced [67]. Stabilisation of the cation produced by the first transmethylation can proceed in two directions (Fig. 8). Loss of a proton from one of the terminal methyl groups produces a 25-methylene compound with a 24β-methyl group, the configuration at C-24 being determined by the stereospecificity of the alkylating enzyme. Reduction of the

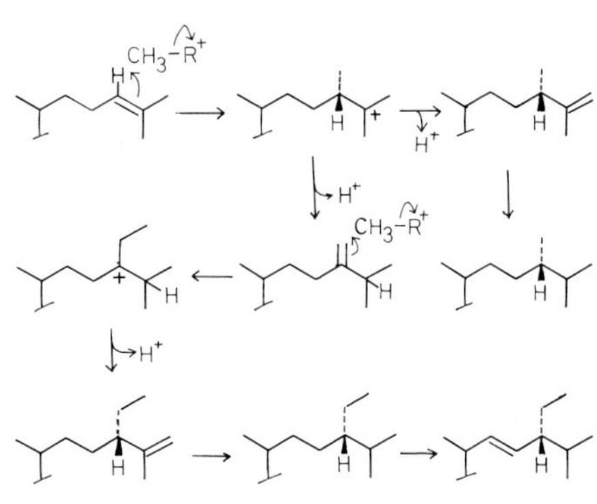

Fig. 8. Transmethylation mechanisms involving 25-methylene intermediates in 24β-alkyl sterol formation

25-methylene group then gives the 24β-methyl saturated side chain found in these algae. Loss of a proton from C-28 of the cation leaves the 24-methylene intermediate encountered previously. By a second transmethylation, this yields a 24-ethyl cation stabilised in these organisms by a 1,2-hydride shift from C-25 to C-24 and loss of a C-26 proton to form a 25-methylene intermediate, which can be reduced to give the saturated side-chain sterols. Again the 24β-configuration is a consequence of the stereochemistry of the transmethylation reaction and not of a double bond reductase, as seen in the previous cases.

Cell-free preparations have been obtained from *Trebouxia* and *Scenedesmus* that catalyse the alkylation of cycloartenol (*10e*) to yield both 24-methylenecycloartanol (*15e*) and cyclolaudenol (*12e*) in amounts approximate to the proportions of 24-ethyl and 24-methyl sterols found in the algae [168]. Other evidence for the operation of the route outlined in Figure 8 is provided by the incorporation by *Trebouxia* of a labelled 24β-methyl-25-methylene compound (*31*-norcyclolaudenol, *12f*) into C_{28} sterol but not C_{29} sterol. Conversely a radioactive 24-methylene sterol (cycloeucalenol, *15f*) was converted only into C_{29} sterol [67]. Recently ^{14}C-labelled clerosterol (*13a*) was synthesised and incubated with *Trebuxia* to yield labelled clionasterol (*5a*, (24*S*)-24-ethylcholest-5-en-3β-ol) and poriferasterol (*9a*), the former with the higher specific radioactivity in accordance with the precursor–product relationship required by the biosynthetic route [93]. In compliance with these observations 25-methylene sterols (*12e, 13b*) have been identified in *Chlorella* species [118], and codisterol (*12a*) and clerosterol (*13a*) are the main sterols in *Codium fragile* of the order Siphonales, which is believed to have evolved from the Chlorococcalean line [141]. It seems that in the latter alga, the 25-methylene reductase activity is absent.

Another group of organisms that do not operate an alkylation mechanism involving a 24-ethylidene intermediate for C_{29} sterol formation is the slime moulds. Stigmastan-3β-ol (*3i*) and stigmast-22-en-3β-ol (*7i*) each contained five deuteriums when synthesised by *Physarum polycephalum* and *Dictyostelium discoideum* in the presence of [CD_3]-methionine [94, 95]. The mechanism outlined in Figure 9 has been proposed, in which introduction of a Δ^{22}-bond is an integral part of the alkylation mechanism [45]. The mode of biosynthesis of the saturated side chain of the stigmastanol (*3i*) in these organisms is uncertain. Also it should be noted that in other organisms (fungi, algae, and higher plants) a Δ^{22} bond is introduced by abstration of two hydrogens from C-22 and C-23 [65].

In vascular plants the 24α-methyl and 24α-ethyl sterols campesterol (*2a*), stigmasterol (*7a*), and sitosterol (*3a*) are the most frequently reported. The identification of 24(*Z*)-ethylidene sterols (*17a, 17b, 17h*) in many higher plants and their rapid labelling from radioactive acetate or mevalonate suggested that

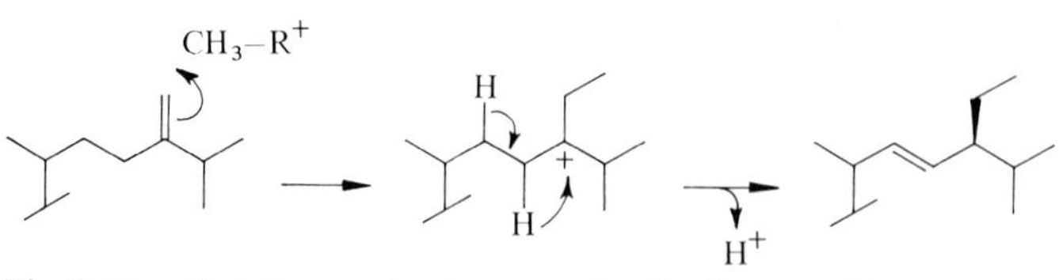

Fig. 9. The alkylation mechanism operative in slime moulds

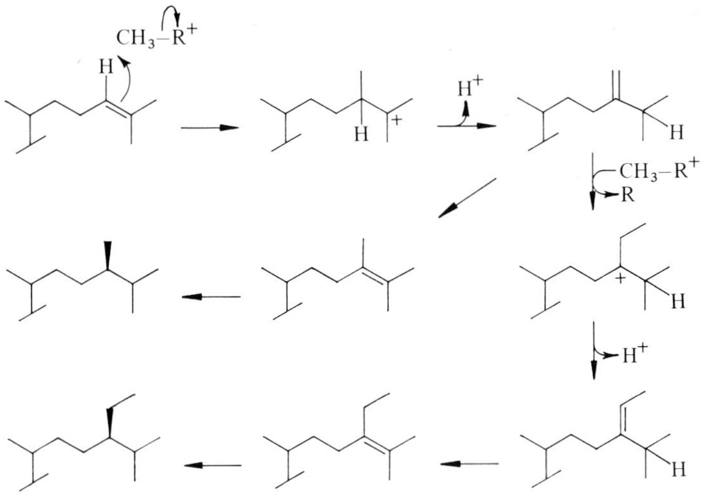

Fig. 10. The higher plant transmethylation reactions leading to 24α-alkyl sterols

these compounds are intermediates to 24α-ethyl sterol production [65]. This has been substantiated by the conversion of labelled 24(*Z*)-ethylidene sterols (*17a, 17h*) into sitosterol (*3a*) [98, 162] and in particular by the incorporation by barley seedlings of only four deuteriums from [CD_3]-methionine into stigmasterol and sitosterol [98]. However, the production of the 24α-ethyl sterol cannot proceed by a direct reduction of the 24(*Z*)-ethylidene double bond as in some algae (see Fig. 7) since it has been established in several plants that the C-25 hydrogen of the 24-methylene and 24-ethylidene intermediates is eliminated in the production of thc 24-ethyl sterols [9, 98, 128, 160].

The biosynthetic route most compatable with these observations is shown in Figure 10. This is essentially the same as the chrysophyte pathway (Fig. 7) up to the production of a 24(*Z*)-ethylidene intermediate, but it is then suggested that isomerisation gives a $\Delta^{24(25)}$ sterol that can be stereospecifically reduced to yield the 24α-ethyl side chain. Since the incorporation of two deuteriums from [CD_3]-methionine into campesterol (*2a*) by barley seedlings implicates a 24-methylene intermediate, a similar route (Fig. 10) is envisaged for 24α-methyl sterol production [68].

The involvement of $\Delta^{24(25)}$-sterols gains some credence from the reports of ergosta-5,24-dien-3β-ol and stigmasta-7,24-dien-3β-ol in two higher plants [81, 100]. However, the incorporation of labelled stigmasta-5,24-dien-3β-ol (*18a*) into sitosterol (*3a*) could not be demonstrated with barley seedlings nor could the incorporation of [2-^{14}C] mevalonic acid into stigmasta-5,24-dien-3β-ol (*18a*) be demonstrated, even in the presence of added unlabelled material (*18a*) to act as a trap [93]. Thus, although all the experimental evidence points to a $\Delta^{24(25)}$-sterol intermediate in some higher plants, this remains to be conclusively established. The preponderance of 24α-alkyl sterols in higher plants can perhaps now be explained by the stereospecifity of a $\Delta^{24(25)}$-reductase favouring production of 24α-methyl or -ethyl compounds, whereas in those fungi and algae producing 24-methylene- and 24-ethylidene intermediates, reduction leads to the characteristic 24β-methyl and -ethyl sterols, respectively.

The co-occurrence of campesterol (*2a*) and brassicasterol (*8a*) in rapeseed [157] and the recent recognition that the campesterol (*2a*) of several plants in fact contains a variable proportion of the 24β-epimer [109, 113] raise the question of how the 24β-methyl sterols are synthesised in higher plants. Evidence is now at hand that indicates that it is produced from cyclolaudenol (*12e*) by the same route as that previously described for some algae (Fig. 8). Cyclolaudenol (*12e*) has been reported as a constituent of a few plants [65] and the C-24 configuration has been assigned as 24β- on the basis of molecular rotation data [79]. Cyclolaudenol side-chain biosynthesis in the fern *Polypodium vulgare* was shown to comply with the mechanism outlined in Figure 8 [54]. With maize, which produces a mixture of the 24β-methyl and 24α-methyl sterols [109], it has now been shown that [2-^{14}C] mevalonic acid and [^{14}C-methyl]-methionine are incorporated into both 24-methylenecycloartanol (*15e*), which can then act as the precursor of 24α-methyl and 24α-ethyl sterols, and into cyclolaudenol (*12e*) from which the 24β-methyl sterols may be obtained [171].

Thus it appears that in some higher plants the stabilisation of the cation produced by the first transmethylation can proceed in two directions to give 24-methylene or 25-methylene compounds in the same way as in some algae. Similarly in plants of the Cucurbitaceae and Verbenaceae the cation produced by the second transmethylation can also lead by the route shown in Figure 8 to 25-methylene sterols such as *13a, 13b, 14a,* and *14b,* which are characteristic of these plant families. In this respect the hydride migrations from C-24 to C-25 and back to C-24 required by the mechanisms (Fig. 8) have been demonstrated during (24*S*)-24-ethylcholesta-5,22,25-trien-3β-ol (*14a*) production by *Clerodendrum campbellii* [25].

A composite alkylation mechanism can be envisaged involving enzyme X^- group participation, which permits the elaboration of the various sterol side chains [56, 68]. Examination of the mechanisms in Figure 11 shows that by

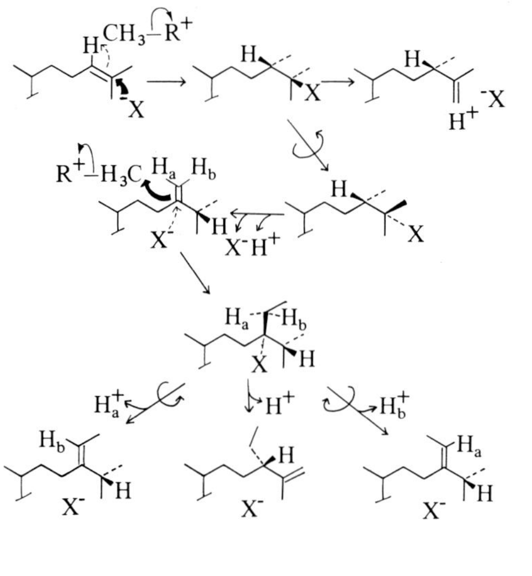

Fig. 11. Theoretical alkylation mechanisms involving X^- group participation

suitable hydride migrations and proton eliminations with, where appropriate, rotation of bonds so that trans-1,2 relationships with the X^- group are maintained, it is possible to obtain 24β-alkyl-25-methylene sterols or the *E* and *Z* 24-ethylidene isomers. Loss of a proton may be facilitated by a second basic group of the enzyme (cf. cyclisation, Fig. 4). Changes in the conformation of the transmethylase enzymes of different organisms may then determine the spatial relationship of the basic group to the sterol side chain, thus favouring the loss of a particular proton to yield the unsaturated side-chain sterol precursor typical of the organisms.

E. Ring Structure Modification in Phytosterol Formation

The conversion of the 4,4-dimethyl sterol produced by squalene-2,3-oxide cyclisation into the various 4-desmethyl sterols requires loss of the C-4 and C-14 methyl groups, formation of the appropriate alkylated side-chain structure, and establishment of the Δ^5, Δ^7, or $\Delta^{5,7}$ ring B double bond system.

In fungi the modification of the ring system of lanosterol (*10d*) to give the $\Delta^{5,7}$-diene system of ergosterol (*8c*) follows a pattern essentially similar to that operating in cholesterol production in animals [51, 53]. Considerable information on the sequencing of the reactions and on the identities of precursors has been derived from the use of yeast mutants blocked at various points in the biosynthetic sequence and also from the use of fungicides that may cause sterol precursors to accumulate (see, e.g., [11, 12, 86, 88, 107, 127, 161]).

The C-14 methyl group is the first to be lost in an oxidative reacton involving elimination of the 15α-hydrogen of lanosterol [5, 28] to give a $\Delta^{8,14}$-diene intermediate in which the Δ^{14} bond is subsequently reduced. As in animals and higher plants, it is likely that the 4α-methyl group is the first to be removed, with concomitant epimerisation of the 4β-methyl to produce a 4α-methyl intermediate (e.g., *15h*) [105].

Isomerisation of the Δ^8-bond to the Δ^7-position at some point in the sequence proceeds with loss of the 7α-hydrogen. This contrasts sharply with animals and photosynthetic plants (excluding *Fucus spiralis*) where the 7β-hydrogen is eliminated [65]. Introduction of the Δ^5-bond proceeds with cis-removal of the 5α- and 6α-hydrogens to leave the $\Delta^{5,7}$-diene sterol, which accumulates in fungi presumably due to the absence of the Δ^7-reductase, which is active in many other organisms.

The conversion of cycloartenol (*10e*) and other 9β,19-cyclopropane sterols into 4-desmethyl sterols in algae and higher plants has been amply demonstrated [59, 65]. Recently, convincing evidence for the intermediacy of cycloartenol (*10e*) has been provided by the demonstration that a deuterium atom is introduced into the C-19 methyl group of sitosterol (*3a*) when pea seeds are germinated after soaking with deuterium oxide [27]. This result is best explained by the opening of the 9β,19-cyclopropane ring of a precursor sterol to produce a C-19 anionic centre to which a deuteron (proton) from the medium is added.

The cation so produced is stabilised by loss of the 8β-hydrogen to give a Δ^8-sterol precursor [66].

A cell-free preparation has been obtained from bramble tissues (*Rubus fruticosus*), which catalysed the opening of the cyclopropane ring of cycloeucalenol (*15f*) to produce obtusifoliol (*15g*) [77]. The enzyme was microsomal and did not require added ATP or NADH for activity. 24-Methylenecycloartanol (*15e*) and cycloartenol (*10e*) were very poor substrates in comparison to cycloeucalenol (*15f*). It is suggested that the presence of a 4β-methyl group hinders the opening of the cyclopropane ring, possibly because of unfavourable 4β-methyl, C-10 methyl interactions when the compound is bound to the enzyme active site. The actual mechanism of cyclopropane opening is envisaged as proceeding by essentially the reverse of the cyclisation mechanism shown in Figure 4. The cyclopropane sterol interacts with an X^- group at the enzyme active site to open the cyclopropane ring; withdrawl of the X^- group and loss of the 8β-hydrogen then gives the Δ^8-sterol.

It appears that only photosynthetic eukaryote organisms or those that have evolved from photosynthetic ancestors (although they may themselves contain no plastids) have the ability to synthesise cycloartenol and to open the cyclopropane ring. The non-photosynthetic parasitic plants *Cuscuta europaea, C. epithymum,* and *Orobanche lutea* biosynthesise sterols via cycloartenol [139]. Both green and etiolated *Euglena gracilis* and also *Astasia longa,* an apoplastid mutant of *Euglena,* utilise the cycloartenol pathway for sterol synthesis [7, 138] as does the non-photosynthetic endosperm of pine seeds [101]. Conversely cycloartenol cannot be utilised for sterol production by either yeast [8] or rat liver preparations [57, 77].

In the conversion of cycloartenol (*10e*) into desmethyl sterols, the removal of the C-14 and C-4 methyls follows a different sequence from that in animals and fungi. The 4α-methyl is the first to be removed with the 4β-methyl epimerising to the 4α-position in the product 4-monomethyl sterol [55, 90]. As mentioned previously, absence of the 4β-methyl group aids opening of the cyclopropane ring to give a Δ^8 sterol and this results in the molecule assuming a more planar structure and so exposing the 14α-methyl group (see Fig. 4). The 14α-methyl group is then removed by an oxidative demethylation reaction that is probably facilitated by the newly generated Δ^8 bond. The reaction involves loss of the 15α-hydrogen, presumably by the same mechanism as that operating in animals and fungi [29, 149].

From this point the modifications of the sterol ring system apparently proceed in much the same manner as in fungi and animals. Loss of the 4α-methyl group from a 4-monomethyl sterol, obtusifoliol (*15g*), proceeds in *Ochromonas malhamensis* with the introduction of a 4β-hydrogen from the medium and inversion of the existing C-4 hydrogen to the 4α-position [92]. The $\Delta^8 \rightarrow \Delta^7$ isomerisation proceeds with elimination of the 7β-hydrogen (cf. fungi) and Δ^5-bond introduction removes the 5α- and 6α-hydro [65, 148].

A general partial biosynthetic sequence is illustrated in Figure 12. From information on sterol distribution and labelling studies in a variety of algae and higher plants it appears that the first side-chain transmethylation usually utilises cycloartenol (*10e*) as the substrate to produce 24-methylenecycloartanol

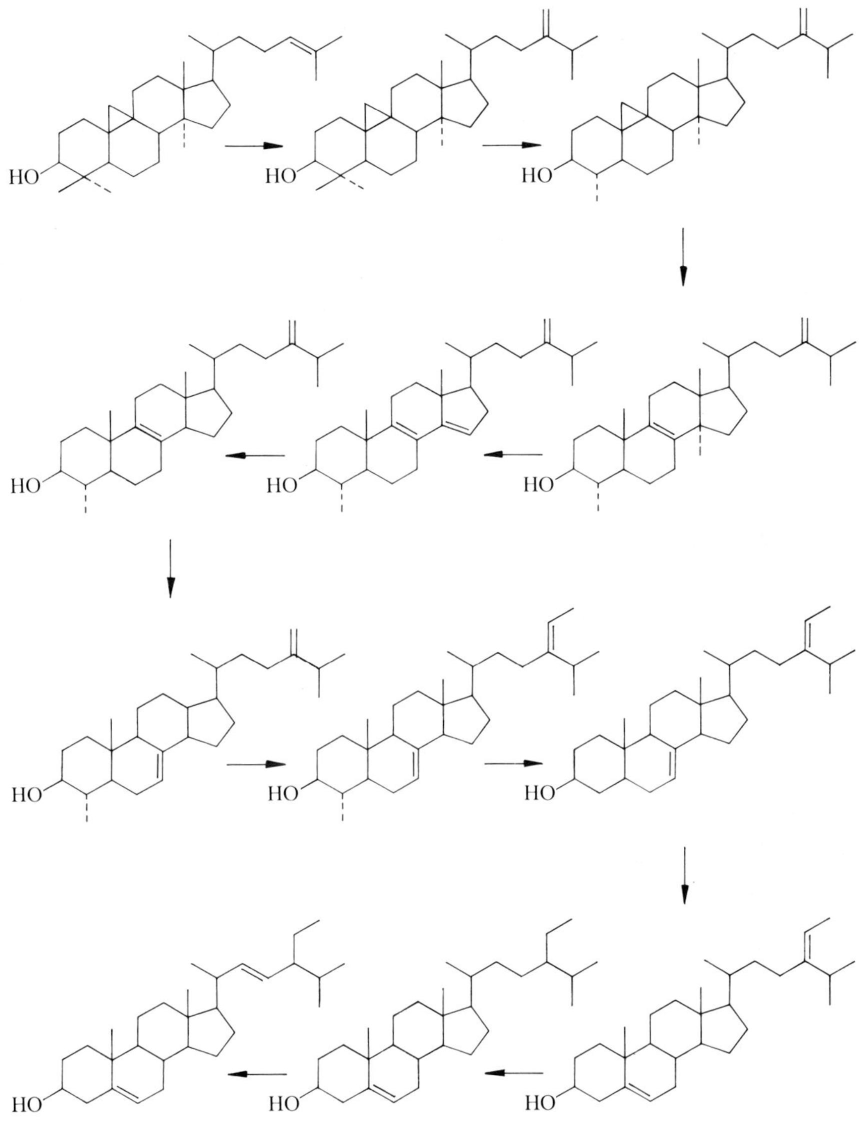

Fig. 12. A generalised route from cycloartenol to 4-desmethyl sterols

(*15e*) or cyclolaudenol (*12e*). The most favoured point for the second transmethylation seems to be at the 4α-methyl sterol level to give for example 24-ethylidenelophenol (*17h*, 4α-methyl-stigmast-5,*Z*-24(28)-dien-3β-ol). It is clear, however, that the number of possible permutations resulting from the several alkylation mechanisms in conjunction with the changes required to the sterol rings allows a myriad of possible routes from cycloartenol to the many 4-desmethyl sterols found in the plant kingdom. A start has been made in determining the principal routes in a few organisms. For example, Patterson and co-workers have conducted comprehensive studies, employing inhibitors of sterol biosynthesis, to gain information on the routes of sterol production in various *Chlorella* species [32, 33, 39, 40, 117, 118].

F. Conclusions

From this account of our present knowledge of the overall patterns of plant sterol biosynthesis it is now possible to see biogenetic and phylogenetic relationships between many of the known plant sterols and to predict biosynthetic routes leading to the novel sterols that are continually being reported. Much of the work to date has employed in vivo studies, but the time now seems appropriate for more extensive work at the enzyme level. Such studies would be particularly appropriate in the case of the transmethylase enzymes in the quest for a better understanding of the factors that determine the nature of the sterol side chains produced in different classes of plants.

Another aspect of plant sterol research that can now perhaps be pursued more fruitfully with the biosynthetic knowledge at hand, is the regulation of plant sterol biosynthesis and the possible role played by HMG-CoA reductase, one of the most important regulatory enzymes in animal sterol production. Such studies should also be aimed at explaining why certain precursor sterols accumulate in some tissues, for example cycloartenol and 4-monomethyl sterols in seeds.

Acknowledgements. I wish to thank sincerely Professor T.W. Goodwin, C.B.E., F.R.S. for his encouragement and all my colleagues for the work they have contributed from our laboratory.

References

1. Aexel, R., Evans, S., Kelly, M., Nicholas, H.J.: Phytochemistry **6**, 511 (1967)
2. Adler, J.H., Patterson, G.W.: Plant Physiol. **53**, S-13 (1974)
3. Akhtar, M., Hunt, P.F., Parvez, M.A.: Biochem. J. **103**, 616 (1967)
4. Akhtar, M., Parvez, M.A., Hunt, P.F.: Biochem. J. **113**, 727 (1969)
5. Alexander, K.T.W., Mitropoulos, K.A., Gibbons, G.F.: Biochem. Biophys. Res. Commun. **60**, 460 (1974)
6. Altman, L.J., Kowerski, R.C., Rilling, H.C.: J. Am. Chem. Soc. **93**, 1782 (1971)
7. Anding, C., Brandt, R.D., Ourisson, G.: Europ. J. Biochem. **24**, 259 (1971)
8. Anding, C., Parks, L.W., Ourisson, G.: Europ. J. Biochem. **4**, 459 (1974)
9. Armarego, W.L.F., Goad, L.J., Goodwin, T.W.: Phytochemistry **12**, 2181 (1973)
10. Barton, D.H.R., Corrie, J.E.T., Marshall, P.J., Widdowson, D.A.: Bio-org. Chem. **2**, 363 (1973)
11. Barton, D.H.R., Corrie, J.E.T., Widdowson, D.A., Bard, M., Woods, R.A.: J. Chem. Soc. Perkin Trans. **I**. 1326 (1974)
12. Barton, D.H.R., Gunatilika, A.A.L., Jarman, T.R., Widdowson, D.A., Bard, M., Woods, R.A.: J. Chem. Soc. Perkin Trans. **I**. 88 (1975)
13. Barton, D.H.R., Jarman, T.R., Watson, K.G., Widdowson, D.A., Boar, R.B., Damps, K.: J. Chem. Soc. Perkin **I**. 1134 (1975)

14. Barton, D.H.R., Kempe, U.M., Widdowson, D.A.: J. Chem. Soc., 513 (1972)
15. Barton, D.H.R., Mellows, G., Widdowson, D.A., Wright, J.J.: J. Chem. Soc. **C**, 1142 (1971)
16. Barton, D.H.R., Shioiri, T., Widdowson, D.A.: Chem. Commun., 939 (1970)
17. Bartnicki-Garcia, S.: In: Phytochemical Phylogeny. Harborne, J.B. (ed.). London: Academic Press, 1970, p. 81
18. Bean, G.A.: Adv. Lipid Res. **11**, 193 (1973)
19. Beastall, G.H., Rees, H.H., Goodwin, T.W.: FEBS Lett. **18**, 175 (1971)
20. Beastall, G.H., Rees, H.H., Goodwin, T.W.: FEBS Lett. **28**, 243 (1972)
21. Benveniste, P., Massy-Westrop, R.A.: Tetrahedron Lett., 3553 (1967)
22. Benveniste, P., Hirth, L., Ourisson, G.: Phytochemistry **5**, 31, 45 (1966)
23. Bergmann, W.: Ann. Rev. Plant Physiol. **4**, 383 (1953)
24. Berti, G., Bottari, F.: Progr. Phytochemistry **1**, 589 (1968)
25. Bolger, L.M., Rees, H.H., Ghisalberti, E.L., Goad, L.J., Goodwin, T.W.: Biochem. J. **118**, 197 (1970)
26. Cahn, R.S., Ingold, C.K., Prelog, V.: Angew. Chem. (Internat. Ed.) **5**, 385 (1966)
27. Caspi, E., Sliwowski, J.: J. Am. Chem. Soc. **97**, 5032 (1975)
28. Caspi, E., Moreau, J.P., Ramm, P.J.: J. Am. Chem. Soc. **96**, 8107 (1974)
29. Caspi, E., Sliwowski, J., Robichaud, C.S.: J. Am. Chem. Soc. **97**, 3820 (1975)
30. Castle, M., Blondin, G.A., Nes, W.R.: J. Chem. Soc. **85**, 3306 (1963)
31. Castle, M., Blondin, G.A., Nes, W.R.: J. Biol. Chem. **242**, 5796 (1967)
32. Chan, J.T., Patterson, G.W.: Plant Physiol. **52**, 246 (1973)
33. Chan, J.T., Patterson, G.W., Dutkey, S.R., Cohen, C.F.: Plant Physiol. **53**, 244 (1974)
34. Clayton, R.B.: Quart. Rev. (London) **19**, 168 (1965)
35. Corey, E.J., Ortiz de Montellano, P.R.: J. Am. Chem. Soc. **89**, 3362 (1967)
36. Corey, E.J., Russey, W.E., Ortiz de Montellano, P.R.: J. Am. Chem. Soc. **88**, 4750 (1966)
37. Cornforth, J.W.: Angew. Chem. (Internat. Ed.) **7**, 903 (1968)
38. Dean, P.D.G.: Steroidologia **2**, 143 (1971)
39. Dickson, L.G., Patterson, G.W.: Lipids **7**, 635 (1972)
40. Dickson, L.G., Patterson, G.W., Cohen, C.F., Dutky, S.R.: Phytochemistry **11**, 3473 (1972)
41. Douglas, L.J., Paleg, L.G.: Plant Physiol. **49**, 417 (1972)
42. Edmond, J., Popják, G., Wong, S.-M., Williams, V.P.: J. Biol. Chem. **246**, 6254 (1971)
43. Ehrhardt, J.D., Hirth, L., Ourisson, G.: Phytochemistry **6**, 815 (1967)
44. Ellouz, R., Lenfant, M.: Tetrahedron Lett., 609 (1969)
45. Ellouz, R., Lenfant, M.: Europ. J. Biochem. **23**, 544 (1971)
46. Eppenberger, U., Hirth, L., Ourisson, G.: Europ. J. Biochem. **8**, 180 (1969)
47. Epstein, W.W., Rilling, H.C.: J. Biol. Chem. **245**, 4597 (1970)
48. Eschenmoser, A., Ruzicka, L., Jeger, O., Arigoni, D.: Helv. Chim. Acta **38**, 1890 (1955)
49. Fieser, L.F., Fieser, M.: Steroids. New York: Reinhold Publishing 1959
50. Fryberg, M., Oehlschlager, A.C., Unrau, A.M.: Biochem. Biophys. Res. Commun. **48**, 593 (1972)
51. Fryberg, M., Oehlschlager, A.C., Unrau, A.M.: J. Am. Chem. Soc. **95**, 5747 (1973)
52. Fryberg, M., Oehlschlager, A.C., Unrau, A.M.: Arch. Biochem. Biophys. **160**, 83 (1974)
53. Fryberg, M., Avruch, L., Oehlschlager, A.C., Unrau, A.M.: Can. J. Biochem. **53**, 881 (1975)
54. Ghisalberti, E.L., de Souza, N.J., Rees, H.H., Goad, L.J., Goodwin, T.W.: Chem. Commun., 1401 (1969)
55. Ghisalberti, E.L., de Souza, N.J., Rees, H.H., Goad, L.J., Goodwin, T.W.: Chem. Commun., 1403 (1969)
56. Gibbons, G.F., Goad, L.J., Goodwin, T.W.: Phytochemistry **7**, 983 (1968)
57. Gibbons, G.F., Goad, L.J., Goodwin, T.W., Nes, W.R.: J. Biol. Chem. **246**, 3967 (1971)
58. Goad, L.J.: In: Terpenoids in Plants. Pridham, J.B. (ed.). London: Academic Press, 1967, p. 159
59. Goad, L.J.: In: Natural Substances Formed Biologically from Mevalonic Acid. Goodwin, T.W. (ed.). London: Academic Press, 1970, p. 45
60. Goad, L.J.: In: Biochemistry of Steroid Hormones. Makin, H.L.J. (ed.). London: Blackwells, 1975, p. 17
61. Goad, L.J.: In: Biochemical and Biophysical Perspectives in Marine Biology. Malins, D.C., Sargent, J.R. (eds.). London: Academic Press, 1976, Vol. 3, p. 213

62. Goad, L.J., Goodwin, T.W.: Biochem. J. **99**, 735 (1966)
63. Goad, L.J., Goodwin, T.W.: Europ. J. Biochem. **1**, 357 (1967)
64. Goad, L.J., Goodwin, T.W.: Europ. J. Biochem. **7**, 502 (1969)
65. Goad, L.J., Goodwin, T.W.: Prog. Phytochemistry **3**, 113 (1972)
66. Goad, L.J., Williams, B.L., Goodwin, T.W.: Europ. J. Biochem **3**, 232 (1967)
67. Goad, L.J., Knapp, F.F., Lenton, J.R., Goodwin, T.W.: Biochem. J. **129**, 219 (1972)
68. Goad, L.J., Lenton, J.R., Knapp, F.F., Goodwin, T.W.: Lipids **9**, 582 (1974)
69. Gonzalez, A.G., Mora, M.C.G.: An. Real. Soc. Esp. Fisc. Quim. **48B**, 475 (1952)
70. Gonzalez, A.G., Toste, A.H.: An. Real. Soc. Esp. Fisc. Quim. **48B**, 487 (1952)
71. Goulston, G., Mercer, E.I., Goad, L.J.: Phytochemistry **14**, 457 (1975)
72. Grunwald, C.: Ann. Rev. Plant Physiol. **26**, 209 (1975)
73. Hartmann, M.A., Ferne, M., Gigot, C., Brandt, R., Benveniste, P.: Physiol. Vég. **11**, 209 (1973)
74. Heftmann, E.: In: Phytochemistry—Organic Metabolites. Miller, L.P. (ed.). New York: Van Nostrand-Reinhold, 1973, Vol. 2, p. 171
75. Heftmann, E.: Phytochemistry **14**, 891 (1975)
76. Heintz, R., Benveniste, P.: Phytochemistry **9**, 1499 (1970)
77. Heintz, R., Benveniste, P.: J. Biol. Chem. **249**, 4267 (1974)
78. Heintz, R., Benveniste, P., Robinson, W.H., Coates, R.M.: Biochem. Biophys. Res. Commun. **49**, 1547 (1972)
79. Henry, J.A., Irvine, D.S., Spring, F.S.: J. Chem. Soc. 1607 (1955)
80. Hewlins, M.J.E., Ehrhardt, J.D., Hirth, L., Ourisson, G.: Europ. J. Biochem. **8**, 184 (1969)
81. Homberg, E.E., Schiller, H.P.K.: Phytochemistry **12**, 1767 (1973)
82. Immer, H., Huber, K.: Helv. Chim Acta **54**, 1346 (1971)
83. Itoh, T., Tamura, T., Matsumoto, T.: Lipids **9**, 173 (1974)
84. Jacobsohn, G.M.: In: Recent Adv. in Phytochemistry. Steelink, C., Runeckles, U.C. (eds.). New York: Appleton-Century-Crofts, 1970, Vol. 3, p. 229
85. Jarman, T.R., Gunatilika, A.A.L., Widdowson, D.A.: Bio-org. Chem. **4**, 202 (1975)
86. Kato, T., Tanaka, S., Ueda, M., Kawase, Y.: Agr. Biol. Chem. (Tokyo) **39**, 169 (1975)
87. Kawaguchi, A., Kobayashi, H., Okuda, S.: Chem. Pharm. Bull. (Japan) **21**, 577 (1973)
88. Kim, S.J., Kwon-Chung, K.J., Milne, G.W.A., Hill, W.B., Patterson, G.: Antimicrob. Agents Chemother. **7**, 99 (1975)
89. Klein, R.M., Cronquist, A.: Quart. Rev. Biol. **42**, 105 (1967)
90. Knapp, F.F., Nicholas, H.J.: Phytochemistry **10**, 97 (1971)
91. Knapp, F.F., Greig, J.B., Goad, L.J., Goodwin, T.W.: Chem. Commun., 707 (1971)
92. Knapp, F.F., Goad, L.J., Goodwin, T.W.: J. Chem. Soc. Chem. Comm., 149 (1973)
93. Largue, C., Goad, L.J., Goodwin, T.W.: Phytochemistry. In Press (1977)
94. Lederer, E.: Quart. Rev. (London) **23**, 453 (1969)
95. Lenfant, M., Ellouz, R., Das, B.C., Zissman, E., Lederer, E.: Europ. J. Biochem. **7**, 159 (1969)
96. Lenfant, M., Lecompte, M.F., Farugia, G.: Phytochemistry **9**, 2529 (1970)
97. Lenton, J.R., Hall, J., Smith, A.R.H., Ghisalberti, E.L., Rees, H.H., Goad, L.J., Goodwin, T.W.: Arch. Biochem. Biophys. **143**, 164 (1971)
98. Lenton, J.R., Goad, L.J., Goodwin, T.W.: Phytochemistry **14**, 1523 (1975)
99. Lin, H.-K., Knoche, H.W.: Phytochemistry **15**, 683 (1976)
100. Lockley, W.J.S., Roberts, D.P., Rees, H.H., Goodwin, T.W.: Tetrahedron Lett., 3773 (1974)
101. Malhotra, H.C., Nes, W.R.: J. Biol. Chem. **247**, 6243 (1972)
102. Mercer, E.I., Carrier, D.J.R.: Phytochemistry **15**, 283 (1976)
103. Mercer, E.I., Harries, W.B.: Phytochemistry **14**, 439 (1975)
104. Mercer, E.I., Johnson, M.W.: Phytochemistry **8**, 2329 (1969)
105. Mercer, E.I., Russell, S.M.: Phytochemistry **14**, 445, 451 (1975)
106. Moore, J.T., Gaylor, J.L.: J. Biol. Chem. **245**, 4684 (1970)
107. Morris, D.C., Safe, S., Subden, R.E.: Biochem. Gen. **12**, 459 (1974)
108. Mulheirn, L.J., Ramm, P.J.: Chem. Soc. Rev. **1**, 259 (1972)
109. Mulheirn, L.J.: Tetrahedron Lett., 3175 (1973)
110. Nes, W.R.: Lipids **9**, 596 (1974)

111. Nes, W.R., Gouinda Malya, P.A., Mallory, F.B., Ferguson, K.A., Landry, J.R., Conner, R.L.: J. Biol. Chem. **246**, 561 (1971)
112. Nes, W.R., Krevitz, K., Behzadan, S., Patterson, G.W., Landrey, J.R., Conner, R.L.: Biochem. Biophys. Res. Commun. **66**, 1462 (1975)
113. Nes, W.R., Krevitz, K., Behzadan, S.: Lipids **11**, 118 (1976)
114. Orcutt, D.M., Patterson, G.W.: Comp. Biochem. Physiol. **50B**, 579 (1975)
115. Palmer, M., Goad, L.J., Boar, R.: unpublished results (1976)
116. Patterson, G.W.: Lipids **6**, 120 (1971)
117. Patterson, G.W., Doyle, P.J., Dickson, L.G., Chan, J.T.: Lipids **9**, 567 (1974a)
118. Patterson, G.W., Thompson, M.J., Dutky, S.R.: Phytochemistry **13**, 191 (1974b)
119. Ponsinet, G., Ourisson, G.: Bull. Soc. Chim. Fr., 3682 (1965)
120. Ponsinet, G., Ourisson, G.: Phytochemistry **6**, 1235 (1967)
121. Ponsinet, G., Ourisson, G.: Phytochemistry **7**, 89 (1968)
122. Ponsinet, G., Ourisson, G.: Phytochemistry **7**, 757 (1968)
123. Popják, G., Cornforth, J.W.: Biochem. J. **101**, 553 (1966)
124. Popják, G., Edmond, J., Wong, S.-M.: J. Am. Chem. Soc. **95**, 2713 (1973)
125. Poppleston, C.R., Unrau, A.M.: Phytochemistry **12**, 1131 (1973)
126. Quershi, A.A., Peytia, E., Porter, J.W.: J. Biol. Chem. **248**, 1848 (1973)
127. Ragsdale, N.N.: Biochim. Biophys. Acta **380**, 81 (1975)
128. Randall, P.J., Rees, H.H., Goodwin, T.W.: Chem. Commun., 1295 (1972)
129. Rees, H.H., Goodwin, T.W.: In: Biosynthesis. Geissman, T.A. (ed.). London: The Chemical Society, 1972, Vol. 1, p. 59
130. Rees, H.H., Goodwin, T.W.: In: Biosynthesis. Geissman, T.A. (ed.). London: The Chemical Society, 1973, Vol. 2, p. 16
131. Rees, H.H., Goodwin, T.W.: In: Biosynthesis. Geissman, T.A. (ed.). London: The Chemical Society, 1975, Vol. 3, p. 14
132. Rees, H.H., Goad, L.J., Goodwin, T.W.: Biochem. J. **107**, 417 (1968)
133. Rees, H.H., Goad, L.J., Goodwin, T.W.: Tetrahedron Lett., 723 (1968)
134. Rees, H.H., Goad, L.J., Goodwin, T.W.: Biochim. Biophys. Acta **176**, 892 (1969)
135. Reid, W.W.: Phytochemistry **7**, 451 (1968)
136. Richards, L., Hendrickson, J.B.: The Biosynthesis of Sterols, Terpenes and Acetogenins. New York: W.A. Benjamin Inc. 1964
137. Rilling, H.C., Poulter, C.D., Epstein, W.W., Larsen, B.: J. Am. Chem. Soc. **93**, 1783 (1971)
138. Rohmer, M., Brandt, R.D.: Europ. J. Biochem. **36**, 446 (1973)
139. Rohmer, M., Ourisson, G., Benveniste, P., Bimpson, T.: Phytochemistry **14**, 727 (1975)
140. Rubinstein, I.: Ph.D. Thesis, University of Liverpool, 1973
141. Rubinstein, I., Goad, L.J.: Phytochemistry **13**, 481 (1974)
142. Rubinstein, I., Goad, L.J.: Phytochemistry **13**, 485 (1974)
143. Rubinstein, I., Goad, L.J., Clague, A.D.H., Mulheirn, L.J.: Phytochemistry **15**, 195 (1976)
144. Schecter, J., Bloch, K.: J. Biol. Chem. **246**, 7690 (1971)
145. Schechter, I., Sweat, F.W., Bloch, K.: Biochim. Biophys. Acta **220**, 463 (1970)
146. Shimizu, I., Nagai, J., Hatamoka, H., Katsuki, H.: Biochim. Biophys. Acta **296**, 310 (1973)
147. Sliwowski, J.: Postepy Biochem. **20**, 281 (1974)
148. Sliwowski, J., Kasprzyk, Z.: Phytochemistry **13**, 1451 (1974)
149. Sliwowski, J.K., Caspi, E.: J. Chem. Soc. Chem. Commun., 196 (1975)
150. Smith, A.R.H., Goad, L.J., Goodwin, T.W., Lederer, E.: Biochem. J. **104**, 56C (1967)
151. Smith, A.R.H., Goad, L.J., Goodwin, T.W.: Phytochemistry **11**, 2775 (1972)
152. Smith, L.L., Dhar, A.K., Gilchrist, J.L., Lin, Y.Y.: Phytochemistry **12**, 2727 (1973)
153. Stoll, A., Jucker, E.: In: Modern Methods of Plant Analysis. Paech, K., Tracey, M.V. (eds.). Berlin: Springer Verlag, 1955, Vol. 3, p. 141
154. Suga, T., Shishibori, T.: Phytochemistry **14**, 2411 (1975)
155. Tai, H.-H., Bloch, K.: J. Biol. Chem. **247**, 3767 (1972)
156. Thompson, E.D., Bailey, R.B., Parks, L.W.: Biochim. Biophys. Acta **334**, 116 (1974)
157. Thompson, M.J., Dutky, S.R., Patterson, E.L., Gooden, E.L.: Phytochemistry **11**, 1781 (1972)
158. Tomita, Y., Uomori, A., Minato, H.: Phytochemistry **9**, 555 (1970)
159. Tomita, Y., Uomori, A., Sakurai, E.: Phytochemistry **10**, 573 (1971)

160. Tomita, Y., Uomori, A.: J. Chem. Soc. Perkin Trans. **1**, 2656 (1973)
161. Trocha, P.J., Jasne, S.J., Sprinson, D.B.: Biochem. Biophys. Res. Commun. **59**, 666 (1974)
162. Van Aller, R.T., Chikamatsu, H., de Souza, N.J., John, J.P., Nes, W.R.: Biochem. Biophys. Res. Commun. **31**, 842 (1968)
163. Van Tamelen, E.E., Willett, J.D., Clayton, R.B., Lord, K.E.: J. Am. Chem. Soc. **88**, 4752 (1966)
164. Von Ardenne, M., Osske, G., Schreiber, K., Steinfelder, K., Tummler, R.: Kulturpflanze **13**, 115 (1965)
165. Walker, W., Goad, L.J.: Unpublished results (1976)
166. Weete, J.D.: Phytochemistry **12**, 1843 (1973)
167. Weete, J.D., Laseter, J.L.: Lipids **9**, 575 (1974)
168. Wojciechowski, Z.A., Goad, L.J., Goodwin, T.W.: Biochem. J. **136**, 405 (1973)
169. Yamamoto, S., Bloch, K.: In: Natural Substances Formed Biologically from Mevalonic Acid. Goodwin, T.W. (ed.). London: Academic Press, 1970, p. 35
170. Yokoyama, A., Natori, S., Aoshima, K.: Phytochemistry **14**, 487 (1975)
171. Zakelj, M., Goad, L.J., Goodwin, T.W.: Unpublished (1976)

CHAPTER 9

Steryl Glycosides and Acylated Steryl Glycosides

W. EICHENBERGER

A. Introduction

Steryl glycosides and acylated steryl glycosides have attracted the attention of both plant physiologists and biochemists, primarily because of the similarity between their properties and those of typical membrane lipids of the plant cell. Because of these similarities, both components, which are widely distributed in higher plants, are possibly involved in membrane function. Although a lot of information has accumulated about these plant lipids in the last few years, their significance for the plant itself is poorly understood. While many details about their occurrence and biosynthesis have been compiled, there exists a broad gap in respect to the metabolism and the physiologic function of steryl glycosides and their acylated forms. The field has been partially reviewed in relation to plant sterols [38] and the biosynthesis of glycolipids [60].

B. Structure and Distribution

A steryl glycoside (SG) was first identified by hydrolysis in 1913 [69] and an acylated (or esterified) steryl glycoside (ASG) was first detected in potato tubers in 1964 [57]. Since then, the structure of both compounds has repeatedly been confirmed [4, 26, 53, 55, 80]. The sterol part consists of the type of phytosterols

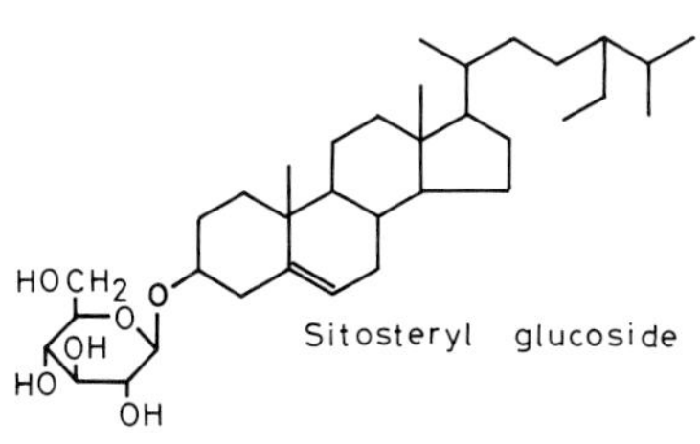

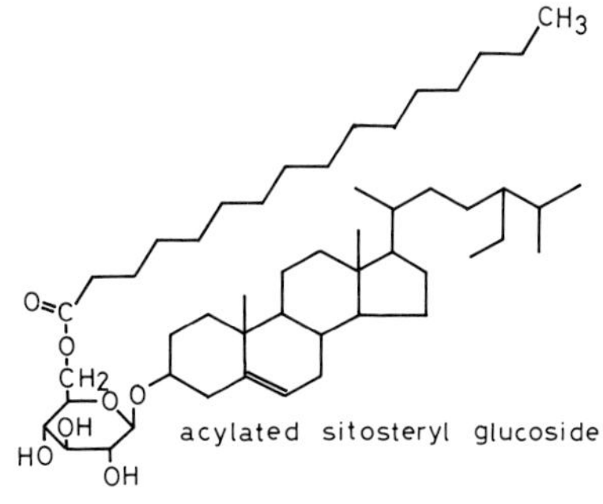

also present in the same plant in the form of free and esterified sterols [1]. The sterol pattern of SG may be quantitatively almost identical to [36] or different [1] from that of free sterols. The sugar part consists mostly of glucose, although galactose [54, 64, 76], mannose [23], xylose [77], and gentiobiose [51]

Table 1. Properties of plant steryl glycosides and steroid glycosides

	Steryl glycosides	Steroid glycosides (Cardiac glycosides, saponins, steroid alkaloids)
Occurrence	Flowering plants Ferns, mosses Fungi? Mycoplasma	Few plant families
Aglycon	Phytosterol (3β-OH)	Steroid with one or more O-functions ring system with N- or O-heteroatoms
Sugar	Glucose Monohexoside and its monoester	Rare sugars Mono- or oligosaccharides
Solubility	Slightly soluble in H_2O	Soluble in H_2O, often surface-active
Physiologic effects	Not reported	Various

have also been found in single cases. Other oligosaccharides are suggested, but their structure has not definitely been established [71]. The linkage to the 3-position of the sterol is in most cases a β-glycosidic bond, but an α-glycosidic bond has also been reported [39]. In the acylated form, a fatty acid is attached to the 6-position of the sugar [4, 53] and consists of one of the fatty acids present in the lipids of the same plant.

Steryl glycosides differ in several respects from the group of plant steroid glycosides, which are well-known for their physiologic effects in animals, as summarized in Table 1. Steryl glycosides are present in all flowering plants thus far tested and are probably widely distributed among ferns [70] and mosses [15]. While their presence in fungi has been reported in individual cases only [6, 27, 58], they have never been reported in algae [62]. Among prokaryotes, however, certain strains of *Mycoplasma* synthesize the free glycoside, not the acylated form [74]. For steryl glycosides no effects on animals have been reported so far. This may be the reason why their investigation has been rather neglected for a long time.

Representative data on the percentage of SG and ASG in the total lipid of different plants, organs, and subcellular fractions are summarized in Table 2. On a total lipid base, photosynthetic tissue contains 1–2%, storage tissue, 3–10% SG and ASG. Even higher amounts occur in embryonic tissue and tissue cultures. Within the cell, ASG and SG have been found in all kinds of subcellular particles, the highest concentration being present in microsomal fractions [11, 20, 36]. In microsomal lipids the percentage increases further if chloroplast lipids are subtracted by calculation [16]. Minor amounts of SG and ASG were found in chloroplast envelopes [68], while lamellae are free of them [3, 68]. Many quantitative data not given here are based on total dry weight rather than total lipid. As a general rule, steryl glycosides, like free and esterified sterols, seem to be concentrated in the microsomal fraction, which is composed mainly of membraneous material from the endoplasmic reticulum, Golgi vesicles,

Table 2. Steryl glycosides (SG) and acylated steryl glycosides (ASG) in different plants, organs, and subcellular fractions

		% Total lipid			References
		SG	ASG	Total	
Radish	leaves	0.7	0.3	1.0	
	roots	1.1	1.7	2.8	
Carrot	leaves	0.8	0.9	1.7	[20]
	roots	0.8	3.6	4.4	
Potato	leaves	0.8	1.2	2.0	
	tubers	1.0	3.6	4.6	
		2.0	6.4	8.4	[34]
Apple	pulp	5.8	1.6	7.4	[35]
Barley	embryo				
	coleoptile		5.8–17.1[a]		
	coleorhiza		4.1–12.3		66]
	scutellum		3.4– 8.6		
Artemisia absinthum					
	callus cultures			ca. 50	[71]
Spinach	chloroplasts				
	envelope	1.1	0.8	1.9	[68]
	lamellae	0	0	0	
Lettuce	crude chloroplasts			0.9	
	crude microsomes			6.7	[16]
	microsomes (calculated)			8.7	

[a] % of total acyl lipids

and the plasmalemma [61]. Chloroplasts and mitochondria [14] contain minor portions only. The portion of total sterols combined as glycosides ranges from 2 to 86% [34, 36, 49].

C. Biosynthesis

The enzymatic synthesis of SG was first demonstrated with homogenates from soybeans [47] and observed later in different tissues and subcellular fractions of several plants [17–19, 24, 31, 48, 50, 56, 63, 66, 67, 83]. All these preparations utilized exogenous UDP-^{14}C-glucose and unlabeled endogenous sterols (and fatty acids) as substrates to produce steryl-^{14}C-glucoside and acylated steryl-^{14}C-glucoside. The same reactions were observed when labeled sterols were administred [9].

Therefore, two reactions were taking place in these preparations: (1) a glycosylation step, and (2) an acylation step. The two enzyme activities were shown to be associated with two different protein fractions isolated from a cotton fiber homogenate and separated by gel chromatography [26].

I. Glycosylation

The glycosyltransferase involved in formation of SG has been characterized in several plants [29, 32, 66, 84]. In all cases the enzyme is particle-bound but may be solubilized with Triton X-100 [32, 66, 84]. The pH-optimum of the activity ranges from pH 6.5 to 8.5 [9, 32, 48, 50, 63, 84].

As a hexose donor UDP-glucose is the most effective nucleotide. TDP-glucose is less effective, while ADP-, CDP-, GDP-, IDP-glucose and UDP-glucuronic acid are almost ineffective [48, 84]. The specificity for UDP-galactose varies [24, 56, 84], while GDP-mannose [56] and UDP-mannose [31] are not utilized. The hexose incorporated from UDP-galactose is always found to be glucose [24, 63, 84], indicating the presence of an effective epimerase in these systems. Values of the apparent K_m for UDP-glucose range from 0.028 to 0.22 mM [48, 56, 66].

As a sterol substrate, sitosterol is always preferred. Stigmasterol, α-spinasterol, ergosterol, and cholesterol are also effective [32, 56, 84], while triterpene alcohols are not accepted [84]. This specificity agrees exactly with analytical data from *Sorghum vulgare*, where only 4-demethyl-sterols saturated in the 24 [36] position are found as glycosides [65]. The apparent K_m for sitosterol ranges from 0.04 to 0.3 mM [32, 66].

The reaction is often stimulated by ATP [9, 17, 29, 48, 83], Mg^{2+} and Ca^{2+} [29, 31, 56, 84], and phosphoglycerides [26, 29, 31, 32]. UDP and UTP [48, 56, 84] as well as phosphate buffer [32, 63] inhibit glycosylation. Many results concerning the influence on the enzyme preparation of acetone and butanol extraction and of different cofactors are contradictory and need further experiments under comparable conditions.

Activity of SG-forming glycosyltransferase has also been established in the prokaryote *Mycoplasma gallinarum* [72, 74] as well as in the yeast *Candida bogoriensis* [27]. A similar enzyme seems to occur also in animal tissue [10].

II. Acylation

Two different routes are possible for the formation of ASG. Route 1 leads from a free sterol to SG, which is further esterified to ASG. This path would be expected to be more convenient for the plant, because it requires only one additional step. Route 2 leads to an acylated hexosyl intermediate, which is then further combined with a free sterol to give ASG. This way, although

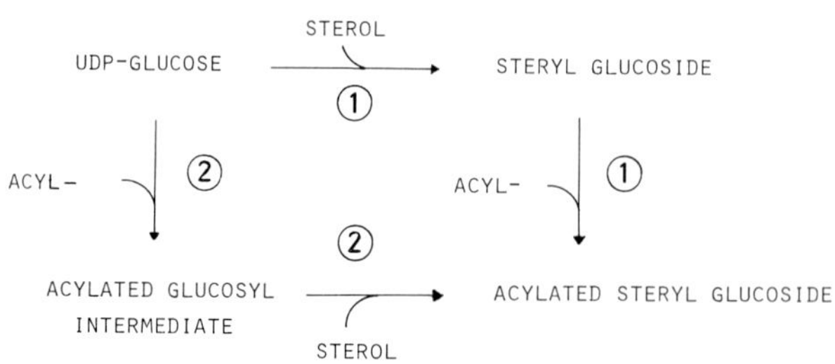

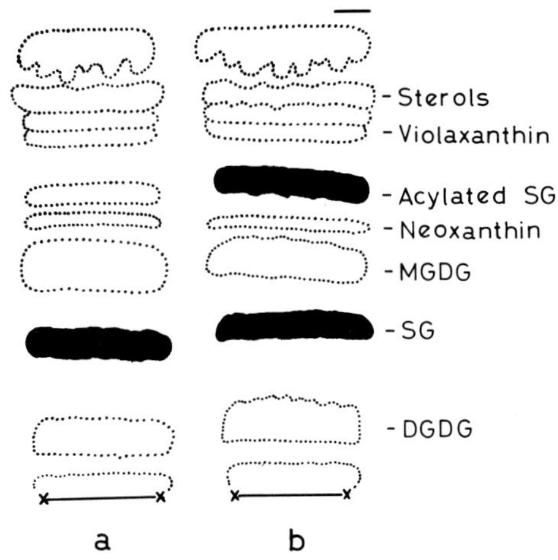

Fig. 1a and b. Acylation of steryl-^{14}C-glucoside by a particulate fraction from lettuce leaf homogenate as demonstrated by TLC and autoradiography [19]. Enzyme preparation (a), heat-inactivated; (b), untreated

requiring two additional steps, has to be considered, because acylated sugars have been found in microorganisms [7, 81], not in plants. Evidence for route 1 was obtained both from direct esterification of labeled SG and from kinetic data.

On incubation of a particulate preparation from lettuce with steryl ^{14}C-glucoside, the formation of acylated steryl ^{14}C-glucoside was demonstrated by thin-layer chromatography (TLC) and autoradiography, as shown in Figure 1. Similar results were obtained with cotton fibers [26].

Incubation of a pea root homogenate with UDP-^{14}C-galactose yielded both labeled SG and ASG [63]. However, the appearence of the label in the latter was delayed, as is shown in Figure 2.

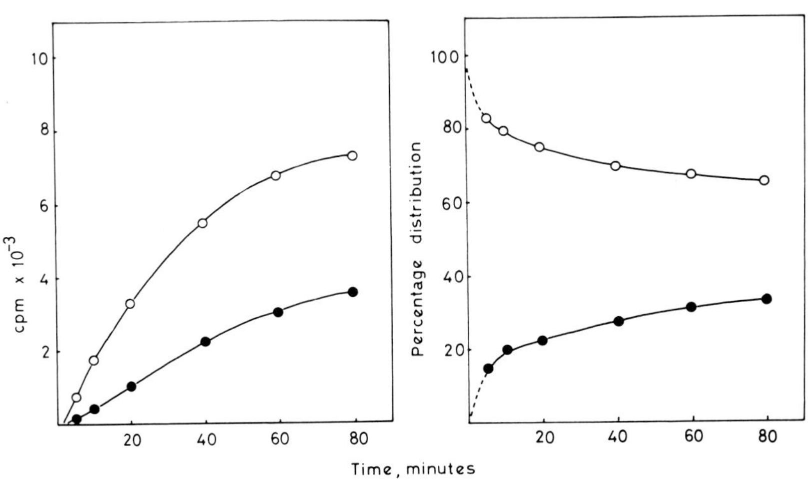

Fig. 2. Biosynthesis of steryl glucoside (○) and acylated steryl glucoside (●) by particles (18,000 g) from pea roots, adapted from [63]

This would be expected if SG were the precursor of ASG. Very similar results were also obtained with wheat seedlings [66, 67] and *Calendula* plants [83].

Although experiments with cell-free systems without exception favor route 1, results obtained with intact tobacco plants fed with ^{14}C-cholesterol suggest the existence of a second way, which is independent of SG as a precursor and which might be identical to route 2 [9]. Based on these results, the formation of ASG from SG in the plant seems very likely. An alternative way cannot, however, be excluded.

The acylation step, which needs SG and an appropriate acyl donor as substrates, is of special interest, because it connects biosynthesis of steryl glycosides with the metabolism of fatty acids and polar lipids of the plant.

Acyltransferase, like glycosyltransferase, is generally particle-bound, its solubility varying between different plants. The acyltransferase from carrot roots [21, 25] and bean leaves [45] is soluble in buffer solutions. The enzymes from wheat seedlings [67], *Calendula* [85], cotton fibers [33], and leaves of lettuce, spinach, radish, and carrot [17, 19, 24, 25] need to be solubilized with Triton X-100. The pH-optimum ranges from 4.8 to 7.0 [25, 33, 45, 63, 83]. A molecular weight of 60,000 daltons has been determined for the acyltransferase from carrot roots [25]. The apparent K_m value for SG as substrate varies between 0.014 and 0.16 mM [25, 33, 67]. The specificity for the glycoside substrate varies from plant to plant. The enzyme from beans transfers fatty acids from monogalactosyldiglyceride (MGDG) in a similar reaction mainly to another molecule of MGDG to give acylgalactosyldiglyceride (AGDG) [45]. In contrast, the enzyme from carrots esterifies 22 times more SG than MGDG under comparable conditions [73] and seems therefore specific for SG. The acylation step is stimulated by different lipids acting as acyl donors. The transfer of acyl groups has been proved using acyl-labeled substrates [33, 73, 85]. The acyltransferase from wheat utilizes phosphatidylethanolamine (PE) [67]. The enzymes from both *Calendula* [85] and cotton fibers [33] prefer PE, but accept also phosphatidylcholine (PC). Triacylglycerol and galactosyldiglycerides are ineffective [85]. The acyltransferase from carrots, on the other hand, prefers digalactosyldiglyceride (DGDG), while both MGDG and phosphoglycerides are less effective [25]. Finally, the enzyme preparation from beans accepts galactolipids but not phospholipids as acyl donors [45].

Stimulation of the reaction by CoA and ATP or acyl-CoA [25, 26] suggests a mechanism involving a CoA derivative as an intermediate. There is also evidence, however, for a CoA-independent path of acyl transfer, since dialyzed enzyme preparations retain considerable activity [25]. An apparent K_m value of 0.019 mM was obtained for DGDG in carrots [25]. The K_m for PE was estimated to be about 0.03 mM in cotton fibers [33] and 5 mM in wheat seedlings [67].

An important question in fatty acid transfer is the origin of acyl groups transferred to SG in the intact plant. The preference for galactolipids of the acyl transferase from both carrots and beans is obvious [25, 45] and could mean that fatty acids of ASG are transferred from chloroplasts, which are generally rich in galactolipids. If this is the case, the acyl pattern of ASG

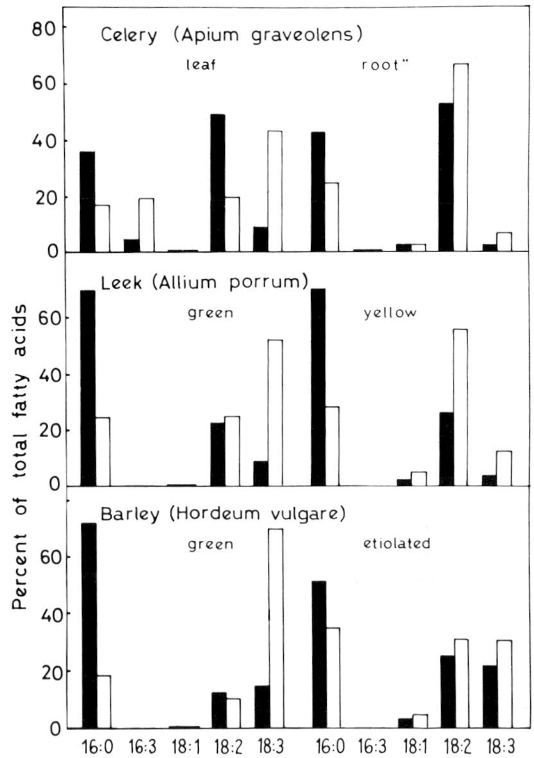

Fig. 3. Fatty acid pattern of ASG (■) and total lipid (□) from green and nongreen tissue from different plants

should contain considerable amounts of typical chloroplast fatty acids and therefore reflect the presence of chloroplasts in the tissue. Experiments comparing the acyl pattern of both ASG and total lipids have been carried out with plants forming both green and nongreen tissue [22].

As shown in Figure 3, ASG always contains large amounts of 16:0 and 18:2 acids in both breen and nongreen tissue, although 18:3 predominates in the total lipid of green tissue. Also, ASG in bleached parts of the stem or in etiolated leaves contains the same fatty acid pattern as ASG from green tissue of the same plant. Experiments with *Calendula* plants demonstrate that the acyl pattern of ASG is almost identical in flowers, leaves, and roots [85]. As a general rule, the fatty acid pattern of ASG seems to be almost independent of the function of tissue, and chloroplast lipids can therefore be eliminated as major acyl donors. In vitro experiments with particulate enzyme preparations from cotton fibers indicate that a phospholipid is very likely the acyl donor, since the acylation is inhibited by phospholipase [33]. The real origin of fatty acids in ASG has, however, to be cleared up. Further experiments should also consider a possible different specificity of acyltransferase for different types of fatty acids, as has been suggested by experiments using different PE preparations varying both in their acyl pattern and in their efficiency as acyl donors [67].

III. Intracellular Localization

Glycosyltransferase and acyltransferase activities have been established in microsomal [47, 48, 67, 83], chloroplast [63], etioplast [29], starch grain [56], and mitochondrial fractions [29, 63]. For the identification of these subcellular fractions, only sedimentation properties have been used. More reliable results were produced by experiments based on the activity of marker enzymes. In maize coleoptiles, the formation of SG is associated with the same particles as is the activity of β-glucan synthetase and is therefore attributed to the plasma membrane [40] or the Golgi vesicles [29, 32]. In *C. officinalis* the activities of both acyltransferase and glycosyltransferase are associated with the crude mitochondrial fraction [85]. After a further separation on a sucrose density gradient, part of glycosyltransferase activity coincides with glucose-6-phosphatase, but not with succinate dehydrogenase or chlorophyll [84], as shown in Figure 4. At least part of the enzyme activity may therefore be attributed to the endoplasmic reticulum, while it is suggested that the major part of the activity is localized in Golgi vesicles [84]. In any case the intracellular sites of both enzymes need to be further clarified by morphologic and enzymatic criteria.

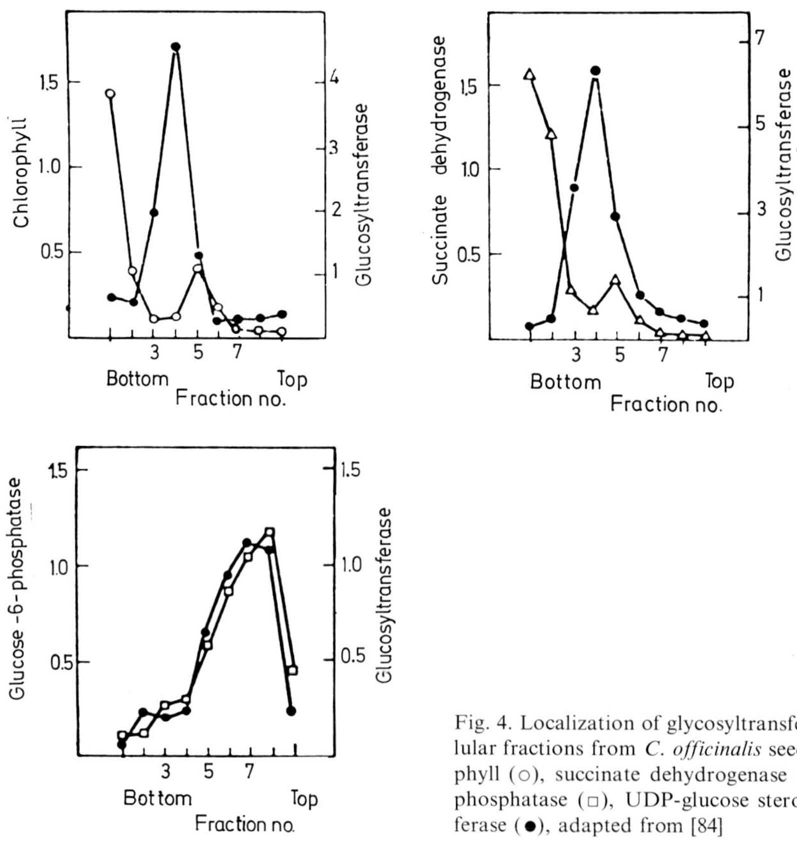

Fig. 4. Localization of glycosyltransferase in subcellular fractions from *C. officinalis* seedlings. Chlorophyll (○), succinate dehydrogenase (△), glucose-6-phosphatase (□), UDP-glucose sterolglycosyltransferase (●), adapted from [84]

Based on the results available at present, the following generalized concept may be suggested: SG and ASG are formed primarily in the endoplasmic reticulum and/or the Golgi vesicles and the plasma membrane. Minor sites of synthesis exist also in other compartments such as mitochondria and chloroplasts. This situation would explain the distribution of enzyme activities and steryl glycosides among different subcellular fractions. It seems likely that the acyl parts of ASG originate directly from lipids of the membranes involved in synthesis. This would explain both the predominance of fatty acids typical for microsomal lipids and the existence of minor amounts of typical chloroplast fatty acids in the ASG of the total cell. This concept, which serves at least as a useful working hypothesis, is supported by results obtained with *Calendula*. ASG isolated from different subcellular fractions of this plant have their individual acyl patterns [85], as expected, when the synthesis of ASG is distributed over different cell compartments.

D. Metabolism

Very little is known about the metabolism of SG and ASG. Evidence for a degradation of steryl glycosides in the plant comes from several experiments.

In potato tubers the amount of ASG decreases during storage [13], while the amount of SG increases moderately [12]. Labeled SG administered to leaves of *Pelargonium hortorum* was found to be partially hydrolyzed after resorption [5]. In germinating tobacco seeds a decrease of steryl glycosides was observed, while free sterols were produced [8]. Indirect evidence for a breakdown was obtained from pulse-labeling experiments with soybeans fed with $^{14}CO_2$. The time-course of total radioactivity of SG reflects a turnover and suggests a role as a carrier for hexose [30]. On the other hand, no decrease of radioactivity incorporated in SG was observed in a pulse-chase experiment with pea seedlings [50]. In this case, degradation could have been overlooked, because the labeled SG formed was diluted by endogenous SG.

It should be mentioned that a degradation reaction has in no case been characterized, although, with regard to the physiologic function of SG and ASG, an insight into their metabolic fate would be very important.

E. Possible Functions

Little is known about the physiologic function of sterols in plants. The present state of knowledge has recently been reviewed [38, 42]. If plant sterols are assigned a function similar to that of sterols in animal cells, the following roles have to be considered: (1) constituents of cellular membranes and (2) hormones or precursors thereof. Both functions could theoretically be attributed also to steryl glycosides. In addition, based on their hexose and acyl content,

SG and ASG could have a (3) metabolic function and act as storage compounds or as a transportable form of sterols.

Basically, suggestions of the involvement of steryl glycosides in the formation of cellular membranes come from the fact that SG and ASG are present in all kinds of organs and subcellular fractions within a single plant [20, 78, 82, 85]. A broad distribution within a single plant was also established for the activity of enzymes involved in the synthesis [2, 83]. For example, glycosyltransferase activity has been found in all plant organs and subcellular fractions checked [29, 86]. An important argument for their role in membranes is the fact that steryl glycosides and their acylated forms are actively synthesized during germination. In *Calendula,* glycosyltransferase is found to be most active in this stage of development [86]. Also in seedlings of radish a distinct increase of SG and ASG was observed during germination [59]. In roots of *Digitalis* seedlings fed with ^{14}C-mevalonate, radioactivity was more rapidly incorporated in SG than in free sterols [28]. An influence on membrane permeability has, however, been demonstrated for free sterols only and not for the glycosides [37].

Only individual observations exist as to their possible roles, as hormones or precursors thereof. Stigmasteryl-β-D-glucoside enhances the growth-promoting effect of auxin in *Avena* coleoptiles [52]. An intermediate hydrolysis to free sterol has, however, not been properly excluded in these experiments. In *Digitalis* plants ^{14}C-cholesteryl-glucoside is transformed to spirostanols [79]. This means that steryl glycosides may act as precursors of different plant steroids, the physiologic importance of which has been discussed elsewhere [41, 43, 44].

Based on their sugar and acyl contents, steryl glycosides have also been considered to have a storage function [48]. Indeed, during maturation an increase of the SG content is observed in soybeans [49]. In germinating tobacco seeds, the amount of SG decreased, while free sterols were formed [8]. In any case, it should be mentioned that because of the relatively low amounts of SG and ASG, they can only have quantitatively minor importance as storage compounds.

From the observation that in *Digitalis* plants fed with ^{14}C-mevalonate, SG is labeled more rapidly than free sterols [28], it was suggested that SG acts as a transportable form of sterols [28, 74]. If this is the case, transport of steryl glycosides within the plant should be detectable. Indeed, labeled SG could be isolated from the stem and the root of *Pelargonium* plants to which steryl-^{14}C-glucoside had been administered through the leaf [5]. Whether SG is transported unchanged or hydrolyzed and resynthesized is not known.

F. Conclusions

Many investigations have been carried out and much information accumulated, primarily on the occurrence and the biosynthesis of steryl glycosides and their acylated forms. Although single steps have been thoroughly studied and enzymes have been partially characterized, there remain several questions to be resolved.

In order to elucidate the mechanisms of both glycosylation and acylation, the effects of cofactors and inhibitors have to be investigated. The origin of the acyl groups needed for the formation of ASG, as well as the question to what extent soluble and particle-bound portions of acyltransferase are responsible for ASG formation, needs further investigation. Also, it will be necessary to check whether the pathways established for cell-free systems are also valid for intact plants. As to the intracellular localization, there is good evidence that enzymes and products are distributed among different cell compartments, with the major site in the endoplasmic reticulum. In further attempts to prove this suggestion, more emphasis will have to be placed on morphologic and enzymatic criteria for the characterization of subcellular fractions.

Concerning metabolism, only a few results are available. Further experiments are necessary to characterize enzymes involved in the degradation of SG and ASG. In this connection, more information on the developmental and seasonal changes of SG and ASG contents of steryl glycosides within the plant need to be considered further.

As to the physiologic role, the ubiquitous presence of SG and ASG, in meristematic tissue as well, suggests that they play a role in cellular membranes. Information on their mode of action at the molecular level is not available and should be expected from more extensive experiments on the influence of steryl glycosides on artificial and natural membranes. As to their hormonal action or relation to steroid hormones, no definitive conclusion is possible. First more details are needed concerning their transformation into steroid derivatives and their hormonal effects. Whatever the function of steryl glycosides may be, it would be difficult to understand why the plant should operate a machinery of this specificity in every cell simply to produce waste products. For further progress in this field, generally more attention should be paid to comparative aspects, by working with different plants under comparable conditions.

References

1. Adler, G., Kasprzyk, Z.: Free sterols, steryl esters, glucosides, acylated glucosides and water-soluble complexes in *Calendula officinalis*. Phytochemistry **14**, 627–631 (1975)
2. Adler, G., Kasprzyk, Z.: The incorporation of mevalonate [2-^{14}C] into the sterol fractions of *Calendula officinalis*. Phytochemistry **14**, 723–726 (1975)
3. Allen, C.F., Good, P., Davis, H.E., Chisum, P., Fowler, S.D.: Methodology for the separation of plant lipids and application to spinach leaf and chloroplast lamellae. J. Am. Oil Chem. Soc. **43**, 223–231 (1966)
4. Aneja, R., Harries, P.C.: Sitosteryl-β-D-(6-0-fattyacyl)-glucopyranosides from the crude phosphatides of Maize. Chem. Phys. Lipids **12**, 351–362 (1971)
5. Attallah, A.M., Aexel, R.T., Ramsay, R.B., Threlkeld, S., Nicholas, H.J.: Translocation of sitosterol and related compounds in *Pelargonium hortorum* and *Helianthus annuus*. Phytochemistry **14**, 1927–1932 (1975)
6. Baraud, J., Maurice, A., Napias, C.: Composition et répartition des lipides au sein des cellules de *Saccharomyces cerevisiae*. Bull. Soc. Chim. Biol. **52**, 421–432 (1970)
7. Brennan, P.J., Lehane, D.P.: Acylglucoses in corynebacteria. Biochim. Biophys. Acta **176**, 675–677 (1969)

8. Bush, P.B., Grunwald, C.: Sterol changes during germination of *Nicotiana tabacum* seeds. Plant Physiol. **50**, 69–72 (1972)
9. Bush, P.B., Grunwald, C.: Steryl glycoside formation in seedlings of *Nicotiana tabacum* L. Plant Physiol. **53**, 131–135 (1974)
10. Collins, D.C., Williamson, D.G., Layne, D.S.: Steroid glucosides. Enzymatic synthesis by a partially purified transferase from rabbit liver microsomes. J. Biol. Chem. **245**, 873–876 (1970)
11. Dupéron, R., Brillard, M., Dupéron, P.: Localisation intracellulaire des différents composés stéroliques dans le tubercule de pomme de terre avant sa germination. C.R. hébd. Séanc. Acad. Sci. **274**, 2321–2324 (1972)
12. Dupéron, R., Dupéron, P., Thiersault, M.: Etude de l'évolution des différents groupes de composés stéroliques dans les tubercules de pomme de terre, pendant leur entreposage et leur germination à l'obscurité. C.R. Acad. Sci. Sér. D **273**, 580–583 (1971)
13. Dupéron, P., Dupéron, R., Thiersault, M.: Biosynthèse des composés stéroliques dans le tubercule de pomme de terre durant son entreposage et sa "germination" à l'obscurité. Origine des stérides apparaissant au cours de cette période. C.R. Acad. Sci. Sér. D **276**, 1169–1172 (1973)
14. Dupéron, R., Méance, J., Dupéron, P., Thiersault, M.: Les composés stéroliques présents dans les mitochondries et les "microsomes" de cellules végétales. C.R. hébd. Séanc. Acad. Sci. Sér. D **280**, 205–208 (1975)
15. Eichenberger, W.: unpublished
16. Eichenberger, W.: Über den Steringehalt pflanzlicher Mikrosomen. Chimia **29**, 132–133 (1975)
17. Eichenberger, W., Grob, E.C.: Zur enzymatischen Bildung von Steringlycosiden durch Partikelfraktionen aus Blättern von Lattich und Spinat. Chimia **22**, 46–48 (1968)
18. Eichenberger, W., Grob, E.C.: Enzymatic formation of steryl glycosides by homogenates and chloroplast preparations from lettuce and spinach leaves. In: Progress in Photosynthesis Research. Metzner, H. (ed.). München: Lichtenstein, 1969, Vol. I, pp. 338–344
19. Eichenberger, W., Grob, E.C.: Zur enzymatischen Bildung von Acyl-Steringlycosid in Pflanzen. Chimia **23**, 368–369 (1969)
20. Eichenberger, W., Grob, E.C.: Über die quantitative Bestimmung von Sterinderivaten in Pflanzen und die intrazelluläre Verteilung der Steringlycoside in Blättern. FEBS Lett. **11**, 177–180 (1970)
21. Eichenberger, W., Grob, E.C.: Enzymatische Acylierung von Sterin-^{14}C-glucosid durch ein lösliches Enzym aus Karotten. Chimia **24**, 394–396 (1970)
22. Eichenberger, W., Grob, E.C.: Fettsäuremuster im pflanzlichen Acylsteringlycosid. FEBS Lett. **20**, 245–248 (1972)
23. Eichenberger, W., Menke, W.: Sterole in Blättern und Chloroplasten. Z. Naturforsch. **21b**, 859–867 (1966)
24. Eichenberger, W., Newman, D.W.: Hexose transfer from UDP-hexose in the formation of steryl glycoside and esterified steryl glycoside in leaves. Biochem. Biophys. Res. Commun. **32**, 366–374 (1968)
25. Eichenberger, W., Siegrist, H.P.: Steryl glycoside acyltransferase from carrots. FEBS Lett. **52**, 153–156 (1975)
26. Elbein, A.D., Forsee, W.T.: Biosynthesis and structure of glycosyl diglycerides, steryl glucosides and acylated steryl glucosides. Lipids **10**, 427–436 (1975)
27. Esders, T.W., Light, R.J.: Occurrence of a Uridine Diphosphate Glucose: Sterol Glucosyltransferase in *Candida bogoriensis*. J. Biol. Chem. **247**, 7494–7497 (1972)
28. Evans, F.J.: Uptake of [2-^{14}C] mevalonic acid by lipids and glycoside sterols. Planta **116**, 99–104 (1974)
29. Fang, T.-Y., Baisted, D.J.: UDPG: sterol glucosyltransferase in etiolated pea seedlings. Phytochemistry **15**, 273–278 (1976)
30. Fisher, D.B.: Kinectics of C-14 translocation in soybean. II. Kinetics in the leaf. Plant Physiol. **45**, 114–118 (1970)
31. Forsee, W.T., Laine, R.A., Chambers, J.P., Elbein, A.D.: UDP-glucose: β-sitosterol glucosyltransferase of developping cotton seeds. In: Methods in Enzymology. Ginsburg, V. (ed.). New York: Academic Press, 1972, Vol. 28, Part B, pp. 478–482
32. Forsee, W.T., Laine, R.A., Elbein, A.D.: Solubilization of a particulate UDP-glucose: sterol β-glucosyl-transferase in developping cotton fibers and seeds and characterization of steryl 6-acyl-D-glucosides. Arch. Biochem. Biophys. **161**, 248–259 (1974)
33. Forsee, W.T., Valkovitch, G., Elbein, A.D.: Acylation of steryl glucosides by phospholipids.

Solubilization and properties of the acyl transferase. Arch. Biochem. Biophys. **172**, 410–418 (1976)
34. Galliard, T.: Aspects of lipid metabolism in higher plants. I. Identification and quantitative determination of the lipids in potato tubers. Phytochemistry **7**, 1907–1914 (1968)
35. Galliard, T.: Aspects of lipid metabolism in higher plants. II. The identification and quantitative analysis of lipids from the pulp of pre- and postclimacteric apples. Phytochemistry **7**, 1915–1922 (1968)
36. Grunwald, C.: Sterol distribution in intracellular organelles isolated from tobacco leaves. Plant Physiol. **45**, 663–666 (1970)
37. Grunwald, C.: Effects of free sterols, steryl ester and steryl glycoside on membrane permeability. Plant Physiol. **48**, 653–655 (1971)
38. Grunwald, C.: Plant sterols. Ann. Rev. Plant Physiol. **26**, 209–236 (1975)
39. Gupta, O.C.D., Rizvi, S.A.I., Gupta, P.C.: Chemical examination of a phytosterolin from the seeds of *Ipomoea fistulosa*. Planta Med. **20**, 172–177 (1971)
40. Hartmann, M.A., Fonteneau, P., Benveniste, P.: Subcellular localization of sterol synthesizing enzymes in maize coleoptiles. Plant Sci. Lett. **8**, 45–51 (1977)
41. Heftmann, E.: Biochemistry of steroidal saponins and glycoalkaloids. Lloydia **30**, 209–230 (1967)
42. Heftmann, E.: Functions of sterols in plants. Lipids **6**, 128–133 (1971)
43. Heftmann, E.: Recent progress in the biochemistry of plant steroids other than sterols (saponins, glycoalkaloids, pregnane derivatives, cardiac glycosides, and sex hormones). Lipids **9**, 626–639 (1974)
44. Heftmann, E.: Functions of steroids in plants. Phytochemistry **14**, 891–901 (1975)
45. Heinz, E., Dieler, H.P., Rullkötter, J.: Enzymatic acylation of steryl glycoside. Z. Pflanzenphysiol. **75**, 78–87 (1975)
46. Hølmer, G., Ory, R.L., Høy, C.-E.: Changes in lipid composition of germinating barley embryo. Lipids **8**, 277–283 (1973)
47. Hou, C.T., Umemura, Y., Nakamura, M., Funahashi, S.: Enzymatic synthesis of steryl glucoside by a particulate preparation from immature soybean seeds. J. Biochem. (Tokyo) **62**, 389–391 (1967)
48. Hou, C.T., Umemura, Y., Nakamura, M., Funahashi, S.: Enzymatic synthesis of steryl glucoside by a particulate preparation from immature soybean seeds. J. Biochem. (Tokyo) **63**, 351–360 (1968)
49. Katayama, M., Katoh, M.: Soybean sterols during maturation of seeds. I. Accumulation of sterols in the free form, fatty acid esters, acylated glucosides and non-acylated glucosides. Plant Cell Physiol. **14**, 681–688 (1973)
50. Kauss, H.: Enzymatische Glucosylierung von pflanzlichen Sterinen. Z. Naturforsch. **23b**, 1522–1526 (1968)
51. Khanna, I., Seshadri, R., Seshadri, T.R.: Sterol and lipid components of green *Thea sinensis*. Phytochemistry **13**, 199–202 (1974)
52. Kimura, Y., Tietz, A., Tamura, S.: Stigmasteryl-β-D-glucoside as an auxin synergist. Planta **126**, 289–292 (1975)
53. Kiribuchi, T., Mizunaga, T., Funahashi, S.: Separation of soybean sterols by Florisil chromatography and characterization of acylated steryl glucoside. Agr. Biol. Chem. (Tokyo) **30**, 770–778 (1966)
54. Köstens, J., Willuhn, G.: Steringlycoside und Acyl-Steringlycoside in Blättern von *Solanum dulcamara*. Planta Med. **24**, 278–285 (1973)
55. Laine, R.A., Elbein, A.D.: Steryl glucosides in *Phaseolus aureus*. Use of Gas-Liquid Chromatography and Mass Spectrometry for structural identification. Biochemistry **10**, 2547–2553 (1971)
56. Lavintman, N., Cardini, C.E.: Biosynthesis of glycolipid in starch grains from sweet corn. Biochim. Biophys. Acta **201**, 508–510 (1970)
57. Lepage, M.: Isolation and characterization of an esterified form of steryl glucoside. J. Lipid Res. **5**, 587–592 (1964)
58. Lopez, A., Burgos, J.: Lipid composition of *Sporendonema epizoum*. Phytochemistry **15**, 971–975 (1976)
59. Méance, J., Dupéron, R.: Présence de glycosides stéroliques et de glycosides stéroliques acylés dans les semences. Evolution de ces substances au cours de la germination du radis. C.R. Acad. Sci. Sér. D **277**, 849–852 (1973)
60. Mudd, J.B., Garcia, R.E.: Biosynthesis of glycolipids. In: Recent Adv. in the Chemistry and Biochemistry of Plant Lipids. Galliard, T., Mercer, E.I. (eds.). London: Academic Press, 1975, pp. 161–201

61. Musgrave, A., Walters, J., Sargent, J.A., Powell, G.: Microsomal membrane lipids from pea shoots and *Ranunculus sceleratus* petioles. Phytochemistry **15**, 1219–1223 (1976)
62. Nichols, B.W.: Comparative lipid biochemistry of photosynthetic organisms. In: Phytochemical Phylogeny. Harborne, J.B. (ed.). New York: Academic Press, 1970, pp. 105–118
63. Ongun, A., Mudd, J.B.: The biosynthesis of steryl glucosides in plants. Plant Physiol. **45**, 255–262 (1970)
64. Osman, A.M., El-Garby Younes, M., Mokhtar, A.: Sitosterol-β-D-galactoside from *Hibiscus sabdariffa*. Phytochemistry **14**, 829–830 (1975)
65. Palmer, M.A., Bowden, B.N.: The pentacyclic triterpene esters and the free, esterified and glycosylated sterols of *Sorghum vulgare* grain. Phytochemistry **14**, 1813–1815 (1975)
66. Péaud-Lenoël, C., Axelos, M.: Uridine diphosphate glucose: stérol transglucosylase. Purification et activité de préparations particulaires et solubles extraites du blé. C.R. hébd. Séanc. Acad. Sci. Sér. D. **273**, 1057–1060 (1971)
67. Péaud-Lenoël, C., Axelos, P.: D-glucosylation des phytostérols et acylation des stéryl-D-glucosides en présence d'enzymes de plantes. Carbohydrate Res. **24**, 247–262 (1972)
68. Poincelot, R.P.: Isolation and lipid composition of spinach chloroplast envelope membranes. Arch. Biochem. Biophys. **159**, 134–142 (1973)
69. Power, F.B., Salway, A.H.: The identification of ipuranol and some allied compounds as phytosterol glucosides. J. Chem. Soc. **103**, 399–406 (1913)
70. Radunz, A.: Über die Lipide der Pteridophyten, II. Die Fettsäuren einiger Lipide aus Blättern von *Dryopteris filix-mas*. Hoppe-Seylers Z. Physiol. Chem. **349**, 303–309 (1968)
71. Radwan, S.S., Spener, F., Mangold, H.K.: Lipids in plant tissue cultures. IV. The characteristic pattern of lipid classes in callus cultures and suspension cultures. Chem. Phys. Lipids **14**, 72–80 (1975)
72. Rothblat, G.H., Smith, P.F.: Nonsaponifiable lipids of representative pleuropneumonia-like organisms. J. Bacteriol. **82**, 479–491 (1961)
73. Siegrist, H.P.: Über die enzymatische Acylierung von Steringlucosid in Karotten. Thesis, Bern (1975)
74. Smith, P.F.: Biosynthesis of Cholesteryl Glucoside by *Mycoplasma gallinarum*. J. Bacteriol. **108**, 986–991 (1971)
75. Sucrow, W.: Inhaltsstoffe von *Momordica charantia* L. $\Delta^{5,25}$-Stigmastadienol-(3β) und sein β-D-Glucosid. Chem. Ber. **99**, 2765–2777 (1966)
76. Thompson, C., Henson, R.C., Gueldner, R.C., Hedin, P.A.: Sterol galactosides and sterol esters of the cotton bud. Lipids **5**, 283–284 (1970)
77. Tin-Wa, M., Farnsworth, N.R., Fong, H.H.S., Blomster, R.N., Trojanek, J., Abraham, D.J., Persinos, D.J., Dokosi, O.B.: Biological and phytochemical evaluation of plants. IX. Antitumor activity of *Maytenus senegalensis* (Celastraceae) and a preliminary phytochemical investigation. Lloydia **34**, 79–87 (1971)
78. Torres, J.V., Carbonero, P., García-Olmedo, F.: Endosperm sterol phenotype and germination in wheat. Phytochemistry **15**, 677–680 (1976)
79. Tschesche, R., Hulpke, H.: Zur Biosynthese von Steroid-derivaten im Pflanzenreich. 3. Spirostanol-biogenese aus Cholesterin-glucosid. Z. Naturforsch. **21b**, 494–495 (1966)
80. Tunmann, P., Frank, W.: Steringlucosidfettsäureester in der Frucht von *Cucumis sativus* L. Arch. Pharm. **305**, 469–471 (1972)
81. Welsh, K., Shaw, N., Baddiley, J.: The occurrence of acylated sugar derivatives in the lipids of bacteria. Biochem. J. **107**, 313–314 (1968)
82. Willuhn, G., Köstens, J.: Die quantitative Verteilung der Sterine und Sterinderivate in Organen von *Solanum dulcamara*. Phytochemistry **14**, 2055–2058 (1975)
83. Wojciechowski, Z.: Biosynthesis of sterol glycosides in cell-free preparations from *Calendula officinalis* L. Acta Biochim. Polon. **19**, 43–49 (1972)
84. Wojciechowski, Z.A., Nguyen van Uon: Intracellular localization and some properties of UDPG: sterol glucosyltransferase from *Calendula officinalis*. Acta Biochim. Polon. **22**, 25–37 (1975)
85. Wojciechowski, Z.A., Zimowski, J.: Acyl composition and biosynthesis of acylated steryl glucosides in *Calendula officinalis*. Biochim. Biophys. Acta **398**, 111–117 (1975)
86. Wojciechowski, Z.A.: Changes in UDPG: sterol glucosyltransferase activity in *Calendula officinalis*. Phytochemistry **13**, 2091–2094 (1974)

CHAPTER 10

The Biosynthesis and Biological Significance of Prenols and Their Phosphorylated Derivatives

F.W. HEMMING

A. Biosynthesis

I. General

The biosynthesis of all prenyl lipids involves the intermediary formation of one or more prenyl pyrophosphates. The basic principles of the pathways were first established by the studies of Bloch, Cornforth, Lynen, Popjak, and others on squalene biosynthesis in preparations of mammalian liver and of yeast. In higher plants the general skeleton of the pathways appears to be much the same, but more detailed aspects are complicated by a large range of specialised products that are often restricted to a few species. Elucidation of biosynthetic pathways in plants has also frequently been beset by technical difficulties peculiar to this source of tissue. Several reviews relating to earlier aspects of isoprenoid biosynthesis in general and including plant work are recommended to the reader [14, 20, 59]. This review summarises briefly this earlier material with reference to some recent relevant original papers and then deals with a few special aspects of the subject matter.

IIa. Formation of Isopentenyl Pyrophosphate from Acetyl-CoA

The route from acetyl-CoA via 3-hydroxy, 3-methylglutaryl-CoA (HMG-CoA) and mevalonate (MVA) has now been demonstrated as the major pathway for formation of isopentenyl pyrophosphate (IPP) in several higher plants (Fig. 1) [37, 39]. The enzymes for most of the steps have been isolated and some have been purified and studied in detail. Recent reports include studies on HMG-CoA reductase (which catalyses step 4, Fig. 1) of *Pisum sativum* seedlings [16, 17, 18], which closely resembles the rat liver enzyme. It appears that the conversion of acetyl-CoA to MVA is catalysed by particulate cell fractions (microsomal) whereas the subsequent formation of IPP involves only soluble enzymes (see Fig. 1).

Alternative pathways have been considered, including the use of malonyl-CoA instead of acetyl-CoA. Indeed, Fimognari and Rodwell [28] suggested that the malonyl-CoA pathway may be used for non-steroid isoprenoids. However, the work of Higgins and Kekwick [44] showed that in animal systems malonyl-CoA is almost certainly decarboxylated to acetyl-CoA and then incorporated into polyisoprenoids, according to Figure 1. A similar situation probably holds in plant systems also.

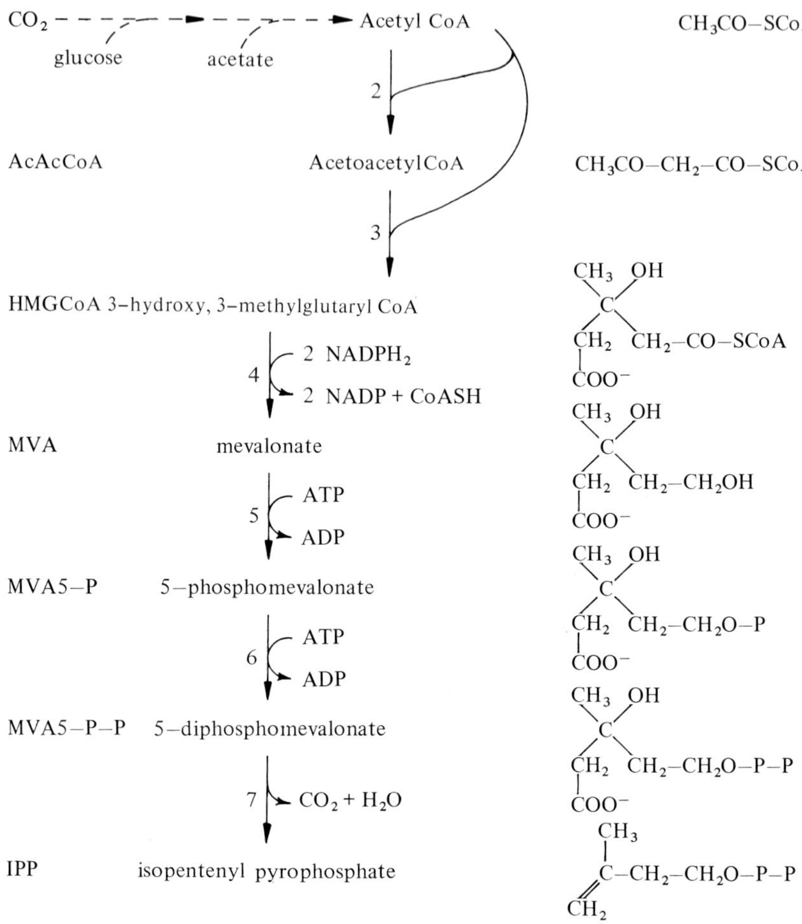

Fig. 1. The biosynthesis of isopentenyl pyrophosphate

IIb. Subcellular Distribution of the Pathway from Acetyl-CoA to Isopentenyl Pyrophosphate

The multiplicity of prenyl lipid products and of their subcellular location has raised the possibility of compartmentalization of the corresponding synthetase systems. Several years ago the relatively low incorporation of [2-^{14}C] MVA into chloroplastidic terpenoids, compared with extraplastidic terpenoids, when administered via the stem to whole leaves and the reverse situation when $^{14}CO_2$ was administered was clear evidence that terpenoid synthetase systems exist both inside and outside the chloroplast [31]. The presence of enzymic activity for a number of the early steps (Fig. 1) in preparations of isolated chloroplasts supported this conclusion. Recently, Brooker and Russell [18] have reported that in green leaves of *P. sativum* seedlings 16% of the cell's HMG-CoA reductase activity (step 4, Fig. 1) is located in the chloroplast with around 10% being present in mitochondria and the rest (74%) associated with microsomal membranes. The separation of the sub-cellular particles was carried out by discontinuous density gradient centrifugation monitored with marker enzymes.

Because of the considerable technical difficulties involved in isolating chloroplasts with negligible contamination by other plant cell fractions, careful interpretation of the results of distribution studies of enzymic activities is required. A detailed study [45, 61] of MVA kinase (which catalyses step 5, Fig. 1) isolated from chloroplasts and from extraplastidic preparations indicated that one was an isoenzyme of the other – differing in pH optimum. This evidence suggested that the enzymic activity recovered in chloroplasts was not due to contamination. However, more recent reports [33] have thrown some doubt on the existence of isoenzymic forms of this enzyme in plant leaves. Nevertheless the balance of evidence favours, early parts of the isoprenoid synthetase systems existing in chloroplasts, and possibly also in mitochondria, as well as in endoplasmic reticulum plus cytoplasm.

IIIa. Monoterpenes—Formation

Monoterpenes are formed (Fig. 2) from diprenyl pyrophosphates which are derived by condensation of IPP with dimethylallyl pyrophosphate (DMAPP), the latter being formed by a stereospecific isomerization of IPP (Section A.VII.1.). DMAPP gives rise to the α-isoprene residue only. In most plant tissues the condensation produces geranyl pyrophosphate (GPP), the trans isomer, which is then used as an acceptor for further isoprene residues (Section A.IV.) or, as a result of phosphatase activity, gives rise to geraniol. The glycosylation of geraniol has been described in a number of tissues, and geranyl β-glucoside is a major terpenoid product in petals of mature rose flowers [30].

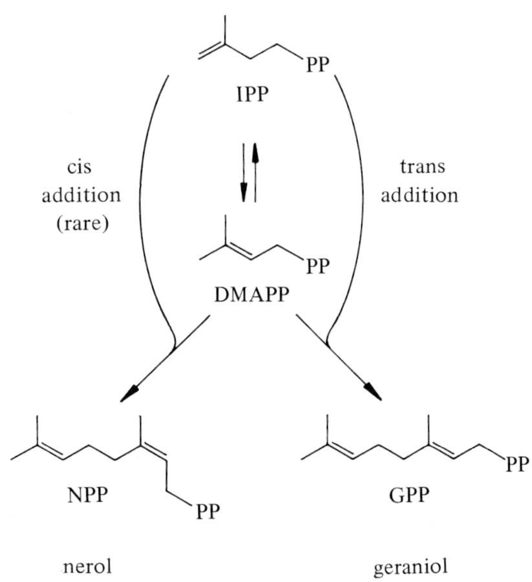

Fig. 2. Formation of dimethylallyl pyrophosphate, geranyl pyrophosphate, and neryl pyrophosphate from isopentenyl pyrophosphate

Nerol, the cis isomer of geraniol, is usually formed from geraniol by an indirect 'isomerization' (Section A.VII.1.). A less common source of the compound is neryl pyrophosphate (NPP), formed by cis addition of IPP to DMAPP [9, 13]. The formation of cyclic monoterpenes is dependent upon the neryl skeleton because the stereochemistry of the trans residue in geraniol does not allow cyclization [10].

IIIb. Different Pools of Soluble Precursors

The rate of biosynthesis of monoterpenes is slow relative to that of sterols and other polyterpenoids in the same tissue. This led Benthorpe et al. [9] to suggest that in *Tanacetum vulgare, Pelargonium graveoleus,* and *Mentha pulegium* there are two soluble pools of monoterpene precursors. They suggested that pool A (Fig. 3) was small and free of precursors whereas pool B was larger and consisted of precursors irreversibly bound to protein. Both pools could be used as a source of GPP. Pool A equilibrated rapidly with exogenous MVA or IPP. The two pools were not separated by membranes. Evidence for these conclusions included the observation that incubation of exogenous [^{14}C] IPP with membrane-free preparations (or of [2-^{14}C] MVA with whole leaves) gave rise to monoterpenes containing in their α-residue (or α-residue-derived parts of cyclic derivatives) far more radioactivity than the ω-residues.

Squalene is labeled symmetrically when formed from exogenous [^{14}C] IPP [8, 24]. If squalene and monoterpenes are being formed at the same time, the two synthetase systems must be kept separate—a situation that obviates the need to propose a pool B in the squalene synthetase system.

The situation is a good example of the complications that the existence of different pools of precursors may bring into interpretations of results obtained by adding exogenous intermediates as potential precursors of terpenoids.

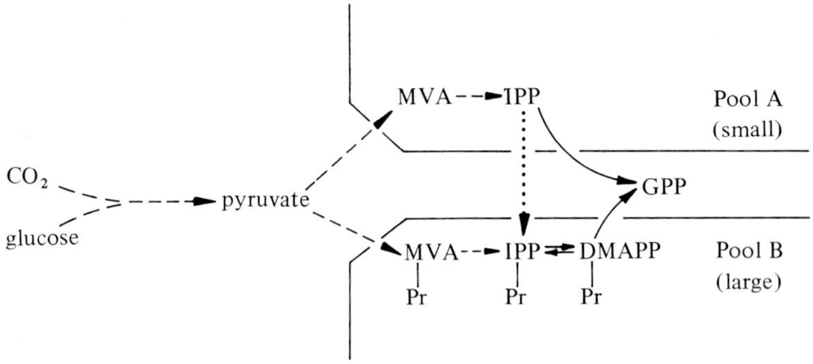

Fig. 3. Two pools of precursors for monoterpene biosynthesis. *Pr:* protein

IV. Sesquiterpenes—Formation

The formation of most sesquiterpenes involves the trans addition of IPP to GPP to form 2-trans-6-trans-farnesyl pyrophosphate (ttFPP), although the addition of IPP to NPP to form both 2-trans-6-cis-farnesyl pyrophosphate (tcFPP) and 2-cis-6-cis-farnesyl pyrophosphate (ccFPP) and hence gossypol has been demonstrated in preparations of cotton roots (Fig. 4; Ref. [1]). The incorporation of MVA into limonene has been reported to occur via linaloyl pyrophosphate (tertiary alcoholic isomer of ttFPP or ctFPP) in citrus fruits [58]. Free 2-trans-6-trans-farnesol (ttF) can be converted in several tissues by a reversible oxido-reducing isomerase to 2-cis-6-trans-farnesol (ctF) via 6-trans-farnesaldehyde (see Section A.VII.1. [19, 27, 56]). This isomerism appears to be a prerequisite for the formation of many cyclic sesquiterpenes.

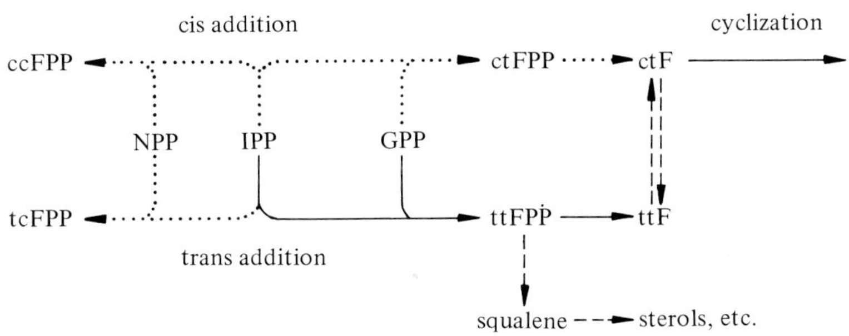

Fig. 4. Formation of sesquiterpenes by condensation of IPP with geranyl and neryl pyrophosphates

It is not yet certain whether caryophyllene (a cyclic sesquiterpene) is formed via ctFPP (formed by cis addition of IPP to GPP) or via ctF which is formed from ttFPP and ttF. However, the work of Croteau and Loomis [23, 24] indicated that it is formed in peppermint leaves from a pool of intermediates that does not equilibrate with the pool giving rise to squalene. This view is based on the observation that IPP is incorporated symmetrically into squalene but asymmetrically into caryophyllene (see also Section A.III.2.).

As in animal tissues the major fate of ttFPP is a head-to-head reductive condensation followed by cyclization to form sterols and other terpenoids.

V. Diterpenes—Formation

Diterpenes are more numerous and more important in higher plants than in animals, and chain lengthening of ttFPP by addition of IPP is an essential process (Fig. 5). Trans addition gives rise to geranylgeranyl pyrophosphate (GGPP) while cis addition forms geranyl neryl pyrophosphate (GNPP). GGPP is presumably used as the source of the diterpenoid side chains of tocopherols, vitamin K, and of the phytol of chlorophyll [67]. A non-reductive head-to-head condensation gives rise to precursors of the tetraprenoid carotenoids [48]. In some tissues, notably *Echinocystis macrocarpus,* the cyclization of GGPP to form kaurene and hence gibberellins has been described [68].

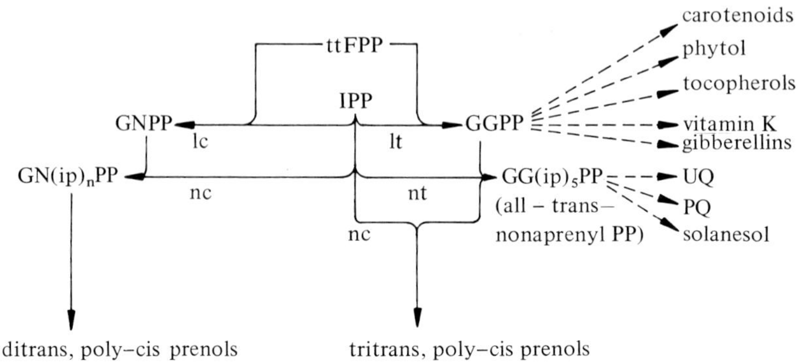

Fig. 5. Formation of di- and polyterpenes by condensation of IPP with geranylgeranyl and geranyl neryl pyrophosphates. *t:* trans additions of IPP; *c:* cis additions of IPP

VI. Polyterpenes—Formation

Chain lengthening of GGPP and of GNPP results in the formation of polyisoprenoid alcohol pyrophosphates (polyprenyl pyrophosphates) containing long hydrophobic chains. In all tissues trans addition of IPP to GGPP is the most likely route to solanesyl pyrophosphate (all-trans-nonaprenyl pyrophosphate), which becomes the side chain of plastoquinone (PQ_9) in chloroplasts or ubiquinones (UQ_9) in mitochondria (Fig. 5). In some plant tissues all-trans-decaprenyl pyrophosphate is the preferred precursor of the ubiquinone side chain, in which case UQ_{10} is the final product. Solanesol (all-trans-nonaprenol) and spadicol (all-trans-decaprenol) accumulate in some tissues [43, 65]. Presumably this reflects the level of phosphatase activity as well as synthetase activity of the tissue.

Repeated cis addition of IPP to GNPP is the means of synthesis of the ditrans-poly-cis prenols (Fig. 5) described in silver birch wood [32]. The mechanism of the synthesis allows flexibility in the size of the final product such that a family of ditrans-, poly-cis prenols is formed in which the number of *cis* residues varies between 3 and 6 [52, 71]. The major part of the mixture of these alcohols is esterified to fatty acids.

Leaves of many plants contain a family of tri*trans-*, poly-*cis* prenols formed by cis addition of IPP to GGPP and in which the number of cis residues varies between 6 and 9 [64, 72].

Mammalian tissues contain as their main polyprenols 2,3 dihydro-, ditrans-, poly-cis prenols (dolichols), which contain between 12 and 18 cis residues. Similar alcohols have been described in plants [43] and more recent evidence suggests that, as in animal systems, dolichyl monophosphate functions in glycosyl transfer reactions (see Section B.I.).

Pinus and *Picea* species contain in their needles high levels (1% dry weight) of long-chain (C_{60}–C_{65}) polyprenols that are predominantly cis [38, 73]. No further information is available regarding the number of cis and trans residues, but it is clear that none of the residues is saturated and, surprisingly, that these polyprenols are present in the needles as acetates.

The longest polyisoprenoid chains are those found in the rubbers. The biosynthesis of poly-cis rubber by *Hevea* latex has been studied extensively [7]. Chain-lengthening occurs by cis addition of IPP to existing rubber particles by a soluble serum fraction of the latex. Whether de novo biosynthesis of poly-cis rubber requires a primer of ttFPP (or GGPP) or simply DMAPP, has not yet been established unequivocally. However, the stimulation of incorporation of ^{14}C-IPP into rubber by DMAPP and NPP, but not by GPP, indicates that the latter is the more likely and that the formation of poly-*cis* rubber is not simply the biosynthesis of polyisoprenoid alcohols that has got out of control. Presumably the biosynthesis of poly-trans rubber (gutta-percha) occurs by extensive trans addition of IPP to GGPP.

VII. Stereochemistry of the Substituted Double Bond in Polyprenols

1. General

In the previous sections it was implied that the presence of cis and trans isoprene residues in polyprenyl pyrophosphates is dependent upon the cis and trans addition, especially, of IPP at the appropriate stage of the chain elongation process. This situation requires that there is no subsequent isomerization of cis to trans or trans to cis isoprene residues.

In the condensation of isopentenyl pyrophosphate with a prenyl pyrophosphate precursor, only one of the hydrogens on carbon 2 of the IPP is retained by the new prenyl pyrophosphate, the other being lost to the medium as a proton (Fig. 6). Similarly the product of IPP-isomerase activity on IPP, namely DMAPP, retains only one of the two hydrogens on C2 of the IPP (Fig. 6). The hydrogen atoms on carbon 2 of IPP originate from position 4 of MVA without any stereochemical change.

In the formation of trans-isoprene residues the 4 pro-R H atom of MVA is retained and the 4 pro-S lost, and vice versa, in the formation of cis-isoprene residues [6, 21].

This approach is particularly useful, for example, in studying the biosynthesis of saturated isoprene residues. It is clear for example, that the saturated residues of phytol are biogenetically trans, indicating that phytol is probably synthesized via geranylgeranyl pyrophosphate [32, 41].

2. Stereochemistry of the α-Isoprene Residue of Mono- and Sesquiterpenes

It was mentioned earlier in this review (Sections A.III. and A.IV.) that the cyclization of mono- and sesquiterpenoid alcohols requires the α-isoprene residue to have the cis configuration. In the light of the foregoing conclusions, it was noted with some initial surprise that in a number of systems nerol and 2-cis, 6-trans-farnesol incorporated tritium into the α-residue from 4R-MVA but not from 4S-MVA [11, 47]. It is now established that an isomerisation occurs via aldehyde intermediates [11, 19, 56] as indicated in Figure 7. The isomerization of farnesol has now been demonstrated in cell-free preparations from seedlings

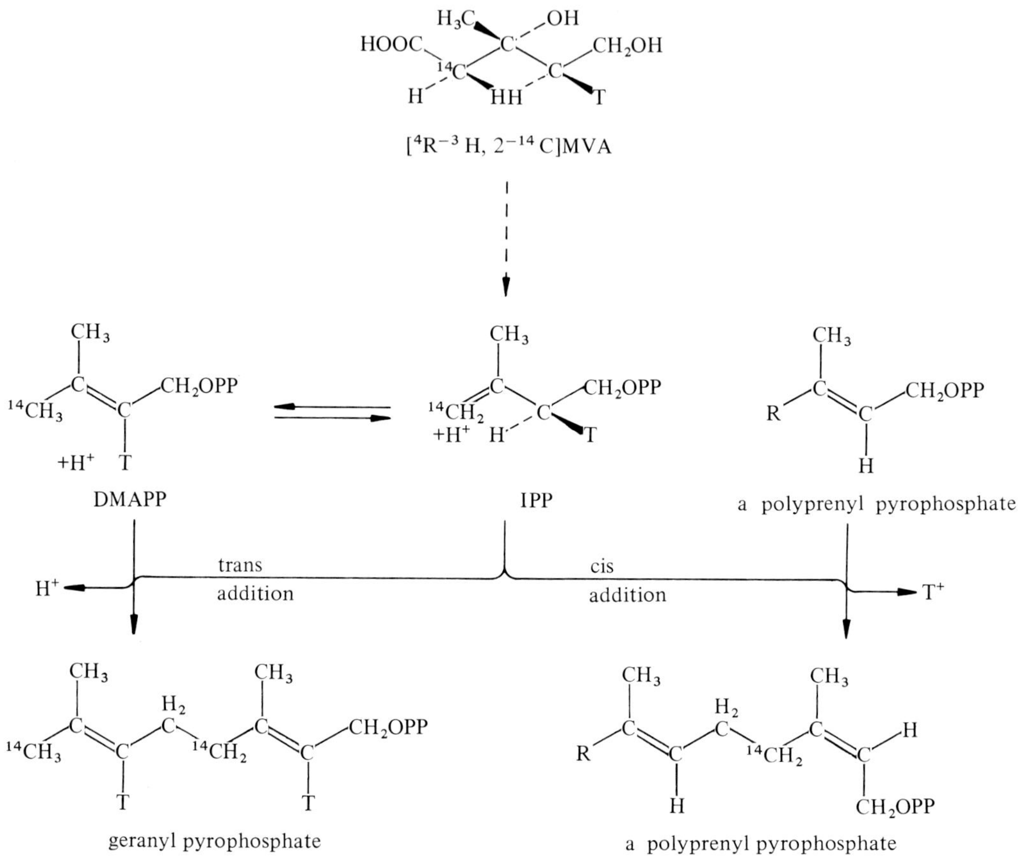

Fig. 6. The fate of the tritium of 4R-MVA during the formation of isoprene residues

of *Pinus radiata* and *Citrus sinensis* [19], from cultures of *Andrographis paniculata* [56], from *Trichothecium roseum* [27], and in *Helminthosporum sativum* [66]. Cofactor requirements (NADPH, NADH) suggest a redox reaction, as shown in Figure 7. In each case it was shown that one of the hydrogens on C-1 of farnesol was lost during the isomerization, and in the work with *A. paniculata,* that this was the pro-1R hydrogen of trans, trans-farnesol but the pro-1S hydrogen of cis, trans-farnesol. This is in fact different from the action of liver alcohol dehydrogenase on farnesol, for this removes the pro-IR hydrogen from both forms of farnesol [56]. Whether or not the isomerization in the plant system uses a different enzyme to convert trans, trans-farnesol to the aldehyde from that acting on cis, trans-farnesol has not yet been settled.

The isomerization of geraniol to nerol according to Figure 7 has been demonstrated in *Pelargonium* leaves [10, 11] and in rose petals [10, 11, 25, 26].

This isomerization of α-isoprene residue has been described only at the mono- and sesquiterpene level and not at the polyprenyl level. Also it clearly can only occur with the unphosphorylated prenols. It thus seems unlikely that

Fig. 7. The isomerization of geraniol and farnesol via their respective aldehydes

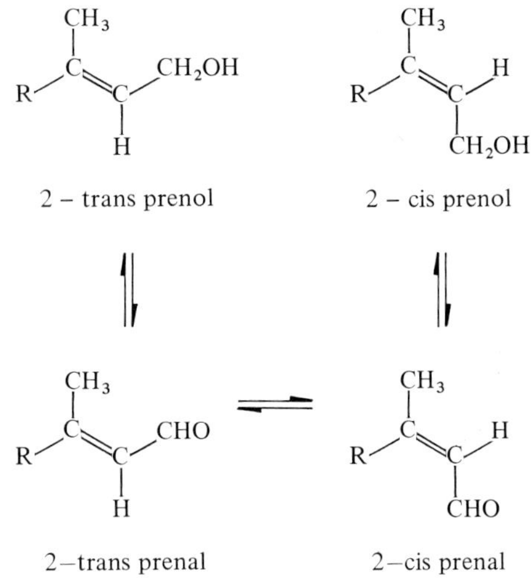

this phenomenon will complicate studies on the biosynthesis of polyisoprenoid chains.

VIII. Phosphorylation of Prenols

Although the major source of prenyl pyrophosphates is undoubtedly the condensation of IPP with a shorter prenyl pyrophosphate catalysed by prenyl transferase, it should not be forgotten that several demonstrations of the phosphorylation of prenols by plant systems are described in the literature. Geraniol kinase activity is present in several plant tissues [4, 5, 53, 54] and orange-juice vesicles catalyse the pyrophosphorylation of linalool (isogeraniol, a tertiary alcohol) [58]. In addition the incorporation of radioactive geranylgeraniol into carotenoids by maize, presumably via GGPP, has been reported by Costes [22], and Gough and Hemming [32] observed in *Betula verrucosa* the incorporation of radioactive geranyl nerol into ditrans, poly-*cis* prenols, presumably via GNPP. The degree of phosphorylation in each cases was small but could become significant in special circumstances.

IX. Control of the Rate of Biosynthesis

1. General

While it is possible that some mono- and sesquiterpene derivatives may be formed in specialised cells, there is little doubt that many of the end products mentioned in Figures 5 and 6 are formed within the same cell. This complexity immediately raises the question of how the situation is controlled.

2. Compartmentation of Early Steps

Compartmentation of synthetase systems is obviously relevant here, although it is clear that this does not simplify the situation very much. For example it has been known for some time that mammalian mitochondria are capable of forming ubiquinone from IPP and appropriate aromatic precursors [63], and it appears that this also takes place in plant mitochondria (Threlfall, private communication, 1976). On the other hand sterols are formed primarily from IPP by enzymes of the cytoplasm and endoplasmic reticulum. The problem then centres around what dictates whether or not IPP, formed by enzymes of the endoplasmic reticulum and cytoplasm, is used for mitochondrial or cytoplasmic biosynthesis. What are the rate-limiting steps and are they subject to control mechanisms?

Within the chloroplast the situation is obviously complex, for here are located carotenoids, phytol, α-tocopherol, and vitamin K_1—apparently all formed via GGPP by fairly direct routes. In addition plastoquinone, and solanesol, formed by involving trans additions of IPP to GGPP are also present, in the chloroplast, as are the tritrans, poly-cis prenols formed by cis addition of IPP to GGPP. Whether or not all of these compounds are formed from the same pool of GGPP is not known. Neither is it understood what dictates the amount of GGPP being used by any synthetase system and how this may vary in response to different physiological situations. There is also a similar dearth of evidence relating to the control of earlier stages of isoprenoid biosynthesis.

3. Key Enzymes?

In mammalian hepatic synthesis of cholesterol, there is a great deal of evidence in favour of HMG-CoA reductase (Step 4, Fig. 1) being a key enzyme in the control of the process. As yet there is no evidence for a similar situation in plants. Alternative enzymes have been suggested. Perhaps the enzyme favoured most for control in plants is DMAPP:IPP prenyl transferase. This enzyme has been studied in *P. sativum* [34, 35], and results suggested that it may well be the rate-limiting enzyme between MVA and squalene.

Another enzyme that may be important in this respect is HMG-CoA synthetase (Step 3, Fig. 1). When sweet potato root tissue is infected by the fungus *Ceratocystis fimbriata,* to produce eventually black rot, large quantities of terpenoids such as ipomearone are produced, and an early event is a sixfold increase in the activity of HMG-CoA synthetase [55].

4. Hormonal

No firm effects of plant hormones on enzyme activities have been published.

5. Light Effects

Light can also be an important parameter. In the chloroplast thylakoids, 25–30% of the lipid present is isoprenoid in character, this portion being made up primarily of chlorophyll, plastoquinone, vitamin K_1, α-tocopherol (and α-tocophenyl quinone), and carotenoids. In the process of stimulating thylakoid development in etiolated leaves, far-red light stimulates a steady increase of C_{20} (mainly of vitamin K_1), C_{40}, and C_{45} isoprenoid chains such that the proportions of these change little from those in etiolated leaves, namely, 40% C_{20}, 25% C_{40}, 35% C_{45} by weight (in *Raphanus sativus*, Lichtenthaler and Becker [51]). White light stimulates a much larger increase in isoprenoids, including a disproportionate rise in chlorophyll, with the result that C_{20} side chains account for 65% of the total isoprenoid chains. This and other evidence led these authors to suggest that far-red light may stimulate, through phytochrome (P_{fr}), all steps between acetyl-CoA and solanesyl pyrophosphate but that it has no effect on the phytylation of α-tocopherol and chlorophyll. They suggested further that the chlorophyll phytylation enzyme is induced by stimulation of P_{fr} but requires light of shorter wavelength for its activation. Since vitamin K_1 did rise it appeared probable that phytyl precursors were in adequate supply. A similar situation has been described in *Hordeum vulgare* except that the phytylation of α-tocopherol as well as that of vitamin K_1 appears to be stimulated by far-red light [50].

Light has also been reported to increase the rate of sterol biosynthesis [12, 36].

It is not at all clear which biosynthetic steps are accelerated under the influence of light. In a negative sense it is of interest that the effect of red or white light on etiolated *P. sativum* is to reduce the activity, without any lag, of HMGCoA reductase (step 4, Fig. 1 [16]), although the same authors report that the inhibition is partly reversed by far-red light.

6. *Compartmentation of Polyisoprenoid Biosynthesis*

It was mentioned earlier (Section A.IX.1) that in a biosynthetic pathway in which precursors and final products are separated by membranes the permeability of the membrane to intermediates is a potentially important factor in controlling the rate of synthesis. Using exogenous substrates, the relative impermeability of the chloroplast membrane to MVA was demonstrated several years ago [31, 62]. However, it is still not understood to what extent this is a problem in the plant with endogenous precursors. Recent results [46] suggest that in *Calendula officinalis* the C_{45} side chain of plastoquinone is formed in the cytoplasm and transferred to the plastid for condensation with the aromatic precursor of the plastoquinone nucleus.

B. Function of Isoprenoid Alcohols and Their Phosphates

I. A Precursor Role

It will be clear from Section A that a major function of the pyrophosphates of prenols is to act as intermediates in the formation of other terpenoid compounds. A feature of those products containing four or more isoprene residues is their localization in membranes (or in lipid globules). Almost certainly the hydrophobic nature of the polyisoprenoid chains is an important factor in this localization. It is not known if the correct localization of PQ and UQ is dependent upon a C_{45} side chain or if once the chain has reached a certain minimal length, and hence hydrophobicity, any further modest increase in size is inconsequential. Likewise the significance of the stereochemistry is not clear. It is conceivable that the evolutionary pressure resulting always in all-trans side chains in PQ and UQ may not be related to optimising their function but rather to optimising control of their synthesis independently of the synthesis of prenyl pyrophosphates of mixed stereochemistry, which appear to have a very different function (see Section B.I.).

II. A Role in Glycosylation

It has been established for a number of years now that the monophosphate of ditrans, octacis-prenol (undecaprenyl monophosphates) is an essential component of the bacterial plasma membrane. In this location it acts as a membrane-bound lipid coenzyme in glycosyl transfer—accepting sugars from nucleotide diphosphate sugars at the cytoplasmic side of the membrane and presenting them to the membrane-bound glycosyl transferase catalysing their transfer to one of a number of cell wall glycan polymers on the other side (Fig. 8). Sometimes there is assembly of a small glycan on the undecaprenyl monophosphate before this is transferred en bloc to the polymer being constructed in the wall. An essential feature of the 'intermediates' is that the transfer potential of the sugar-l-phosphate bond seen in nucleoside diphosphate sugars is also present in polyprenyl monophosphate or diphosphate sugars. This area has been reviewed several times [42, 49].

It is ironic that although the tritrans, poly-cis prenols of plant leaves were some of the first polyprenols to be characterised, a function for them still has not been established. This is not to say that the transfer of sugar to a membrane lipid and its subsequent transfer to glycoprotein has not been described in plants. Indeed this was shown to be the case for the transfer of mannose in preparations from *Phaseolus aureus* by Villemez's group [69, 70] and by Hemming's group [2, 3], but it was not possible to establish unequivocally the nature of the lipid. The transfer of N-acetyl glucosamine from UDP GlcNAc to an unidentified lipid and to protein in the same plant system has also been reported by Roberts and Pollard [60]. In fact if this is a prenyl phosphate-

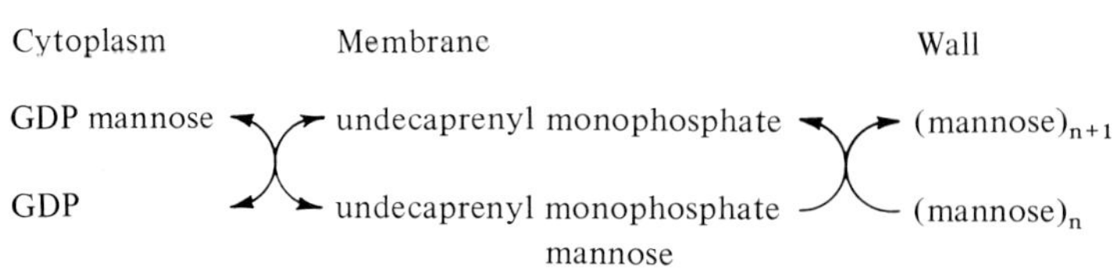

Fig. 8. Mannosylation of mannan in *M. lysodeikticus* according to Lennarz [49]

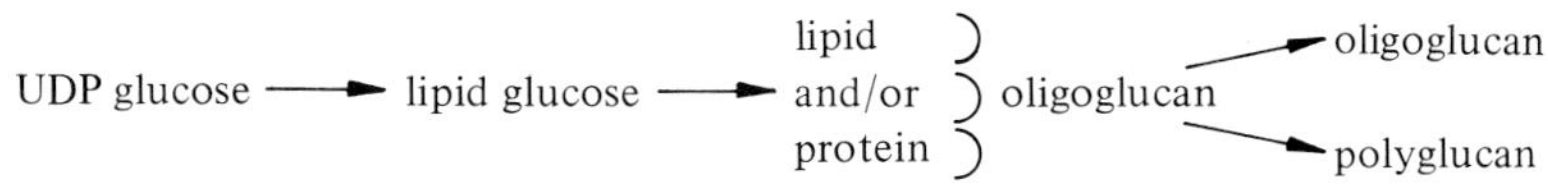

Fig. 9. The transfer of glucose from UDP glucose to glucans in particulate preparations of *P. sativum* roots according to Brett and Northcote [15]

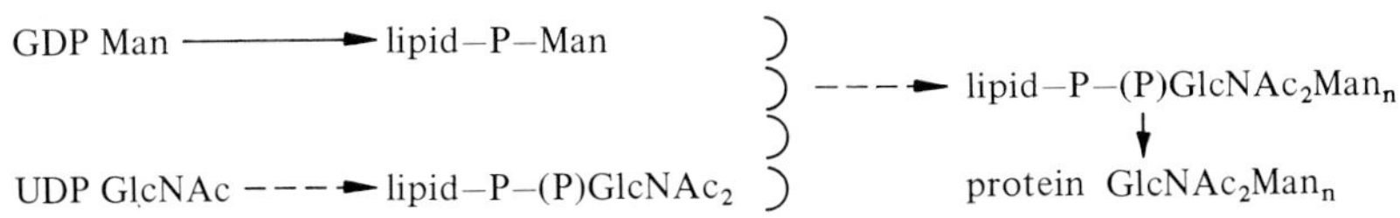

Fig. 10. The transfer of mannose and N-acetylglucosamine to glycoprotein in particulate preparations of cotton bolls according to Forsee and Elbein [29]

mediated phenomenon, it is more likely to involve dolichols or the ditrans, poly-cis prenols than the tritrans, poly-cis prenols, because the plant system used was one of germinating seedlings of *P. aureus* whereas the tritrans, poly-cis prenols are usually associated with photosynthetic tissue.

More recently a role in glucosyl transfer to oligoglucans and polyglucans (linked β1–3 and β1–4) by dictyosome preparations of pea roots has been reported ([15]; Fig. 9). It is suggested that control of synthesis of β1–3 or β1–4 glucans may occur at the lipid glucose level. Also, particulate preparations of cotton-boll fibres have been reported to catalyse the transfer of mannose and N-acetyl glucosamine to a lipid and hence to protein according to the scheme illustrated in Figure 10 [29].

In most of these reports the hydrolytic and chromatographic properties of the lipid-linked sugars are consistent with their being prenyl phosphate derivatives. Pont-Lezica et al. [57] have shown that soybean, wheat germ, and pea seedlings produce a lipid that will function as an acceptor in glucosyl transfer. The properties (chromatographic, hydrolytic, and hydrogenolytic) of the glucolipid so formed indicate that it is unlikely to be a phosphate derivative of a polyprenol containing a normal unsaturated α-residue. These authors point out that if the lipid is a polyprenol it is likely to be of the dolichol type (i.e., a very long chain with a saturated α-residue).

References

1. Adams, S.R., Heinstein, P.F.: Evidence for *trans-trans* and *cis-cis* farnesyl pyrophosphate synthesis in *Gossipium hirsutum*. Phytochemistry **12**, 2167–2172 (1973)
2. Alam, S., Hemming, F.W.: Betulaprenol phosphate as an acceptor of mannose from GDP-mannose in *Phaseolus aureus* preparations. FEBS Lett. **19**, 60–62 (1971)
3. Alam, S., Hemming, F.W.: Polyprenol phosphates and mannosyl transferases in *Phaseolus aureus*. Phytochemistry **12**, 1641–1649 (1973)
4. Allen, R.T. van, Nes, W.R.: The phosphorylation of geraniol in germinating peas. Phytochemistry **5**, 85–88 (1966)
5. Attaway, J.A., Pieringer, A.P., Barabas, L.T.: The origin of citrus flavor components—III.

A study of the percentage variations in peel and leaf oil terpenes during one season. Phytochemistry **6**, 25–32 (1967)

6. Archer, B.L., Barnard, D., Cockbain, E.G., Cornforth, J.W., Cornforth, R.H., Popjak, G.: The stereospecificity of the biosynthesis of rubber by the latex of *Hevea brasiliensis*. Proc. Roy. Soc. (London) Ser. B. **163**, 519–523 (1966)
7. Audley, B.G.: Biosynthesis of rubber. Adv. Enzymol. **29**, 221–257 (1967)
8. Armarego, W.L.F., Goad, L.J., Goodwin, T.W.: Biosynthesis of α-spinasterol from [2-^{14}C, (4R)-4-$^{3}H_1$] mevalonic acid by *Spinacea oleracea* and *Medicago satira*. Phytochemistry **12**, 2181–2187 (1973)
9. Banthorpe, D.V., Bucknall, G.A., Doonan, H.J., Doonan, S., Rowan, M.G.: The biosynthesis of geraniol and nerol in cell free extracts of *Tanacetum vulgare*. Phytochemistry **15**, 91–100 (1976)
10. Banthorpe, D.V., Charlwood, B.V., Francis, M.J.O.: The biosynthesis of monoterpenes. Chem. Revs. **72**, 115–155 (1972a)
11. Banthorpe, D.V., Le Patourel, G.N.J., Francis, M.J.O.: Biosynthesis of geraniol and nerol and their β-glucosides in *Pelargonium graveolus* and *Rosa dilecta*. Biochem. J. **130**, 1045–1054 (1972b)
12. Benveniste, P., Hartmann, M.A., Durst, F.: Biosynthesis of sterols in Jerusalem Artichoke tuber tissue. Phytochemistry **11**, 3003–3005 (1972)
13. Beytia, E., Valenzuela, P., Cori, O.: Terpene biosynthesis: formation of nerol, geraniol and other prenols by an enzyme system from *Pinus radiata* seedlings. Arch. Biochem. Biophys. **129**, 345–351 (1969)
14. Bloch, K.: The biological synthesis of cholesterol. Science **150**, 19–23 (1965)
15. Brett, C.T., Northcote, D.H.: The formation of oligoglucans linked to a lipid during synthesis of a β-glucan by characterised membrane fractions isolated from peas. Biochem. J. **148**, 107–117 (1975)
16. Brooker, J.D., Russell, D.W.: Mechanisms of regulation of plant growth. Proc. Roy. Soc. New Zealand 365–370 (1974)
17. Brooker, J.D., Russell, D.W.: Properties of a microsomal 3-hydroxy, 3-methyl glutaryl CoA reductase from *Pisum sativum* seedlings. Arch. Biochem. Biophys. **167**, 723–729 (1975a)
18. Brooker, J.D., Russell, D.W.: Subcellular localization of 3-hydroxy, 3-methyl glutaryl CoA reductase in *Pisum sativum* seedlings. Arch. Biochem. Biophys. **167**, 730–737 (1975b)
19. Chayet, L., Pont Lezica, R., George-Nascimento, C., Cori, O.: Biosynthesis of sesquiterpene alcohols and aldehydes by cell-free extracts from orange flavedo. Phytochemistry **12**, 95–101 (1973)
20. Clayton, R.B.: Biosynthesis of sterols, steroids and terpenoids. Part I. Biogenesis of cholesterol and the fundamental steps in terpenoid biosynthesis. Quart. Rev. **19**, 168–200 (1965)
21. Cornforth, J.W., Cornforth, R.H., Donninger, C., Popjak, G.: Studies on the biosynthesis of cholesterol. XIX. Steric course of hydrogen eliminations and of C-C bond formations in squalene biosynthesis. Proc. Roy. Soc. B. **163**, 492–514 (1966)
22. Costes, C.: Biosynthese du phytol des chlorophylles et du squelette tetraterpenique des carotenoides dans les feuilles vertes. Phytochem. **5**, 311–324 (1966)
23. Croteau, R., Loomis, W.D.: Biosynthesis of mono- and sesquiterpenes in peppermint from mevalonate-2-^{14}C. Phytochemistry **11**, 1055–1064 (1972)
24. Croteau, R., Loomis, W.D.: Biosynthesis of squalene and other triterpenes in *Mentha piperita* from mevalonate-2-^{14}C. Phytochemistry **12**, 1957–1965 (1973)
25. Dunphy, P.J.: Biosynthesis if monoterpenes in rose petals. Phytochemistry **12**, 1512 (1973)
26. Dunphy, P.J., Allcock, C.: Isolation and properties of a monoterpene reductase from rose petals. Phytochemistry **11**, 1887–1891 (1972)
27. Evans, R., Holton, A.M., Hanson, J.R.: Biosynthesis of 2-*cis*-farnesol. J. Chem. Soc. Commun. 465 (1973)
28. Fimognari, G.M., Rodwell, V.W.: Mevalonate biosynthesis in rat liver. Lipids **5**, 104–108 (1970)
29. Forsee, W.T., Elbein, A.D.: Glycoprotein biosynthesis in plants. Demonstration of lipid-linked oligosaccharides of mannose and N-acetylglucosamine. J. Biol. Chem. **250**, 9283–9293 (1975)
30. Francis, M.J.O., Allcock, C.: Geraniol β-D-glucoside; occurrence and synthesis in rose flowers. Phytochemistry **8**, 1339–1347 (1969)

31. Goodwin, T.W.: The biosynthesis of carotenoids. In: Biosynthetic Pathways in Higher Plants. Pridham, J.B., Swain, T. (eds.). London-New York: Academic Press, 1965, pp. 37–71
32. Gough, D.P., Hemming, F.W.: The stereochemistry of betulaprenol biosynthesis. Biochem. J. **117**, 309–317 (1970)
33. Gray, J.C., Kekwick, R.G.O.: Mevalonate kinase in green leaves and etiolated cotyledons of the French bean *Phaseolus vulgaris*. Biochem. J. **133**, 335–347 (1973)
34. Green, T.R., Baisted, D.J.: Development of the squalene synthesizing system during early stages of pea seed germination. Biochem. J. **125**, 1145–1147 (1971)
35. Green, T.R., Baisted, D.J.: Development of the activities of enzymes of the isoprenoid pathway during early stages of pea-seed germination. Biochem. J. **130**, 983–995 (1972)
36. Grunwald, C.: Sterol changes during germination of *Nicotiana tabacum* seeds. Plant Physiol. **50**, 69–72 (1972)
37. Grunwald, C.: Plant sterols. Ann. Rev. Plant Physiol. **26**, 209–236 (1975)
38. Hannus, K., Pensar, G.: Polyisoprenols in *Pinus sylvestris* needles. Phytochemistry **13**, 2563–2566 (1974)
39. Hanson, J.R.: The biosynthesis of C_5–C_{20} terpenoid compounds. Chem. Soc. Biosynth. **3**, 1–10 (1975)
40. Hartmann, M.A., Benveniste, P., Durst, F.: Biosynthesis of sterols in Jerusalem artichoke tuber tissue. Phytochemistry **11**, 3003–3005 (1972)
41. Hemming, F.W.: Polyprenols. Biochem. Soc. Symp. **29**, 105–117 (1970)
42. Hemming, F.W.: Lipids in glycan biosynthesis. M.T.P. Internat. Rev. Sci. Biochem. Ser. One **4**, 39–97 (1974)
43. Hemming, F.W., Morton, R.A., Pennock, J.F.: Constituents of the unsaponifiable lipid from the spadix of *Arum muculatum*. Proc. Roy. Soc. (London) Ser. B. **158**, 291–310 (1963)
44. Higgins, M.J., Kekwick, R.G.O.: An investigation into the role of malonyl-coenzyme A in isoprenoid biosynthesis. Biochem. J. **134**, 295–310 (1973)
45. Hill, H.M., Rogers, L.D.: Mevalonate activating enzymes and phosphatases in higher plants. Phytochemistry **13**, 763–777 (1974)
46. Janiszowska, W., Michalski, W., Kasprzyk, Z.: Polyprenyl quinones and α-tocopherol in *Calendula officinalis*. Phytochemistry **15**, 125–127 (1976)
47. Jedlicki, E., Jacob, G., Frini, F., Cori, O.: Stereospecificity of isopentenylpyrophosphate isomerase and prenyl transferase from *Pinus* and *Citrus*. Arch. Biochem. Biophys. **152**, 590–596 (1972)
48. Jungwala, F.B., Porter, J.W.: Biosynthesis of phytoene from isopentenyl and farnesyl pyrophosphates by a partially purified tomato enzyme system. Arch. Biochem. Biophys. **119**, 209–219 (1967)
49. Lennarz, W.J., Scher, M.G.: Metabolism and function of polyisoprenol sugar intermediates in membrane-associated reactions. Biochim. Biophys. Acta **265**, 417–441 (1972)
50. Lichtenthaler, H.K.: Regulation of prenyl chain synthesis in etiolated *Hordeum* seedlings by far-red and white light. Physiol. Plantarum **33**, 241–244 (1975)
51. Lichtenthaler, H.K., Becker, K.: The influence of continuous far-red and white light on prenyl chain synthesis in plastids of *Raphanus* seedlings. Planta **122**, 255–258 (1975)
52. Lindgren, B.O.: Homologous aliphatic C_{30}–C_{45} terpenols in birch wood. Acta Chem. Scand. **19**, 1317–1326 (1965)
53. Loomis, W.D., Battaile, J.: Plant phenolic compounds and the isolation of plant enzymes. Phytochemistry **5**, 423–438 (1966)
54. Madyastha, K.M., Loomis, W.D.: Phosphorylation of geraniol by cell-free enzymes from *Mentha piperita*. Federation Proc. **28**, 665 (1969)
55. Oshina, K., Uritani, I.: Enzymatic synthesis of a β-hydroxy-β-methylglutaric acid-derivative by a cell-free system from sweet potato with black rot. J. Biochem. (Tokyo) **63**, 617–623 (1968)
56. Overton, K.H., Roberts, F.M.: Interconversions of *trans-trans-* and *cis-trans-*farnesol by enzymes from *Andrographis* tissue cultures. Biochem. J. **144**, 585–592 (1974)
57. Pont Lezica, R., Brett, C.T., Martinez, P.R., Dankert, M.A.: A glucose acceptor in plants with the properties of an α-saturated polyprenol monophosphate. Biochem. Biophys. Res. Commun. **66**, 980–987 (1975)
58. Potty, V.H., Breummer, J.H.: The formation of isoprenoid pyrophosphates from mevalonate by orange enzymes. Phytochemistry **9**, 1229–1237 (1970)
59. Pridham, J.B.: Terpenoids in Plants. London-New York: Academic Press 1967

60. Roberts, R.M., Pollard, W.E.: The incorporation of D-glucosamine into glycolipids and glycoproteins of membrane preparations from *Phaseolus aureus* hypocotyls. Plant Physiol. **55**, 431–436 (1975)
61. Rogers, L.J., Shah, S.P.J., Goodwin, T.W.: Intracellular localization of mevalonate-activating enzymes in plant cells. Biochem. J. **99**, 381–388 (1966a)
62. Rogers, L.J., Shah, S.P.J., Goodwin, T.W.: Mevalonate kinase isoenzymes in plant cells. Biochem. J. **100**, 14c–17c (1966b)
63. Rudney, H.: The biosynthesis of terpenoid quinones. Biochem. Soc. Symp. **29**, 89–103 (1970)
64. Sasak, W., Chojnacki, T.: The distribution of polyprenols in sub-tropical plants. Acta Biochem. Polon. **20**, 343–350 (1973)
65. Stevenson, J., Hemming, F.W., Morton, R.A.: The intracellular distribution of solanesol and plastoquinone in green leaves of the tobacco plant. Biochem. J. **88**, 52–56 (1963)
66. Suzuki, Y., Marumo, S.: *Trans*- to -*cis* 2,3-double bond isomerization of epoxyfarnesol and farnesol by fungus. Tetrahedron Lett. 5101–5104 (1972)
67. Threlfall, D.R.: Biosynthesis of terpenoid quinones. In: Terpenoids in Plants. Pridham, J.B. (ed.). London-New York: Academic Press, 1967, pp. 191–222
68. Upper, C.D., West, C.A.: Biosynthesis of gibberelins. II. Enzymic cyclization of geranylfarnesyl pyrophosphate to kaurene. J. Biol. Chem. **242**, 3285–3292 (1967)
69. Villemez, C.L.: Characterization of intermediates in plant cell wall biosynthesis. Biochem. Biophys. Res. Commun. **40**, 636–641 (1970)
70. Villemez, C.L., Clark, A.F.: A particle bound intermediate in the biosynthesis of plant cell wall polysaccharides. Biochem. Biophys. Res. Commun. **36**, 57–63 (1969)
71. Wellburn, A.R., Hemming, F.W.: Polyprenols of wood and leaf tissue of the silver birch, *Betula verrucosa*. Nature (London) **212**, 1364–1366 (1966a)
72. Wellburn, A.R., Hemming, F.W.: The occurrence and seasonal distribution of higher isoprenoid alcohols in the plant kingdom. Phytochemistry **5**, 969–975 (1966b)
73. Zinkel, D.F., Evans, B.B.: Terpenoids of *Pinus strobus* cortex tissue. Phytochemistry **11**, 3387–3389 (1972)

CHAPTER 11

Carotenoids in Higher Plants

B.H. DAVIES

Of nearly 450 naturally occurring carotenoids, only one quarter have been found in higher plants (for list and structures, see Davies, [24]). Higher plant carotenoids were prominent among the carotenoids whose structures were elucidated by the classical chemical approaches of Karrer and of Kuhn [50] and, by 1952, as many as 60 of the 95 carotenoids then known had been isolated from higher plants [38]. This situation resulted not so much from an early intrinsic interest in these particular pigments as from the fact that higher plant tissues were available on a sufficient scale to provide the large amounts of pure carotenoid necessary for structural determinations by chemical methods. Since the advent of small-scale preparative (e.g., thin-layer chromatography) and physico-organic methods (e.g., mass spectrometry) and with the ready availability, through the culture collections, of a wealth of pure strains and mutants of microorganisms, the interests of carotenoid chemists have broadened to include algae, bacteria, and fungi. It is from smaller-scale studies of the pigments of such organisms that many unusual carotenoid structures have emerged in recent years [57]. Nevertheless, the higher plants, because of their fundamental importance to man, are of continuing interest. This is reflected in the many recent studies that have contributed to our current understanding of the distribution, metabolism, and function of higher plant carotenoids.

A. Nomenclature

Because many of the established trivial names of higher plant (and other) carotenoids are related more to their original source (e.g., chrysanthemaxanthin, eschscholtzxanthin) than to their chemical structure, the introduction of a standardized semi-systematic nomenclature [22, 23] is a welcome rationalization. Although nearly 450 carotenoid structures exist in nature, all are based on only 7 different end groups and, of these, only 4 are found in higher plant carotenoids (β, ε, κ and ψ; see Fig. 1).

The end groups at either extremity of the (fully unsaturated all-trans) polyene chain are used to form the C_{40} carotene (hydrocarbon) stem names (e.g., β,β-carotene, β,ε-carotene, ψ,ψ-carotene). These names are modified by appropriate prefixes and suffixes (with locant designation; see Fig. 1) to denote changes in the carbon skeleton (seco-, apo-), hydrogenation level (hydro-, dehydro-, *retro*-), oxygenation (of xanthophylls; epoxy-, -ol, -one, -oic acid), and configuration (cis, trans, *R, S*).

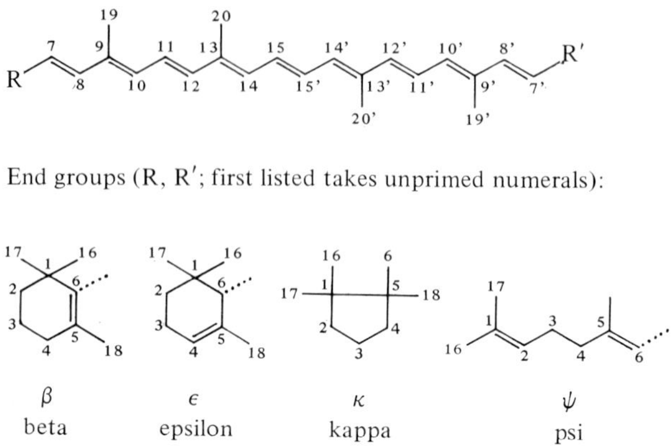

Fig. 1. End groups occurring in higher plant carotenoids and the numbering of the carbons of a carotenoid molecule

These conventions are used, in the Appendix, to define the structures of all the higher plant carotenoids mentioned in this review.

B. Distribution

The structures of the carotenoids occurring in higher plants have been reviewed recently and in detail [42, 43]. Only the general features of their distribution in photosynthetic tissues, fruits, flowers, and other tissues will be considered here.

I. Photosynthetic Tissues

Chloroplasts of actively photosynthetic tissues, apparently irrespective of the type and habitat of the higher plant [81], contain β-carotene and a number of unesterified xanthophylls, lutein, violaxanthin, and neoxanthin. These carotenoids occur in two granal chromoproteins [86] corresponding to photosystems I and II [85]. α-Carotene, β-cryptoxanthin, zeaxanthin, and antheraxanthin may also be present, but in smaller quantities. It is thought that the constancy of the distribution pattern reflects not only the common ancestry of higher plants but also that any mutation that causes a departure from this pattern is lethal [42]. The disintegration of chloroplasts during senescence (see Ref. [79]) is accompanied by esterification of the xanthophylls [40].

II. Fruits

The xanthophylls of ripened fruit, unlike those in the active chloroplast, are esterified. Unripe (green) fruit contain, of course, the same pigments as other photosynthetic tissues. In some ripe fruits, very little carotenoid is present; other ripe fruits contain low concentrations of the chloroplast-type carotenoids. On ripening, and as the chloroplasts become chromoplasts, there is often an increased de novo synthesis of carotenoids [40]. There is also chlorophyll degradation (see [79]), but this may not be complete, and it is not a prerequisite for renewed carotenoid synthesis [97]. The carotenoids formed by de novo synthesis in fruits may be of the following 5 general types [44]:

1. Bicyclic β-carotene and its hydroxylated derivatives, β-cryptoxanthin and zeaxanthin, as in *Physalis alkekengi*.

2. Acyclic lycopene and its acyclic and more saturated precursors, phytoene, phytofluene, ζ-carotene, and neurosporene, as in tomatoes; in some (tangerine) strains of tomato, the more unsaturated of these are in their poly-cis forms (e.g., prolycopene).

3. Virtually genus-specific xanthophylls such as rubixanthin (*Rosa* spp.) or capsanthin (*Capsicum* spp., but also in *Berberis* spp.).

4. Presumed products of the oxidative degradation of C_{40} carotenoids, such as the citrus apocarotenoids, β-apo-8′-carotenal, β-apo-10′-carotenal, and β-citraurin, and the secocarotenoids, β-carotenone, the (α and β) semicarotenones, and triphasiaxanthin.

5. Carotenoid epoxides, often in the furanoid (5,8) form and in high concentrations [39]. Other epoxides, e.g., 1,2-epoxides at acyclic end groups and lycopene 5,6-epoxide, also occur in tomato fruit [9, 15, 17].

III. Flowers

Highly oxidized 5,8-epoxides such as auroxanthin and flavoxanthin are also characteristic of the petals of some carotenoid-forming flowers. Many flower colours are due, of course, to flavonoids rather than to lipid-soluble pigments. When petals do contain carotenoids, however, they are usually of one of a number of distinct types. These are (1) 5,8-epoxides, (2) carotenes, sometimes present in high concentrations (*Mimulus cupreus*, Ref. [45]; *Narcissus poeticus recurvis*, Ref. [11]), (3) ketocarotenoids (*Adonis* spp.; Refs. [33, 77]), and (4) species-specific carotenoids such as gazaniaxanthin (5′-cis rubixanthin; *Gazania rigens*), the *retro* carotenoid eschscholtzxanthin (*Eschscholtzia californica*) or the 8,8′-diapocarotenoic acid, crocetin (*Crocus sativus*). Like fruit carotenoids, petal carotenoids occur in chromoplasts [58] and the xanthophylls are esterified.

IV. Other Tissues

Carotenoids may also occur in other higher plant tissues. The carotenoids of anthers and pollen have been listed [43] and include carotenes and their hydroxy-, epoxy-, and hydroxyepoxy-derivatives. Most seeds contain only traces of carotenoids, so maize is atypical in containing large amounts of β-carotene, β-cryptoxanthin, and zeaxanthin, with some minor components [89]. The seeds of *Bixa orellana* contain bixin, the methyl ester of a 6,6′-diapocarotenoic acid (see [50]).

Few roots have been studied, but the unusual azafrin is the main pigment of *Escobedia scabrifolia* root (see [50]) while the occurrence of β-carotene in the carrot root (*Daucus carota*) is of both commercial and historical significance.

C. Formation

Two general approaches have been used in investigating carotenoid biosynthesis. The first, the kinetic approach, consists of quantitative determinations of each of the carotenoids present in a tissue under different conditions. Analyses may be made (1) at different stages of development, (2) on different strains (mutants), and (3) after growth in the absence and presence, and sometimes after the removal, of an inhibitor. Method (1) is fundamental and can be used with all carotenogenic organisms. The second technique (2) is more readily applicable to microorganisms (see e.g., Ref. [25]) than to higher plants because microbial mutants may be prepared relatively easily; a number of mutants of tomatoes and peppers, however, have been studied in this way. Inhibitors are available that block either carotene desaturation or cyclization; most of these may be used with plant or microbial systems, but others (e.g., diphenylamine, nicotine) are only useful in microbial studies. Biosynthetic pathways may be postulated, but not substantiated, by using this first general approach.

The second approach uses labelling with ^{3}H, ^{14}C, or ^{18}O to prove that individual reactions (or series of reactions) occur. Permeability barriers to added substrates have to be circumvented by preparing cell-free systems, and the water-insolubility and instability of labelled carotene substrates cause problems [26]. Once these are overcome, the use of labelled substrates, especially if labelled stereospecifically, can yield considerable information about reaction sequences and the mechanism of individual reactions [13, 14]. Because of the many techniques available for breaking microbial cells, most carotenogenic microorganisms yield suitable enzyme systems. On the other hand, only two cell-free preparations from higher plants have consistently proved successful. These are a soluble extract of an acetone powder from tomato fruit plastids (referred to subsequently as the tomato system), developed in Madison [49], and the "non-aqueous chloroplast" system developed at Aberystwyth [16, 20] and used so successfully in Liverpool.

Because of these considerations, it is not surprising that many recent biosynthetic studies have used microbial systems. Nevertheless, sufficient metabolic experiments have now been carried out, particularly with the tomato and chloroplast systems, for us to conclude that the general features of carotenoid biosynthesis are essentially the same in higher plants as in algae, fungi, and bacteria. A number of detailed reviews of general carotenoid biosynthesis have appeared recently [13, 14, 25, 26, 28, 41]. The following account will concentrate on experiments utilizing higher plants. Only where the information from such studies is insufficient will reference be made to microbial systems.

I. Early Steps (Up to C_{20})

In carotenoid biosynthesis, as in the formation of other terpenoids (see Hemming, Chapter 10), mevalonic acid (MVA, C_6) is converted via its 5-phosphate and 5-pyrophosphate, and with the loss of C-1 as CO_2, into isopentenyl pyrophosphate (IPP, C_5). Isomerization to dimethylallyl pyrophosphate (DMAPP), the condensation of these two C_5 molecules (IPP and DMAPP) to yield geranyl pyrophosphate (GPP, C_{10}), and subsequent additions of C_5 units (IPP) result in the formation of farnesyl pyrophosphate (FPP, C_{15}; dimerizes to C_{30} squalene via presqualene pyrophosphate (PSPP), [34]) and then geranylgeranyl pyrophosphate (GGPP, C_{20}). The carotenes are products of GGPP dimerization, again with a cyclopropylcarbinyl pyrophosphate, prephytoene pyrophosphate (PPPP) as intermediate [2].

II. Formation of 15-Cis-Phytoene

In squalene formation, the cyclopropylcarbinyl cation formed by rearrangement following the loss of pyrophosphate from PSPP is stabilized by the stereospecific transfer of a hydride ion from NADPH [34]. An analogous reaction for carotene biosynthesis would involve the formation from PPPP of lycopersene, the C_{40} counterpart of squalene. This would then have to lose two hydrogens to yield phytoene, a known precursor of the coloured carotenoids.

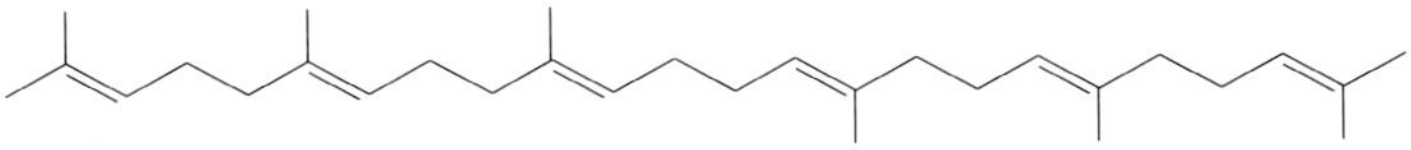

Squalene

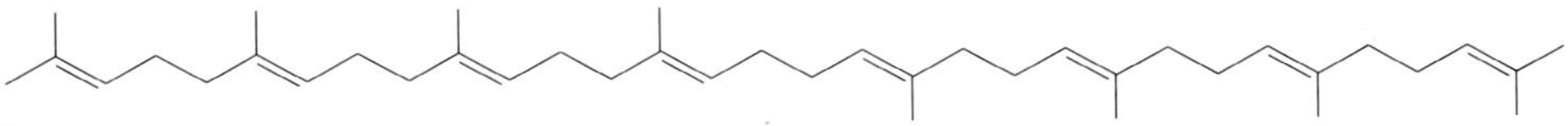

Lycopersene

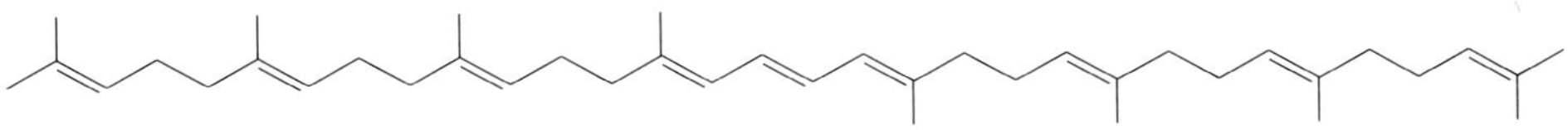

Phytoene

Although there have been a few reports of the presence of lycopersene in carotenogenic organisms (e.g., in some carrot strains; Ref. [66]), most studies failed to reveal it [3, 8, 29, 68]. A particularly rigorous investigation of 2 kg lipid from the green tissues of *Vicia sativa* showed the presence of 42 mg phytoene but no lycopersene, although the detection method was sensitive to 0.05 μg [64]. When carotene production in microorganisms is blocked by mutation or inhibitors, phytoene rather than lycopersene accumulates (see [13]). Higher plants have less potential for similar studies, but an albino maize mutant accumulates

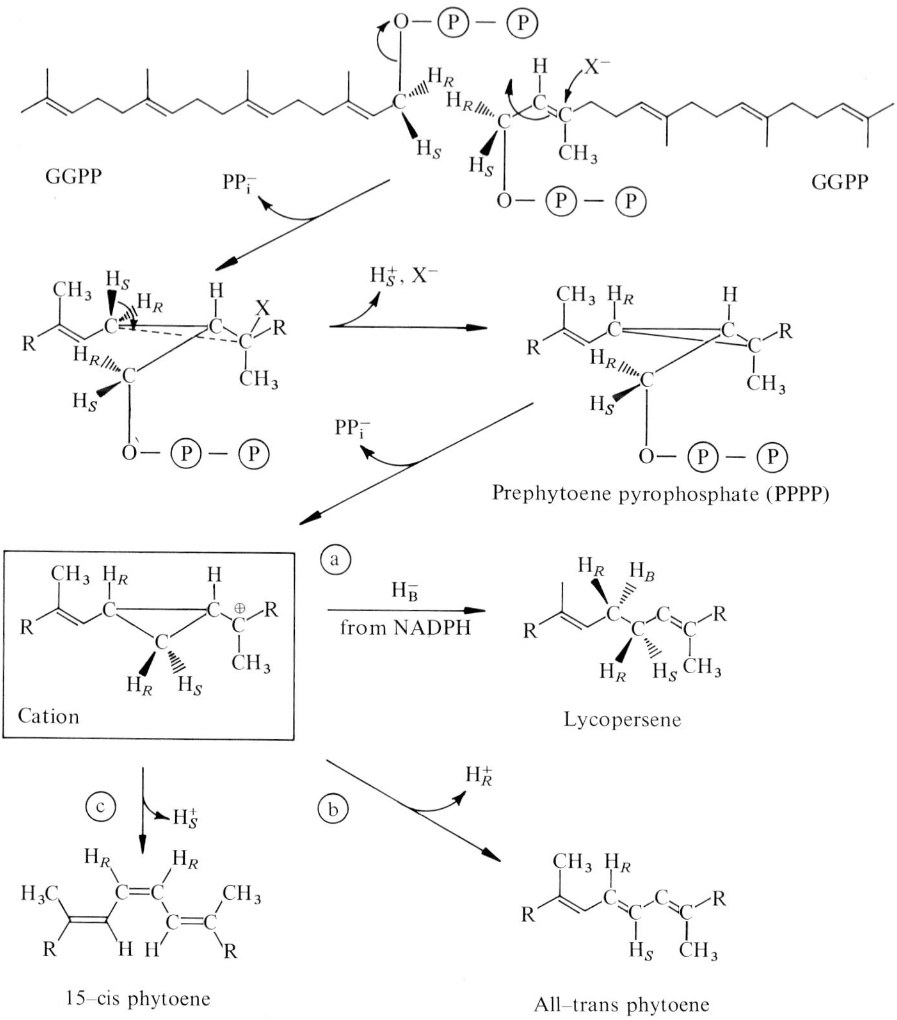

Fig. 2a–c. Formation of prephytoene pyrophosphate from GGPP (2 molecules) and its conversion into (a), lycopersene by a non-specific squalene synthetase; (b), all-trans phytoene as in a *Mycobacterium* system; and (c), 15-cis phytoene as in higher plants and fungi. See text for details

phytoene [4, 87] and the herbicide Sandoz 6706 causes the accumulation of phytoene in wheat seedlings [7, 10].

PPPP is formed from GGPP by purified yeast squalene synthetase [71] and by the tomato system [6] and is converted by both into lycopersene as long as NADPH is present [6, 71]. There is, however, no requirement for NADPH in phytoene formation by chloroplasts [20].

All this evidence suggests that lycopersene is not an obligatory intermediate; its formation from GGPP by yeast squalene synthetase and the tomato system may result from a lack of absolute substrate specificity for FPP by squalene synthetase, presumably a component of the tomato system. Its conversion into phytoene by the latter [6] may also be rationalized in terms of low (dehydrogenase) substrate specificity. Recently, a partially (500-fold) purified "phytoene synthetase complex" from the tomato system has been examined [62]. It converts IPP into phytoene with GGPP as the only demonstrable intermediate; phytoene formation does not require NADPH [63].

A proposal that phytoene could be formed directly from the cyclopropylcarbinyl cation (from PPPP) by proton loss [1] has been substantiated in microbial extracts [46]. Samples of phytoene from higher plants have a predominant 13-trans, 15-cis, 13′-trans configuration [5, 48]. Studies with tomato slices [92] and bean leaf chloroplasts [18] using [2-^{14}C, (5*R*)-5-$^{3}H_1$] MVA and [2-^{14}C, 5-$^{3}H_2$] MVA show that the 1-*pro-S* hydrogen is lost from each GGPP molecule (5-*pro-S* H of MVA) in 15-cis phytoene formation. This has been confirmed in microbial systems, where no tritium is incorporated from [4-^{14}C, (1*S*)-1-$^{3}H_1$] GGPP into 15-cis phytoene (*Phycomyces*) while half the tritium is retained in all-trans phytoene (*Mycobacterium*; [46]).

A current view of the situation [28] is illustrated in Figure 2. The cyclopropylcarbinyl cation formed from PPPP can be stabilized in any of three ways, (1) by gaining a hydride ion from NADPH to yield lycopersene, as in a nonspecific squalene synthetase system, (2) by the stereospecific loss of a *pro-R* proton to yield all-trans phytoene (as in some bacteria), or (3) by the stereospecific loss of the adjacent *pro-S* proton to form 15-cis phytoene, the first C_{40} hydrocarbon precursor of higher plant carotenoids.

III. Carotene Desaturation

As a result of detailed carotenoid analyses on the fruits of mutant tomatoes, Porter and Lincoln [70] proposed that phytoene is sequentially desaturated to lycopene. Determinations of the structures of the intermediates [31] showed that didehydrogenations occur alternately to the left and right of the central chromophore of phytoene to form phytofluene, ζ-carotene, neurosporene, and lycopene (Fig. 3). Presumably, further didehydrogenations lead to the formation of 3,4,3′,4′-tetradehydrolycopene, present in Valencia oranges [94]. Since higher plant phytoene has a 15-cis configuration and ζ-carotene, neurosporene, and lycopene generally have all-trans geometry, a cis-trans isomerization must occur at an early stage in the desaturation sequence. Tomato phytofluene [48] differs from microbial phytofluene (see [28]) in that it is predominantly 15-cis. It is therefore reasonable to postulate that, in higher plants, 15-cis phytoene is dehydrogenated to 15-cis phytofluene, which then isomerizes to all-trans prior to further desaturation.

The conversion of detergent-solubilized 15-cis [^{14}C] phytoene into unsaturated acyclic (including 15-cis phytofluene) and cyclized carotenes has been demonstrated in tomato systems from the fruit of different genetic selections [56]. An extract from red tomato fruit converts cis phytofluene to trans phytofluene, trans phytofluene to trans ζ-carotene, and trans ζ-carotene into neuro-

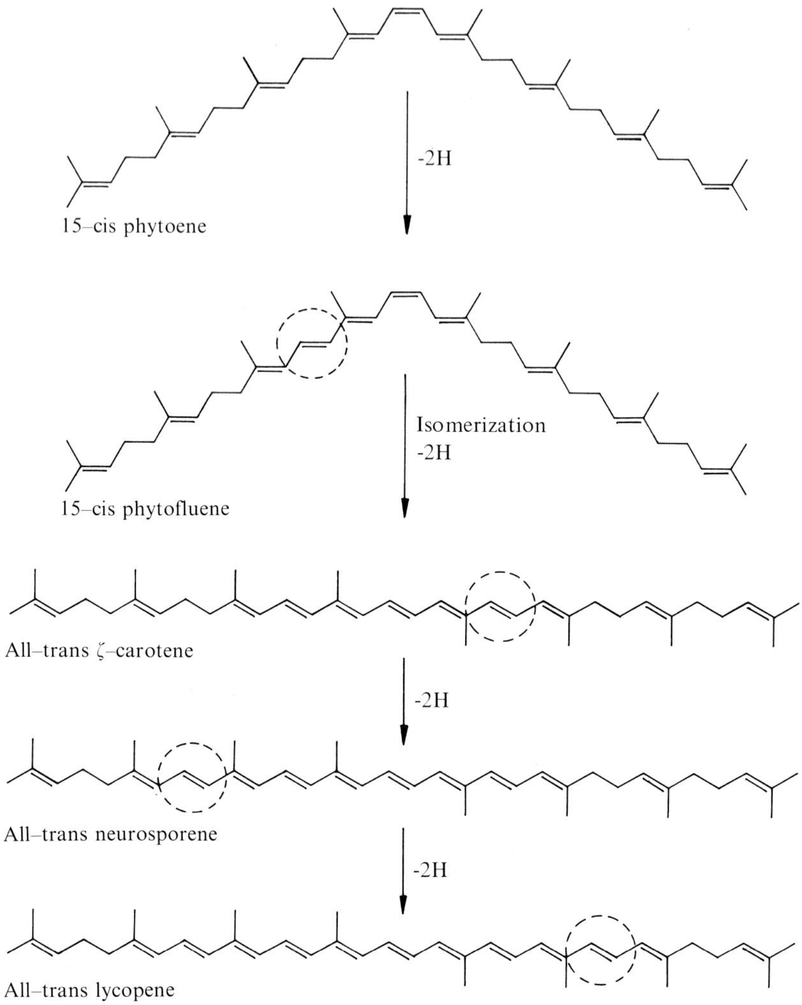

Fig. 3. The Porter-Lincoln sequence of carotene desaturation in higher plants

sporene and lycopene as well as into cyclized carotenes [72]. A similar system from tangerine mutant tomatoes is capable both of converting trans ζ-carotene into more unsaturated all-trans carotenes and their poly-cis isomers (proneurosporene and prolycopene; Ref. [73]) and also of converting cis ζ-carotene [75] into poly-cis and all-trans carotenes. ζ-Carotene appears to be the branch point for poly-cis carotene formation in the tangerine tomato [74] (Fig. 4). The final step of the general desaturation sequence, the conversion of neurosporene into lycopene, has been demonstrated only in a fungal system [12, 25].

The introduction of each new double bond in the desaturation sequence takes place by trans elimination of the 2-*pro-S* and 5-*pro-R* hydrogens originating from MVA [13, 60, 92].

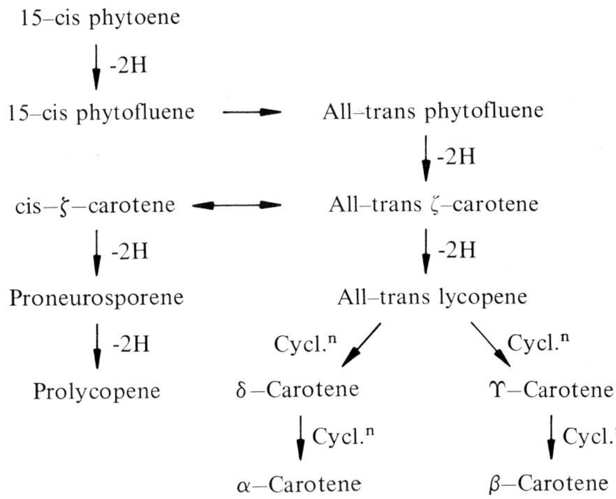

Fig. 4. Formation of carotenes in extracts from red and tangerine tomato plastids [73]

IV. Carotene Cyclization

Although a mechanism is now generally accepted for the conversion of the acyclic (ψ) end group of a carotene into a β- or ε-ring (Refs. [13, 14]; Fig. 5), the precise nature of the cyclization substrate in higher plants is still not clear. There is considerable evidence that lycopene cyclizes (1) via γ-carotene to β-carotene or (2) via δ-carotene to α-carotene; the strongest evidence comes from the conversion of labelled lycopene into the cyclized carotenes by chloroplasts or by the tomato system [32, 47, 55, 56, 67]. What is not known in higher plants is whether lycopene is the sole cyclization substrate. Some inhibitors block cyclization in higher plants, when lycopene accumulates. Both cyocel [52] and CPTA [78] have this effect in pumpkin cotyledons, while CPTA affects citrus [99, 100] and other fruits [21]. None of these inhibitor effects shows that lycopene is the sole cyclization substrate, for a blockage of neurosporene cyclization would presumably result in lycopene accumulation as well.

Neurosporene is certainly a substrate for cyclization, for both α- and β-zeacarotenes ("cyclized neurosporenes") are known plant constituents [69, 93]. By using unlabelled carotenes to dilute out the incorporation by a fungal system

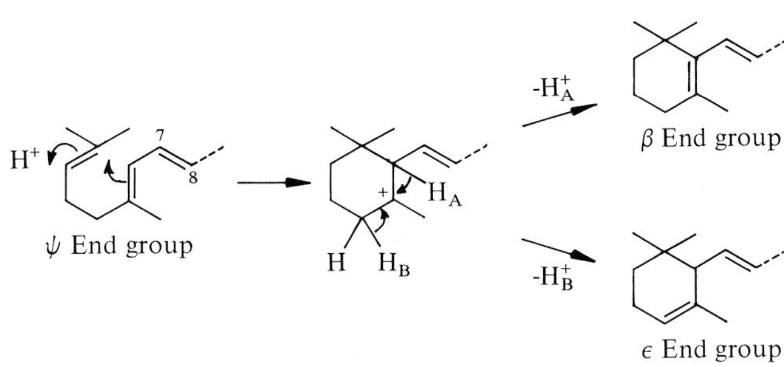

Fig. 5. Mechanisms for β- and ε-ring formation in carotenes

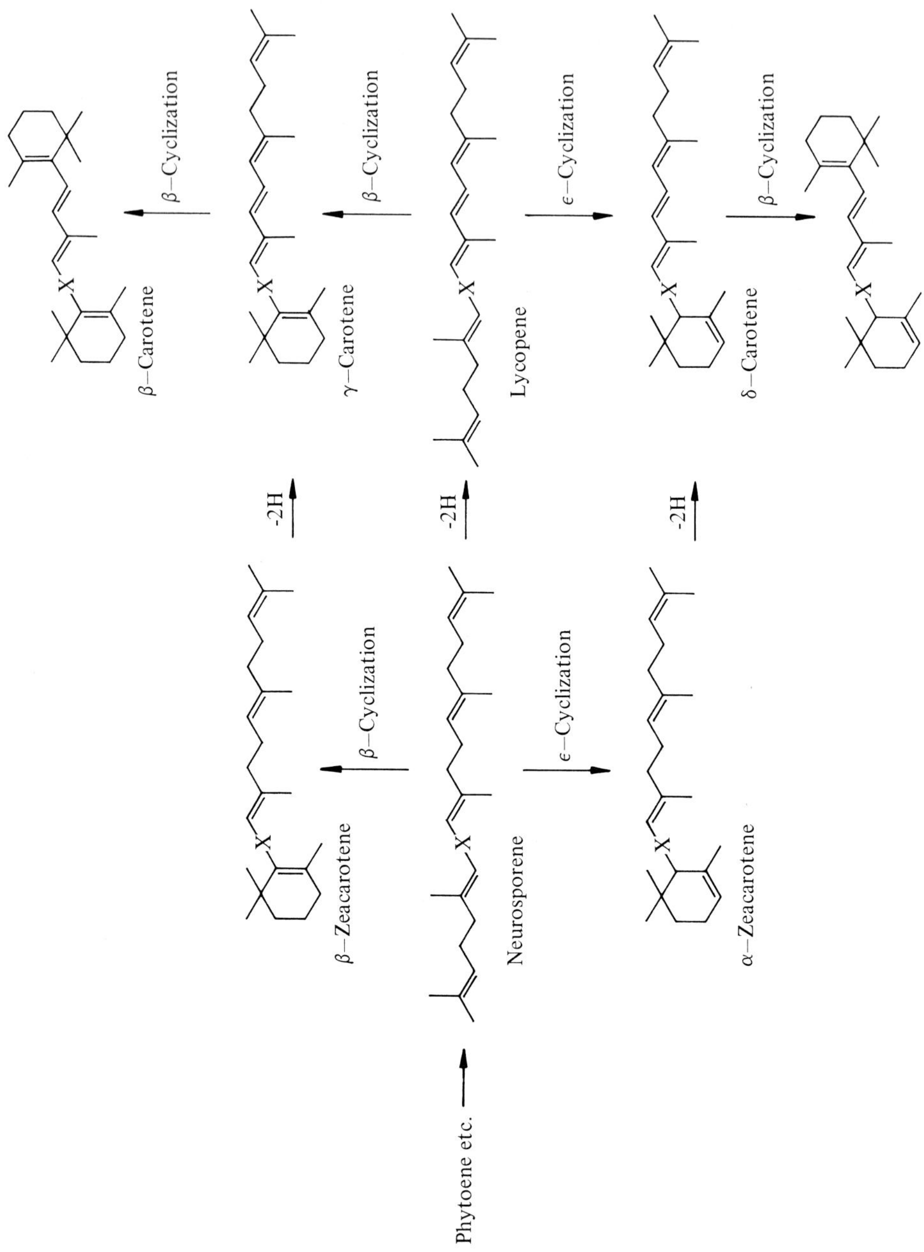

Fig. 6. Possible alternative pathways for carotene cyclization in higher plants

of [^{14}C] neurosporene into β-carotene, it has been established that lycopene and neurosporene (the only two higher plant acyclic carotenes with the presumed structural requirement for cyclization, namely a 7,8 double bond [27]) are both used, approximately equally, as cyclization substrate. It may be that a similar situation obtains in higher plants (Fig. 6).

V. Xanthophyll Formation

It has been assumed for a number of years, and as a result of rather indirect evidence, that xanthophylls are formed from carotenes. A recent study using a *Flavobacterium* sp. [61] has provided direct and conclusive evidence that this is so. The bacteria normally form zeaxanthin (3,3′-dihydroxy-β-carotene), but when grown in the presence of [^{14}C] MVA and nicotine, they accumulated lycopene. When inhibitor and substrate were removed, anaerobic resuspension resulted in the lycopene being converted into labelled β-carotene. When air was admitted, labelled zeaxanthin was formed from the β-carotene.

(a) Conversion of antheraxanthin to capsanthin, violaxanthin to capsanthin 5,6–epoxide, or capsanthin 5,6–epoxide to capsorubin

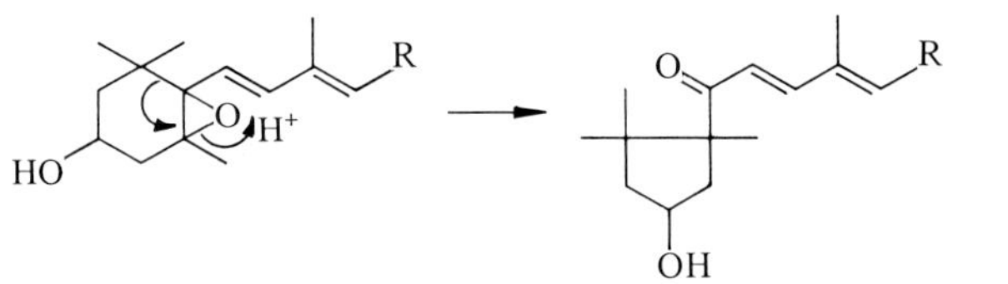

(b) Conversion of violaxanthin into neoxanthin

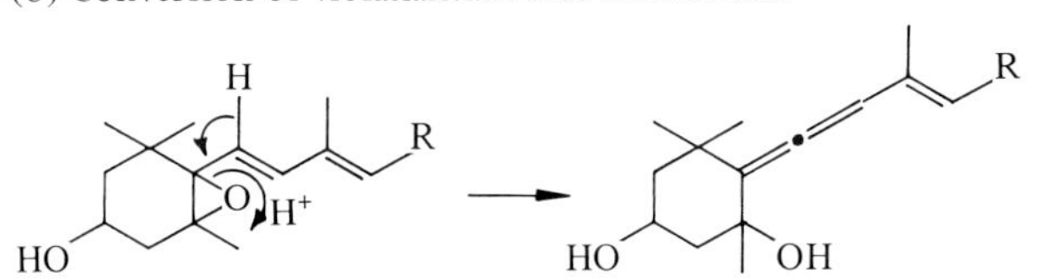

(c) Conversion of antheraxanthin into eschscholtzxanthin

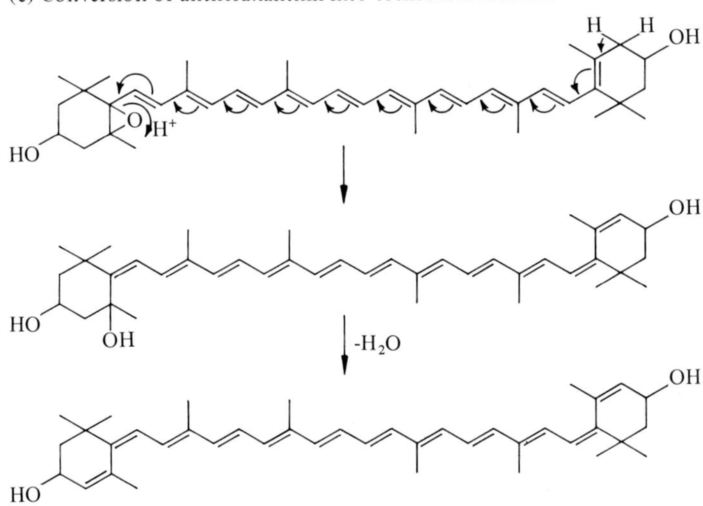

Fig. 7a–c. Possible precursor roles of xanthophylls with 3-hydroxy-5,6-epoxy-β-end groups

The significance of this microbial experiment to higher plant carotenoids is that all the chloroplast xanthophylls are, like zeaxanthin, 3-hydroxy-derivatives of β- or α-carotene. The 3-hydroxyl groups originate in molecular oxygen [95], and are probably inserted by a mixed-function oxygenase, for the hydroxylation is stereospecific and no ketonic intermediate is involved [13].

A second important feature of chloroplast xanthophylls is that some (antheraxanthin, violaxanthin, neoxanthin) are 5,6-epoxides. These oxygen functions also come from molecular oxygen and are inserted stereospecifically and, therefore, enzymically (see Ref. [13]). It is not known whether the formation of the furanoid (5,8) epoxides (of some flowers and fruit) is enzymic, but the fact that a rapid isomerization (5,6 to 5,8) is caused in vitro by acids (even 0.5% citric acid; Ref. [24]) may be significant. Zeaxanthin, its 5,6-epoxide (antheraxanthin), and its 5,6,5′,6′-diepoxide (violaxanthin) all participate in the light-driven xanthophyll (epoxide) cycle of higher plants (see Siefermann-Harms, Chapter 12).

The 3-hydroxy-5,6-epoxy-β-end group (of antheraxanthin and violaxanthin) may have an important role in the formation of other xanthophylls. Studies on the levels of fruit carotenoids of colour varieties of *Capsicum annuum* [30] showed that, in red fruits, capsanthin and capsorubin appear to be formed from antheraxanthin and violaxanthin, respectively, presumably by pinacollic rearrangements (Fig. 7a). Little neoxanthin was present in the red fruits, but the mature yellow fruit of another variety, which lacked capsanthin and capsorubin, formed large amounts of neoxanthin, apparently at the expense of violaxanthin. Thus an alternative transformation of the 3-hydroxy-5,6-epoxy-β-end group may form the allenically linked 3,5-dihydroxy-β-end group of neoxanthin [26] (Fig. 7b), although another view is that an intermediate peroxide may be involved [88]. The *retro* carotenoid eschscholtzxanthin appears to be formed from antheraxanthin, again by a mechanism involving the 3-hydroxy-5,6-epoxy-β-end group (Fig. 7c; Ref. [91]).

The seco- and apocarotenoids of citrus fruit probably represent the first stages of carotenoid degradation by ring-opening and β-oxidation. Yokoyama and White [96] suggested that β-apo-8′-carotenal and β-apo-10′-carotenal are natural degradation products of β-carotene, while β-citraurin arises from zeaxanthin (or β-cryptoxanthin). A ring-opening mechanism involving an intermediate peroxide [88] could lead to secocarotenoid formation (Fig. 8). Thus β-carotene would be converted into semi-β-carotenone and then β-carotenone, α-carotene would yield semi-α-carotenone, and triphasiaxanthin would be formed from β-cryptoxanthin. Such reactions would be consistent with the changes in the carotenoid levels of ripening fruit of *Triphasia trifolia* [98].

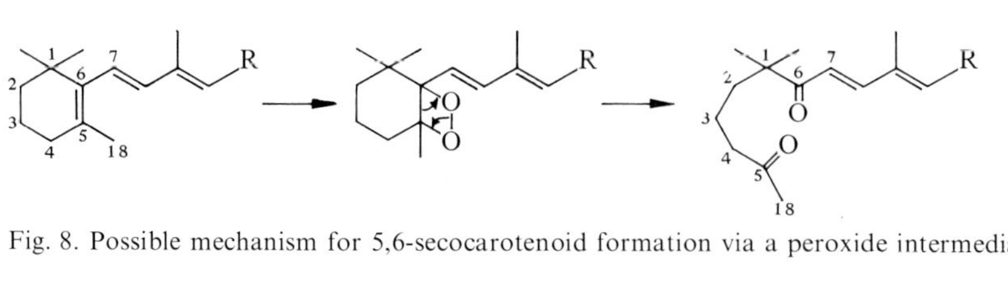

Fig. 8. Possible mechanism for 5,6-secocarotenoid formation via a peroxide intermediate [88]

VI. Factors Affecting Carotenoid Formation

A number of factors affect carotenoid formation in ripening fruit. In the tomato, lycopene synthesis is inhibited at temperatures above 30° C, but β-carotene formation is not. This has been interpreted [13] in terms of the development of a physically separate (temperature-sensitive) pathway of synthesis in the ripening fruit superimposed on the (temperature-insensitive) synthesis of normal chloroplast pigments such as β-carotene. A biosynthetic autonomy of chloroplast and chromoplast carotenoids is also concluded from a recent study of a number of tomato phenotypes in which the partial dependence of carotenoid synthesis on light was demonstrated [76].

Many genetic studies have been carried out on plants that produce commercially important fruit; those on tomatoes have yielded the most information on carotenoids, and a number of genes have been recognized that regulate carotenoid biosynthesis (see Refs. [41, 44]). Red tomatoes have the dominant allele r^+ while mature yellow fruits are homozygous for the recessive allele *r*. The *rr* genotype forms only about 5% of the carotenoid of the r^+r^+ genotype. The r^+/r gene is but one of a number that affect total carotenoid content, for r^+/r, at^+/at, and hp^+/hp all act before phytoene in the biosynthetic sequence. Other genes act elsewhere; gh^+/gh controls phytoene desaturation and b^+/b, the cyclization of lycopene to β-carotene. Another gene, del^+/del, regulates cyclization to form ε-rather than β-rings, while t^+/t determines the degree of desaturation and the extent to which poly-cis carotenes are formed.

Such control of carotenoid formation in other tissues is less well understood. Carotene desaturation and cyclization are both under nuclear control in maize; an albino mutant accumulates phytoene [4, 87] and two other (leaky) mutants accumulate ζ-carotene and lycopene, respectively [36]. This situation is in contrast to that in *Phycomyces blakesleeanus*, in which the entire carotene desaturation sequence is thought to be controlled by one gene (*carB*), with four copies of its product acting as a complex to carry out the four successive dehydrogenation steps between phytoene and lycopene [35].

Etiolated seedlings (germinated in the dark) have no chloroplasts and very little carotenoid, mostly xanthophyll (lutein, antheraxanthin, violaxanthin). Illumination results in the rapid development of chloroplasts and the concomitant synthesis of β-carotene, lutein, violaxanthin, and neoxanthin. This effect is mediated by phytochrome in both mono- [51] and dicotyledonous plants [59].

D. Function

The functions of the carotenoids have been reviewed recently and in much more detail than is possible here [19, 53, 90]. The following is merely a summary of some of their known and presumed functions in higher plants.

Their universal function, exemplified in higher plants by the destruction in bright light and oxygen of chlorophyll in the phytoene-accumulating maize mutant (see Ref. [53]), is protection against photodynamic damage (caused by singlet oxygen, $^1O_2$, formed in aerobes by the transfer of energy from an excited photosensitizer such as chlorophyll or another porphyrin). The possible mechanisms of photoprotection, including the direct quenching of $^1O_2$ by carotenoids, have been reviewed recently [54]. Another photofunction of carotenoids is their role, as components of photosystems I and II, in the light reaction of photosynthesis itself; they are less efficient as accessory pigments in higher plants, however, than in e.g., brown algae [90]. Both carotenoids and flavins have been considered as photoreceptors for phototropism [19]; a recent and largely theoretical study is against carotenoids having such a role [80]. The possible functions of the light-driven xanthophyll cycle, in which 5,6-epoxides participate, are discussed elsewhere in this volume (Siefermann-Harms, Chapter 12).

Another effect of light on carotenoids is that violaxanthin is converted by high intensity light into a mixture of products including xanthoxin, which has abscisic acid (ABA) activity [82–84]. Lipoxygenase converts violaxanthin into a range of products similar to those formed by photolysis [37]. The structural similarities of violaxanthin, 2-cis (+)-xanthoxin (the active isomer), and (+)-ABA are shown in Figure 9. The circumstantial evidence that violaxanthin may also be a precursor of ABA, the evidence for the conversion of xanthoxin into ABA, and also a strong body of evidence from labelling experiments that ABA is formed independently of carotenoids, have all been considered in a review [65].

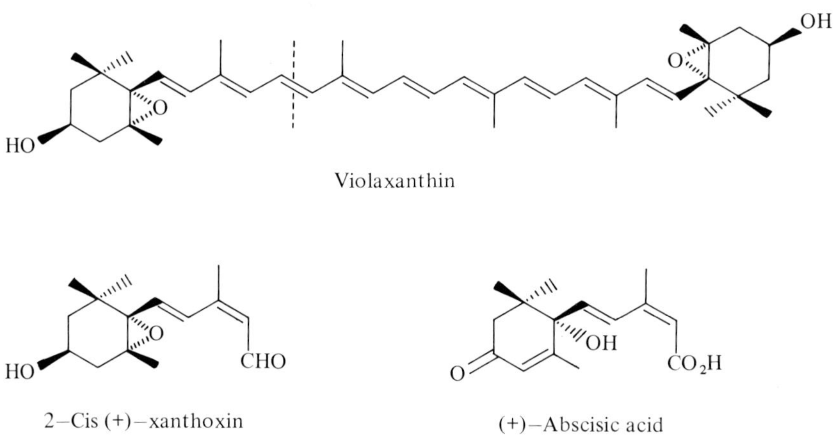

Fig. 9. Structural similarity of violaxanthin, xanthoxin, and abscisic acid

Carotenoids may be precursors of sporopollenin (probably a carotenoid polymer comprising the extremely stable outer layer of both spores and pollen grains; see Ref. [53]). Other possible functions of carotenoids in higher plants—e.g., in facilitating pollination by insects or seed dispersal by animals, because of their colours and very presence in flowers and fruits [42]—are difficult to substantiate experimentally.

E. Appendix

Trivial and semi-systematic names and structures of some higher plant carotenoids[a]

Antheraxanthin	5,6-Epoxy-5,6-dihydro-β,β-carotene-3,3′-diol
β-Apo-8′-carotenal	8′-Apo-β-caroten-8′-al
β-Apo-10′-carotenal	10′-Apo-β-caroten-10′-al
Auroxanthin	5,8,5′,8′-Diepoxy-5,8,5′,8′-tetrahydro-β,β-carotene-3,3′-diol
Azafrin	5,6-Dihydroxy-5,6-dihydro-10′-apo-β-caroten-10′-oic acid
Bixin	Methyl hydrogen 9′-*cis*-6,6′-diapocarotene-6,6′-dioate
Capsanthin	(3*R*,3′*S*,5′*R*)-3,3′-Dihydroxy-β,κ-caroten-6′-one
Capsanthin 5,6-epoxide	5,6-Epoxy-3,3′dihydroxy-5,6-dihydro-β,κ-caroten-6′-one
Capsorubin	(3*S*,5*R*,3′*S*,5′*R*)-3,3′-Dihydroxy-κ,κ-carotene-6,6′-dione
α-Carotene	(6′R)-β,ε-carotene
β-Carotene	β,β-Carotene
γ-Carotene	β,ψ-Carotene
δ-Carotene	(6*R*)-ε,ψ-Carotene
ζ-Carotene	7,8,7′,8′-Tetrahydro-ψ,ψ-carotene
β-Carotenone	5,6,5′,6′-Diseco-β,β-carotene-5,6,5′,6′-dione
Chrysanthemaxanthin	Epimer of flavoxanthin
β-Citraurin	(3*R*)-3-Hydroxy-8′-apo-β-caroten-8′-al
Crocetin	8,8′-Diapocarotene-8,8′-dioic acid
β-Cryptoxanthin	(3*R*)-β,β-Caroten-3-ol
Eschscholtzxanthin	4′,5′-Didehydro-4,5′-*retro*-β,β-carotene-3,3′-diol
Flavoxanthin	5,8-Epoxy-5,8-dihydro-β,ε-carotene-3,3′-diol
Gazaniaxanthin	(3*R*)-5′-Cis-β,ψ-caroten-3-ol
Lutein	(3*R*,3′*R*,6′*R*)-β,ε-Carotene-3,3′-diol
Lycopene	ψ,ψ-Carotene
Lycopene 5,6-epoxide	5,6-Epoxy-5,6-dihydro-ψ,ψ-carotene
Neoxanthin	(3*S*,5*R*,6*R*,3′*S*,5′*R*,6′*S*)-5′,6′-Epoxy-6,7-didehydro-5,6,5′,6′-tetrahydro-β,β-carotene-3,5,3′-triol
Neurosporene	7,8-Dihydro-ψ,ψ-carotene
Phytoene	15-Cis 7,8,11,12,7′,8′,11′,12′-Octahydro-ψ,ψ-carotene
Phytofluene	15-Cis 7,8,11,12,7′,8′-Hexahydro-ψ,ψ-carotene
Rubixanthin	(3*R*)-β,ψ-Caroten-3-ol
Semi-α-carotenone	(6′*R*)-5,6-Seco-β,ε-carotene-5,6-dione
Semi-β-carotenone	5,6-Seco-β,β-carotene-5,6-dione
Tetradehydrolycopene	3,4,3′,4′-Tetradehydro-ψ,ψ-carotene
Triphasiaxanthin	3′-Hydroxy-5,6-seco-β,β-carotene-5,6-dione
Violaxanthin	(3*S*,5*R*,6*S*,3′*S*,5′*R*,6′*S*)-5,6,5′,6′-Diepoxy-5,6,5′,6′-tetrahydro-β,β-carotene-3,3′-diol
α-Zeacarotene	7′,8′-Dihydro-ε,ψ-carotene
β-Zeacarotene	7′,8′-Dihydro-β,ψ-carotene
Zeaxanthin	(3*R*,3′*R*)-β,β-Carotene-3,3′-diol

[a] Figure 1 shows end groups and numbering of carbons throughout the molecule. A fuller list of the names and structures of plant carotenoids has been published elsewhere [24].

References

1. Altman, L.J., Kowerski, R.C., Rilling, H.C.: Synthesis and conversion of presqualene alcohol to squalene. J. Am. Chem. Soc. **93**, 1782–1783 (1971)
2. Altman, L.J., Ash, L., Kowerski, R.C., Epstein, W.W., Larsen, B.R., Rilling, H.C., Muscio, F., Gregonis, D.E.: Prephytoene pyrophosphate. A new intermediate in the biosynthesis of carotenoids. J. Am. Chem. Soc. **94**, 3257–3259 (1972)

3. Anderson, D.G., Porter, J.W.: The biosynthesis of phytoene and other carotenes by enzymes of isolated higher plant plastids. Arch. Biochem. Biophys. **97**, 509–519 (1962)
4. Anderson, I.C., Robertson, D.S.: Role of carotenoids in protecting chlorophyll from photodestruction. Plant Physiol. **35**, 531–534 (1960)
5. Aung Than, Bramley, P.M., Davies, B.H., Rees, A.F.: The stereochemistry of phytoene. Phytochemistry **11**, 3187–3192 (1972)
6. Barnes, F.J., Qureshi, A.A., Semmler, E.J., Porter, J.W.: Prelycopersene pyrophosphate and lycopersene, intermediates in carotene biosynthesis. J. Biol. Chem. **248**, 2768–2773 (1973)
7. Bartels, P.G., McCullough, C.: A new inhibitor of carotenoid synthesis in higher plants: 4-chloro-5-(dimethylamino)-2-α,α,α-(trifluoro-*m*-tolyl)-3(2H)-pyridazinone (Sandoz 6706). Biochem. Biophys. Res. Commun. **48**, 16–22 (1972)
8. Beeler, D.A., Anderson, D.G., Porter, J.W.: The biosynthesis of squalene from mevalonic acid-2-C^{14} and farnesyl pyrophosphate-4,8,12-C^{14} by carrot and tomato enzymes. Arch. Biochem. Biophys. **102**, 26–32 (1963)
9. Ben-Aziz, A., Britton, G., Goodwin, T.W.: Carotene epoxides of *Lycopersicon esculentum*. Phytochemistry **12**, 2759–2764 (1973)
10. Ben-Aziz, A., Koren, E.: Interference in carotenogenesis as a mechanism of action of the pyridazinone herbicide Sandoz 6706. Accumulation of C_{40} carotenoid precursors, inhibition of β-carotene synthesis and enhancement of phytoene epoxidation. Plant Physiol. **54**, 916–920 (1974)
11. Booth, V.H.: β-Carotene in the flowers of *Narcissus*. Biochem. J. **65**, 660–663 (1957)
12. Bramley, P.M., Davies, B.H., Aung Than: Alternative pathways of carotene cyclisation in *Phycomyces blakesleeanus*. Phytochemistry **16**, 235–238 (1977)
13. Britton, G.: Biosynthesis of carotenoids. In: Chemistry and Biochemistry of Plant Pigments, 2nd ed. Goodwin, T.W. (ed.). London-New York-San Francisco: Academic Press, 1976a, Vol. 1, pp. 262–327
14. Britton, G.: Carotene biosynthesis—later reactions. Pure Appl. Chem. **47**, 223–236 (1976)
15. Britton, G., Goodwin, T.W.: The occurrence of phytoene 1,2-epoxide and related compounds in tomatoes. Phytochemistry **8**, 2257–2258 (1969)
16. Britton, G., Goodwin, T.W.: Biosynthesis of carotenoids. Methods Enzymol. **18(C)**, 654–701 (1971)
17. Britton, G., Goodwin, T.W.: Carotene epoxides from the *delta* tomato mutant. Phytochemistry **14**, 2530–2532 (1975)
18. Buggy, M.J., Britton, G., Goodwin, T.W.: Stereochemistry of phytoene biosynthesis by isolated chloroplasts. Biochem. J. **114**, 641–643 (1969)
19. Burnett, J.H.: Function of carotenoids other than in photosynthesis. In: Chemistry and Biochemistry of Plant Pigments, 2nd ed. Goodwin, T.W. (ed.). London-New York-San Francisco: Academic Press, 1976, Vol. 1, pp. 655–679
20. Charlton, J.M., Treharne, K.J., Goodwin, T.W.: Incorporation of [2-^{14}C] mevalonic acid into phytoene by isolated chloroplasts. Biochem. J. **105**, 205–212 (1967)
21. Coggins, C.W., Jr., Henning, G.L., Yokoyama, H.: Lycopene accumulation induced by 2-(4-chlorophenylthio)-triethylamine hydrochloride. Science **168**, 1589–1590 (1970)
22. Commission on Biochemical Nomenclature: IUPAC Commission on the Nomenclature of Organic Chemistry and IUPAC-IUB Commission on Biochemical Nomenclature. Tentative rules for the nomenclature of carotenoids. Biochemistry (U.S.A.) **10**, 4827–4837 (1971)
23. Commission on Biochemical Nomenclature: IUPAC Commission on the Nomenclature of Organic Chemistry and IUPAC-IUB Commission on Biochemical Nomenclature. Nomenclature of carotenoids (Recommendations 1974). Biochemistry (U.S.A.) **14**, 1803–1804 (1975)
24. Davies, B.H.: Carotenoids. In: Chemistry and Biochemistry of Plant Pigments, 2nd ed. Goodwin, T.W. (ed.). London-New York-San Francisco: Academic Press, 1976, Vol. 2 (Analytical Methods), pp. 38–165
25. Davies, B.H.: Carotene biosynthesis in fungi. Pure Appl. Chem. **35**, 1–28 (1973)
26. Davies, B.H.: Carotenoids—Aspects of biosynthesis and enzymology. Ber. Dtsch. Bot. Ges. **88**, 7–25 (1975)
27. Davies, B.H., Rees, A.F.: 7′,8′,11′,12′-Tetrahydro-γ-carotene: a novel carotene from *Phycomyces blakesleeanus*. Phytochemistry **12**, 2745–2750 (1973)
28. Davies, B.H., Taylor, R.F.: Carotenoid biosynthesis—the early steps. Pure Appl. Chem. **47**, 211–221 (1976)

29. Davies, B.H., Mercer, E.I., Goodwin, T.W.: Attempts to detect lycopersene in carotenogenic organisms. Biochem. J. **81**, 40P (1961)
30. Davies, B.H., Matthews, S., Kirk, J.T.O.: The nature and biosynthesis of carotenoids of different colour varieties of *Capsicum annuum*. Phytochemistry **9**, 797–805 (1970)
31. Davis, J.B., Jackman, L.M., Siddons, P.T., Weedon, B.C.L.: Carotenoids and related compounds. Part XIV. The structure and synthesis of phytoene, phytofluene, ζ-carotene and neurosporene. J. Chem. Soc. **(C)**, 2154–2165 (1966)
32. Decker, K., Uehleke, H.: Eine enzymatische Isomerisierung von Lycopin und β-Carotin. Hoppe Seylers Z. Physiol. Chem. **323**, 61–76 (1961)
33. Egger, K., Kleinig, H.: Die Ketocarotinoide in *Adonis annua* L. III. Vergleich mit synthetischen Substanzen. Phytochemistry **6**, 903–905 (1967)
34. Epstein, W.W., Rilling, H.C.: Studies on the mechanism of squalene biosynthesis. The structure of presqualene pyrophosphate. J. Biol. Chem. **245**, 4597–4605 (1970)
35. Eslava, A.P., Cerdá-Olmedo, E.: Genetic control of phytoene dehydrogenation in *Phycomyces*. Plant Sci. Lett. **2**, 9–14 (1974)
36. Faludi-Dániel, A., Láng, F., Fradkin, L.I.: The state of chlorophyll *a* in leaves of carotenoid mutant maize. In: Biochemistry of Chloroplasts. Goodwin, T.W. (ed.). London-New York: Academic Press, 1966, Vol. 1, pp. 269–274
37. Firn, R.D., Friend, J.: Enzymatic production of the plant growth inhibitor, xanthoxin. Planta **103**, 263–266 (1972)
38. Goodwin, T.W.: Comparative Biochemistry of the Carotenoids. London: Chapman and Hall 1952
39. Goodwin, T.W.: Studies in carotenogenesis. 19. A survey of the polyenes in a number of ripe berries. Biochem. J. **62**, 346–352 (1956)
40. Goodwin, T.W.: Studies in carotenogenesis. 24. The changes in carotenoid and chlorophyll pigments in the leaves of deciduous trees during Autumn necrosis. Biochem. J. **68**, 503–511 (1958)
41. Goodwin, T.W.: Biosynthesis. In: Carotenoids. Isler, O. (ed.). Basel: Birkhäuser, 1971, pp. 577–636
42. Goodwin, T.W.: Carotenoids. In: Phytochemistry. Miller, L.P. (ed.). New York-Cincinnati-Toronto-London-Melbourne: Van Nostrand Reinhold, 1973, Vol. 1, pp. 112–142
43. Goodwin, T.W.: Distribution of carotenoids. In: Chemistry and Biochemistry of Plant Pigments, 2nd ed. Goodwin, T.W. (ed.). London-New York-San Francisco: Academic Press, 1976, Vol. 1, pp. 225–261
44. Goodwin, T.W., Goad, L.J.: Carotenoids and triterpenoids. In: The Biochemistry of Fruits and their Products. Hulme, A.C. (ed.). London-New York: Academic Press, 1970, Vol. 1, pp. 305–368
45. Goodwin, T.W., Thomas, D.M.: The carotenoid pigments in the petals of *Mimulus cupreus* and *Mimulus tigrinis*. Phytochemistry **3**, 47–50 (1964)
46. Gregonis, D.E., Rilling, H.C.: The stereochemistry of *trans*-phytoene synthesis. Some observations on lycopersene as a carotene precursor and a mechanism for the synthesis of *cis*- and *trans*-phytoene. Biochemistry **13**, 1538–1542 (1974)
47. Hill, H.M., Calderwood, S.K., Rogers, L.J.: Conversion of lycopene to β-carotene by plastids isolated from higher plants. Phytochemistry **10**, 2051–2058 (1971)
48. Jungalwala, F.B., Porter, J.W.: The configuration of phytoene. Arch. Biochem. Biophys. **110**, 291–299 (1965)
49. Jungalwala, F.B., Porter, J.W.: Enzymatic synthesis of phytoene in tomato. Methods Enzymol. **15**, 454–460 (1969)
50. Karrer, P., Jucker, E.: Carotenoids. New York-Amsterdam-London-Brussels: Elsevier, 1950
51. Kleudgen, H.K., Lichtenthaler, H.K.: Die Wirkung von Phytochrom auf die Bildung von Einzelcarotinoiden in etiolierten Hordeum-Keimlingen. Z. Naturforsch. **30C**, 67–68 (1975)
52. Knypl, J.S.: Accumulation of lycopene in detached cotyledons of pumpkin treated with (2-chloroethyl)-trimethylammonium chloride. Naturwissenschaften **56**, 572 (1969)
53. Krinsky, N.I.: Function. In: Carotenoids. Isler, O. (ed.). Basel: Birkhäuser, 1971, pp. 669–716
54. Krinsky, N.I.: Cellular damage initiated by visible light. In: The Survival of Vegetative Microbes (Soc. Gen. Microbiol., Symp. 26). Gray, T.R.G., Postgate, J.R. (eds.). Cambridge-London-New York-Melbourne: Cambridge Univ. Press, 1976, pp. 209–239
55. Kushwaha, S.C., Subbarayan, C., Beeler, D.A., Porter, J.W.: The conversion of lycopene-15,15′-

^{3}H to cyclic carotenes by soluble extracts of higher plant plastids. J. Biol. Chem. **244**, 3635–3642 (1969)

56. Kushwaha, S.C., Suzue, G., Subbarayan, C., Porter, J.W.: The conversion of phytoene-^{14}C to acyclic, monocyclic and dicyclic carotenes and the conversion of lycopene-15,15′-^{3}H to mono- and dicyclic carotenes by soluble enzyme systems obtained from the plastids of tomato fruit. J. Biol. Chem. **245**, 4708–4717 (1970)
57. Liaaen-Jensen, S., Andrewes, A.G.: Microbial carotenoids. Ann. Rev. Microbiol. **26**, 225–248 (1972)
58. Lichtenthaler, H.K.: Die Lokalisation der Plastidenchinone und Carotinoide in den Chromoplasten der Petalen von *Sarothamnus scoparius* (L.) Wimm ex Koch. Planta **90**, 142–152 (1970)
59. Lichtenthaler, H.K.: Control of light-induced carotenoid synthesis in *Raphanus* seedlings by phytochrome. Physiol. Plantarum **34**, 357–358 (1975)
60. McDermott, J.C.B., Britton, G., Goodwin, T.W.: Carotenoid biosynthesis in a *Flavobacterium* sp. Stereochemistry of hydrogen elimination in the desaturation of phytoene to lycopene, rubixanthin and zeaxanthin. Biochem. J. **134**, 1115–1117 (1973)
61. McDermott, J.C.B., Brown, D.J., Britton, G., Goodwin, T.W.: Alternative pathways of zeaxanthin biosynthesis in a *Flavobacterium* species. Experiments with nicotine as inhibitor. Biochem. J. **144**, 231–243 (1974)
62. Maudinas, B., Bucholtz, M.L., Papastephanou, C., Katigar, S.S., Briedis, A.V., Porter, J.W.: Adenosine-5′-triphosphate stimulation of the activity of a partially purified phytoene synthetase complex. Biochem. Biophys. Res. Commun. **66**, 430–436 (1975a)
63. Maudinas, B., Bucholtz, M.L., Porter, J.W.: The partial purification and properties of a phytoene synthetase complex isolated from tomato fruit plastids. Abstr. 4th Int. Symp. Carotenoids (Berne), 1975b, pp. 41–42
64. Mercer, E.I., Davies, B.H., Goodwin, T.W.: Studies in carotenogenesis. 29. Attempts to detect lycopersene in higher plants. Biochem. J. **87**, 317–325 (1963)
65. Milborrow, B.V.: The chemistry and physiology of abscisic acid. Ann. Rev. Plant Physiol. **25**, 259–307 (1974)
66. Nusbaum-Cassuto, E., Villoutreix, J.: Mise en évidence du lycopersène chez les végétaux supérieurs. C.R. Acad. Sci. **260**, 1013–1015 (1965)
67. Papastephanou, C., Barnes, F.J., Briedis, A.V., Porter, J.W.: Enzymatic synthesis of carotenes by cell-free preparations of fruit of several genetic selections of tomatoes. Arch. Biochem. Biophys. **157**, 415–425 (1973)
68. Pennock, J.F., Hemming, F.W., Morton, R.A.: Some unsaponifiable lipids of the spadix of *Arum maculatum*. Biochem. J. **82**, 11P (1962)
69. Petzold, E.N., Quackenbush, F.W., McQuistan, M.: Zeacarotenes, new provitamins A from corn. Arch. Biochem. Biophys. **82**, 117–124 (1959)
70. Porter, J.W., Lincoln, R.E.: I. *Lycopersicon* selection containing a high content of carotenes and colorless polyenes. II. The mechanism of carotene biosynthesis. Arch. Biochem. **27**, 390–403 (1950)
71. Qureshi, A.A., Barnes, F.J., Semmler, E.J., Porter, J.W.: Biosynthesis of prelycopersene pyrophosphate and lycopersene by squalene synthetase. J. Biol. Chem. **248**, 2755–2767 (1973)
72. Qureshi, A.A., Andrewes, A.G., Qureshi, N., Porter, J.W.: The enzymatic conversion of *cis*-[^{14}C] phytofluene, *trans*-[^{14}C] phytofluene, and *trans*-[^{14}C] ζ-carotene to more unsaturated acyclic, monocyclic and dicyclic carotenes by a cell-free preparation of red tomato fruits. Arch. Biochem. Biophys. **162**, 93–107 (1974a)
73. Qureshi, A.A., Kim, M., Qureshi, N., Porter, J.W.: The enzymatic conversion of *cis*-[^{14}C] phytofluene, *trans*-[^{14}C] phytofluene and *trans*-[^{14}C] ζ-carotene to poly-*cis* acyclic carotenes by a cell-free preparation of tangerine tomato fruit plastids. Arch. Biochem. Biophys. **162**, 108–116 (1974b)
74. Qureshi, A.A., Qureshi, N., Kim, M., Porter, J.W.: The isolation, purification and characterization of *cis*-ζ-carotene and the demonstration of its conversion to acyclic, monocyclic and dicyclic carotenes by a soluble enzyme system obtained from the plastids of tangerine tomato fruit. Arch. Biochem. Biophys. **162**, 117–125 (1974c)
75. Raymundo, L.C., Simpson, K.L.: The isolation of a poly-*cis*-ζ-carotene from the tangerine tomato. Phytochemistry **11**, 397–400 (1972)
76. Raymundo, L.C., Chichester, C.O., Simpson, K.L.: Light dependent carotenoid synthesis in the tomato fruit. J. Agr. Food Chem. **24**, 59–64 (1976)

77. Seybold, A.: Untersuchungen über den Farbwechsel von Blumenblättern, Früchten und Samenschalen. Heidelberg: Springer, 1954, 96 pp.
78. Simpson, D.J., Chichester, C.O., Lee, T.H.: Chemical regulation of plastid development. I. Inhibition of chlorophyll biosynthesis in detached pumpkin cotyledons by CPTA. A pigment and ultrastructural study. Australian J. Plant Physiol. **1**, 135–147 (1974)
79. Simpson, K.L., Lee, T.-C., Rodriguez, D.B., Chichester, C.O.: Metabolism in stored and senescent tissues. In: Chemistry and Biochemistry of Plant Pigments, 2nd ed. Goodwin, T.W. (ed.). London-New York-San Francisco: Academic Press, 1976, Vol. 1, pp. 779–842
80. Song, P.-S., Moore, T.A.: On the photoreceptor pigment for phototropism and phototaxis: Is a carotenoid the most likely candidate? Photochem. Photobiol. **19**, 435–441 (1974)
81. Strain, H.H.: Fat-soluble chloroplast pigments: their identification and distribution in various Australian plants. In: Biochemistry of Chloroplasts. Goodwin, T.W. (ed.). London-New York: Academic Press, 1966, Vol. 1, pp. 387–406
82. Taylor, H.F., Burden, R.S.: Identification of plant growth inhibitors produced by photolysis of violaxanthin. Phytochemistry **9**, 2217–2223 (1970)
83. Taylor, H.F., Burden, R.S.: Xanthoxin, a new naturally occurring plant growth inhibitor. Nature (London) **227**, 302–304 (1970)
84. Taylor, H.F., Burden, R.S.: Xanthoxin, a recently discovered plant growth inhibitor. Proc. Roy. Soc. (London) Ser. B **180**, 317–346 (1972)
85. Tevini, M., Lichtenthaler, H.K.: Untersuchungen über die Pigment- und Lipochinonausstattung der zwei photosynthetischen Pigmentsysteme. Z. Pflanzenphysiol. **62**, 17–32 (1970)
86. Thornber, J.P., Stewart, J.C., Hatton, M.W.C., Bailey, J.L.: Studies on the nature of chloroplast lamellae. II. Chemical composition and further physical properties of two chlorophyll-protein complexes. Biochemistry **6**, 2006–2014 (1976)
87. Treharne, K.J., Mercer, E.I., Goodwin, T.W.: Carotenoid biosynthesis in some maize mutants. Phytochemistry **5**, 581–587 (1966)
88. Weedon, B.C.L.: Allenic and acetylenic carotenoids. Rev. Pure Appl. Chem. (Australia) **20**, 51–66 (1970)
89. White, J.W., Jr., Zscheile, F.P., Brunson, A.M.: The carotenoids of yellow corn grain. J. Am. Chem. Soc. **64**, 2603–2606 (1942)
90. Whittingham, C.P.: Function in photosynthesis. In: Chemistry and Biochemistry of Plant Pigments, 2nd ed. Goodwin, T.W. (ed.). London-New York-San Francisco: Academic Press, 1976, Vol. 1, pp. 624–654
91. Williams, R.J.H., Britton, G., Goodwin, T.W.: A possible mechanism for the biosynthesis of eschscholtzxanthin. Biochim. Biophys. Acta **124**, 200–203 (1966)
92. Williams, R.J.H., Charlton, J.M., Britton, G., Goodwin, T.W.: The stereospecific biosynthesis of phytoene and polyunsaturated carotenes. Biochem. J. **104**, 767–777 (1967a)
93. Williams, R.J.H., Britton, G., Goodwin, T.W.: The biosynthesis of cyclic carotenes. Biochem. J. **105**, 99–105 (1967b)
94. Winterstein, A., Studer, A., Rüegg, R.: Neuere Ergebnisse der Carotinoidforschung. Ber. Deut. Chem. Ges. **93**, 2951–2965 (1960)
95. Yamamoto, H.Y., Chichester, C.O., Nakayama, T.O.M.: Biosynthetic origin of oxygen in the leaf xanthophylls. Arch. Biochem. Biophys. **96**, 645–649 (1962)
96. Yokoyama, H., White, M.J.: Citrus carotenoids—VI. Carotenoid pigments in the flavedo of Sinton citrangequat. Phytochemistry **5**, 1159–1173 (1966)
97. Yokoyama, H., White, M.J.: Carotenoids in the flavedo of Marsh seedless grapefruit. J. Agr. Food Chem. **15**, 693–696 (1967)
98. Yokoyama, H., White, M.J.: Carotenoid formation in *Triphasia trifolia*. Phytochemistry **9**, 1795–1797 (1970)
99. Yokoyama, H., Coggins, C.W., Henning, G.L.: The effect of 2-(4-chlorophenylthio)-triethylamine hydrochloride on the formation of carotenoids of citrus. Phytochemistry **10**, 1831–1834 (1971)
100. Yokoyama, H., de Benedict, C., Coggins, C.W., Henning, G.L.: Induced color changes in grapefruit and orange. Phytochemistry **11**, 1721–1724 (1972)

Chapter 12

The Xanthophyll Cycle in Higher Plants

D. Siefermann-Harms

A. Introduction

Carotenoids are widespread among bacteria, algae, fungi, and higher plants (see Ref. [34] for a review). Because of their lipophilic properties they are membrane components or desolved in lipid vacuoles. Usually their metabolic activity is small, with lifetimes in the range of several days. In photosynthetic membranes, however, fast carotenoid reactions have been observed upon illumination. One type of reaction takes place within a few nanoseconds and consists of the light-induced conversion of carotenoids to their metastable states [36]. Another type takes place within several minutes, is enzymic, and involves a few specific carotenoids in a cyclic pathway. The properties of this pathway will be presented here.

B. General Characteristics of the Xanthophyll Cycle

In 1957 Sapozhnikov et al. [19] observed that in leaves of various higher plants violaxanthin decreased upon illumination, while another carotenoid, then believed to be lutein, increased. The reaction was reversible in the dark, but contrary to the light-dependent reaction, the reversal required molecular oxygen [22]. A few years later, Yamamoto et al. [43] showed that the α-carotene derivative lutein was not involved in these reactions; instead, three β-carotene derivatives took part, namely, the diepoxide violaxanthin, the monoepoxide antheraxanthin, and epoxide-free zeaxanthin (see Fig. 11 for the structures). Thus the light-dependent reaction is a de-epoxidation sequence from violaxanthin through antheraxanthin to zeaxanthin, and the dark reaction is an epoxidation sequence from zeaxanthin to violaxanthin through the same intermediate. Figure 1 illustrates the kinetics of both reaction sequences. As will be shown below, de-epoxidation and epoxidation follow different pathways, thus forming a cyclic system. This cycle has been named xanthophyll cycle [4] or, with respect to its main epoxide compound, violaxanthin cycle [14, 17].

Knowledge of the xanthophyll cycle increased in parallel with improvements in the methods for studying its kinetics. First, samples were analyzed by paper chromatography [19] or by conventional thin-layer chromatography [40], which did not separate zeaxanthin from its isomer lutein; the latter was achieved by column chromatography [43]. Then Hager and Meyer-Bertenrath [7] introduced a thin-layer system that allowed the separation of both isomers. Contin-

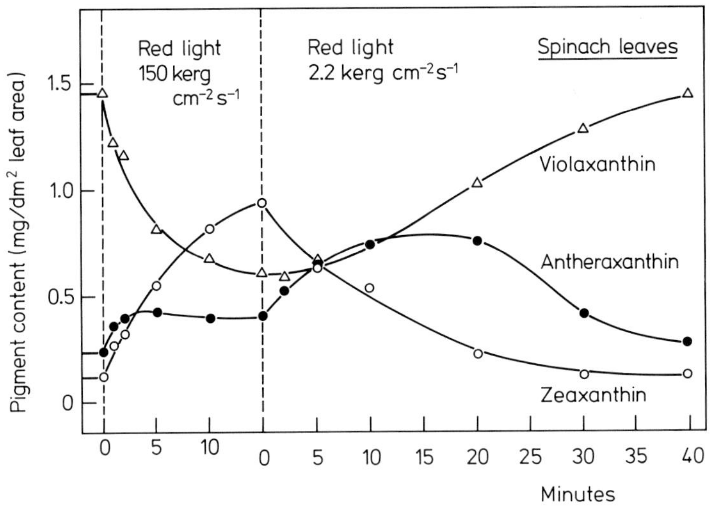

Fig. 1. Kinetics of de-epoxidation (under strong light) and epoxidation (under weak light) in spinach leaves. Weak illumination was required for epoxidation to avoid anaerobiosis by closure of the stomata. From Figs. 1–3 of Ref. [4]

uous measurement of xanthophyll changes in isolated chloroplasts became possible when Yamamoto et al. [42] discovered an absorbance change at 505 nm that was due to violaxanthin de-epoxidation. The development of a stirring arrangement that prevented settling of the chloroplasts [26] enabled long-term monitoring of this absorbance change. Since the absorbance increase was directly correlated to violaxanthin de-epoxidation [26] and the absorbance decrease was correlated to zeaxanthin epoxidation (for further details see Ref. [30]), the 505-nm change allowed detailed kinetic studies of the xanthophyll changes.

C. The De-epoxidation Pathway

I. The Cosubstrates of De-epoxidation

Although violaxanthin is converted to zeaxanthin when whole leaves are illuminated, no such reaction occurs with isolated chloroplasts in the commonly used media. This lack of activity was overcome when Hager [2] discovered that de-epoxidation could be induced by addition of excess ascorbate. The question of whether ascorbate is the endogenous reductant for de-epoxidation has not been settled, but several lines of evidence favor this possibility: (1) ascorbate cannot be replaced by pyridine nucleotides, indicating some specificity of de-epoxidation for its reductant; (2) de-epoxidation activity in isolated chloroplasts with ascorbate was of the same order of magnitude as in whole leaves; (3) ascorbate can be replaced by dehydroascorbate (Fig. 2) under conditions

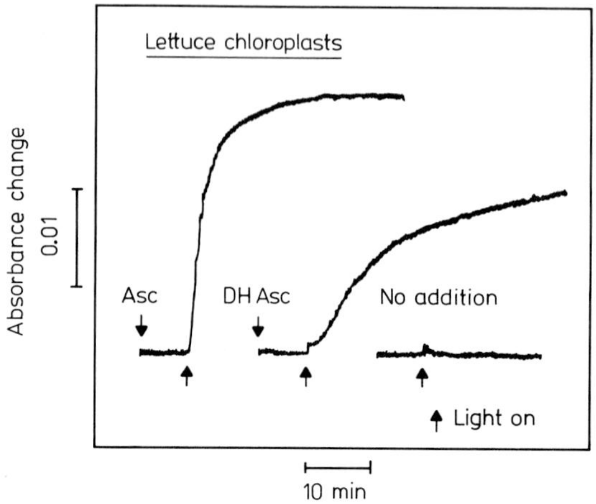

Fig. 2. Kinetics of de-epoxidation in isolated chloroplasts in the presence of ascorbate (*Asc*) and dehydroascorbate (*DHAsc*). Violaxanthin de-epoxidation is shown as absorbance increase at 505 nm with 540 nm as reference. From Fig. 2 of Ref. [28]

where dehydroascorbate is reduced by photosynthetic electron transport [28]. The latter suggests the existence of a pathway from the electron transport chain to ascorbate. This pathway seems to accept electrons after photosystem I [28], and the electrons appear to flow from NADPH through glutathione to the ascorbate-dehydroascorbate couple [6]. The ascorbate requirement indicates that de-epoxidation is a reductive reaction as presented in the following scheme:

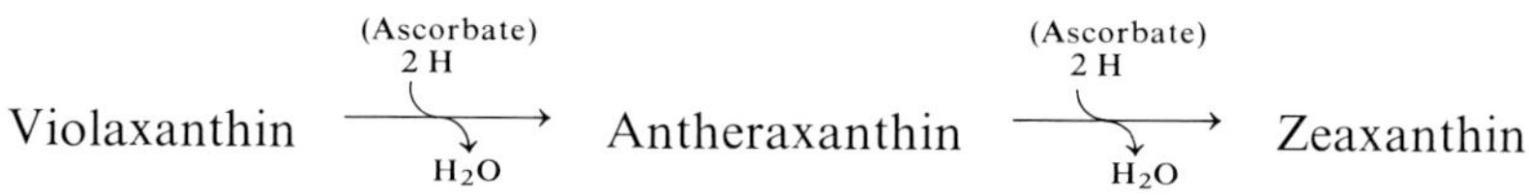

Sapozhnikov [16] concluded from studies with whole cell systems that two types of de-epoxidation exist, a reductive reaction where the epoxide oxygen is converted to water and an oxidative reaction where it is evolved as molecular oxygen. Under anaerobiosis, the reductive reaction was insensitive to diuron or hydroxylamine, while the oxidative reaction was blocked by these inhibitors [21, 23]. The existence of an oxidative reaction may, however, be questioned, since it has not been confirmed by direct experimental evidence; also, de-epoxidation in isolated chloroplasts under oxidative conditions has not been established.

II. The De-epoxidase

De-epoxidation requires not only reducing conditions but also light. The reaction could be induced by red light [2], which suggested that the light acceptor was not a carotenoid but chlorophyll. This was confirmed by the action spectrum of de-epoxidation [14, 18]. Therefore de-epoxidation seemed to be linked to photosynthetic processes. Indeed, photosynthetic electron transport inhibitors inhibited de-epoxidation [2, 3], but uncouplers that enhance electron flow had the same effect [2, 3]. Therefore proton pumping rather than electron transport appeared to affect de-epoxidation. This was confirmed when Hager [2, 5]

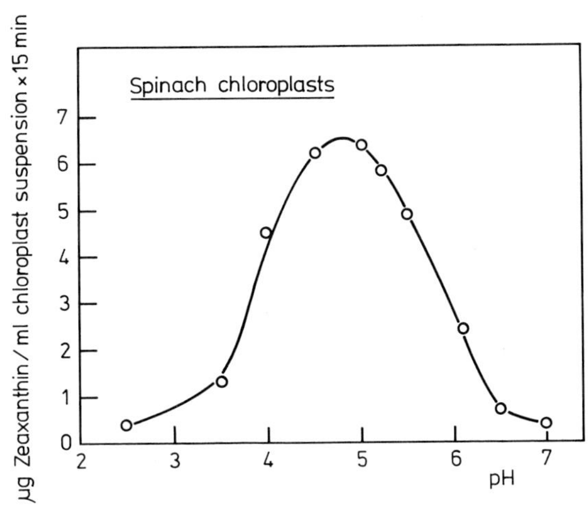

Fig. 3. Effect of medium pH on violaxanthin de-epoxidation in isolated chloroplasts in the dark. From Fig. 5a of Ref. [5]

observed that the reaction also occurred in the dark when the medium was acidic. Figure 3 shows that dark de-epoxidation in isolated chloroplasts has an optimum near pH 5. Since this is the pH of the thylakoid loculus during photosynthetic proton pumping, Hager [5] concluded that the de-epoxidase is located in the loculus and is activated by low pH.

The de-epoxidase was isolated by Perz and Hager [8, 15]. It is a water-soluble enzyme with an optimum at pH 5.2 [8] and a molecular weight of 54,000 [6]. Similar to chloroplast-bound de-epoxidase [38], the isolated enzyme is inhibited by the sulfhydryl reagent dithiothreitol (Chenchin and Yamamoto, personal communication). Yamamoto et al. [41] showed that de-epoxidation of purified violaxanthin required the presence of chloroplast lipids in addition to ascorbate and low pH. Monogalactosyl diglyceride, the main lipid of chloroplast membranes, was most effective in stimulating the reaction. The authors suggested that monogalactosyl diglyceride functions by solubilizing violaxanthin in micelles that meet structural or binding site requirements of the enzyme.

D. The Epoxidation Pathway

I. The Cosubstrates of Epoxidation

The oxygen requirement of epoxidation raised the question of whether oxygen was consumed or had some other effect on the reaction. Yamamoto and coworkers [33, 37, 39] showed that whole cell systems in an $^{18}O_2$-atmosphere incorporated ^{18}O into both antheraxanthin and violaxanthin and that incorporation rates were typical for epoxidation (Fig. 4). These results provided evidence that molecular oxygen is a cosubstrate for epoxidation and is incorporated into the epoxide groups.

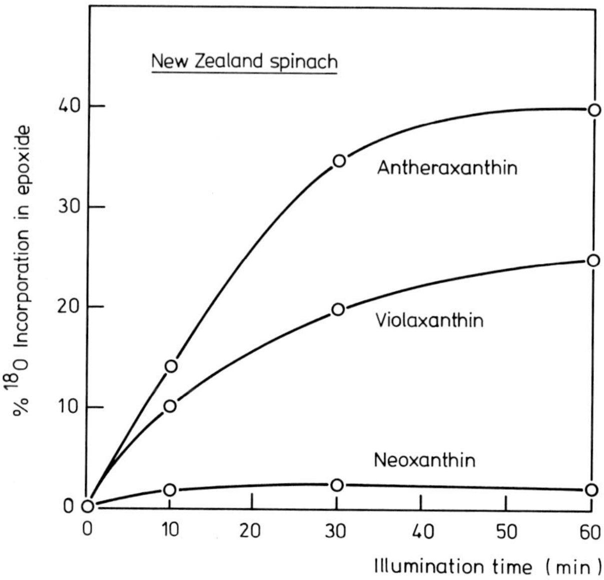

Fig. 4. $^{18}O_2$ incorporation by epoxi xanthophylls in New Zealand spinach leaves. From Fig. 4 of Ref. [33]

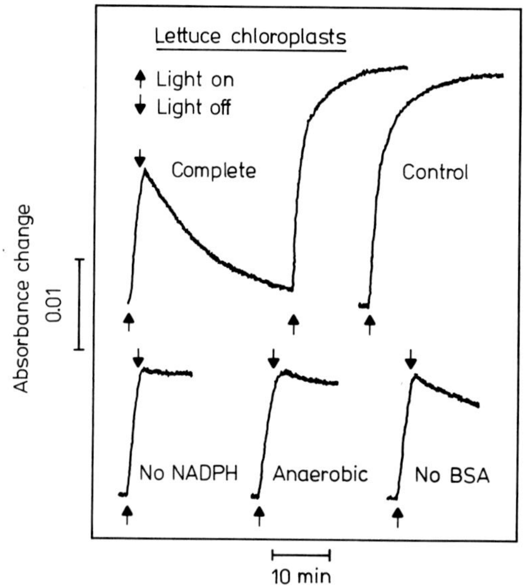

Fig. 5. Requirements of zeaxanthin epoxidation in isolated chloroplasts. Zeaxanthin was formed by illuminating the chloroplasts with red light for 2 min. *BSA*: bovine serum albumin. Violaxanthin de-epoxidation is shown as absorbance increase and zeaxanthin epoxidation as absorbance decrease at 505 nm. From Fig. 1 of Ref. [29]

Epoxidation studies in isolated chloroplasts showed that this reaction was dependent not only on molecular oxygen but also on NADPH [6, 29, 30]. Figure 5 illustrates this NADPH requirement and further shows the effect of bovine serum albumin; this protein protects the epoxidation system from inhibition by fatty acids [30] and enables epoxidation to occur at rates comparable to those in whole cell systems. The NADPH and O_2 dependency suggest that epoxydation is an external mono-oxygenation as described in the following scheme:

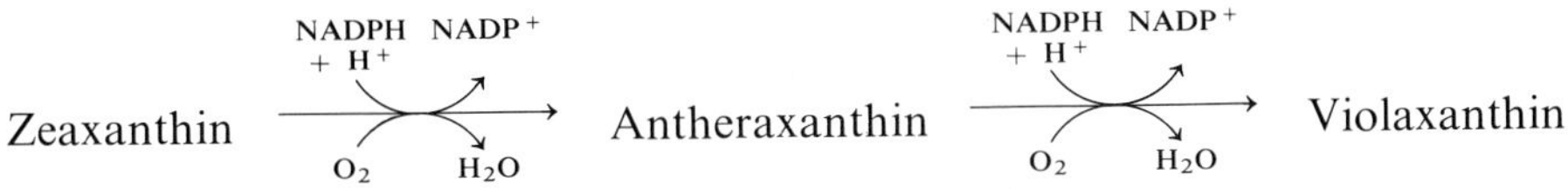

The origin of the epoxide oxygen has also been studied by Sapozhnikov et al. [20] using $H_2{}^{18}O$. This group also obtained ^{18}O-incorporation into violaxanthin, which led Sapozhnikov to conclude that the epoxide could originate from both O_2 and H_2O [17] and, for stoichiometric reasons, to postulate that reducing equivalents are produced. This proposed mechanism for epoxidation, however, does not seem probable since the ^{18}O-incorporation kinetics from $H_2{}^{18}O$ did not show a correlation with epoxidation kinetics [20] and since epoxidation requires reduced instead of oxidized pyridine nucleotides [6, 29].

II. The Epoxidase

In contrast with the location of the de-epoxidase in the loculus of the thylakoid, it appears that the epoxidase is bound on the stroma side of the membrane [6, 30]. This conclusion is based on the fact that, as shown in Figure 6, epoxidation is optimal near the pH of the stroma [10] but is inactive at the pH of the loculus under light. Since epoxidation occurs in light, the epoxidation system should be on the stroma side where it is shielded from loculus pH.

The preceding view is also supported by experiments in which NADPH was generated from NADP through photosynthetic electron transport [30, 35]. Epoxidation and NADP reduction seemed to be closely coupled because epoxidation occurred even though NADPH production was low [30] or undetectable [35]. Consequently, the epoxidase should be close to the ferredoxin-NADPH-reductase, which is known to be bound to the stroma side of the thylakoid membrane [1].

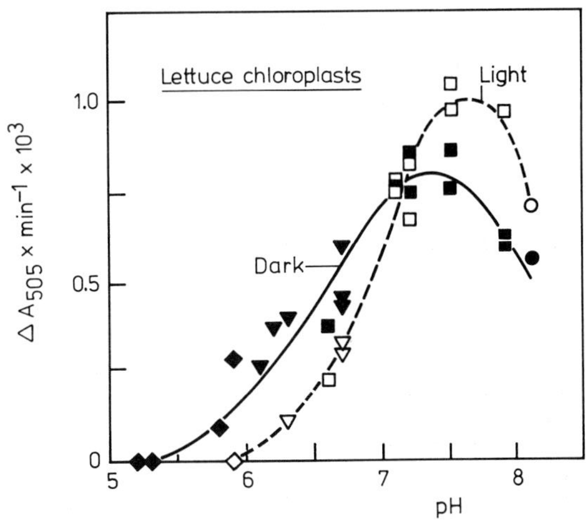

Fig. 6. Effect of medium pH on zeaxanthin epoxidation in isolated chloroplasts in light and dark. From Fig. 3 of Ref. [29]

The epoxidase has not been isolated. Washing experiments indicate that it is membrane-bound [30]. Contrary to the de-epoxidase, the epoxidase is not inhibited by dithiothreitol [29].

E. A Transmembrane Model of the Xanthophyll Cycle

The properties of de-epoxidation and epoxidation given above enable an overall view of the xanthophyll cycle [6, 30] as presented in Figure 7. The cycle is depicted as a transmembrane system wherein de-epoxidation takes place on the loculus side and epoxidation on the stroma side of the thylakoid membrane [1]. According to this scheme, turnover of the cycle requires migration of the carotenoids from one side to the other in a type of shuttle within the lipophilic membrane core.

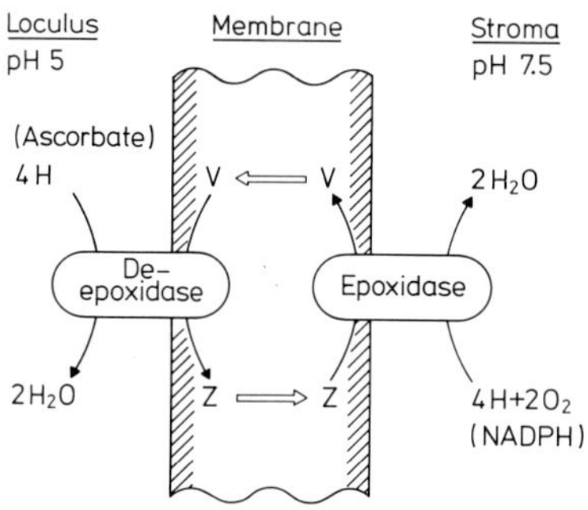

Fig. 7. Transmembrane model of the xanthophyll cycle. *V*: violaxanthin; *Z*: zeaxanthin; the intermediate antheraxanthin is not shown. From Fig. 7 of Ref. [30]

A single turnover of the cycle results in the reduction of two oxygen molecules to water with accompanying oxidation of NADPH and ascorbate. Turnover of the cycle is dependent on the loculus pH established by proton pumping as well as on the levels of NADPH, oxygen, and ascorbate, produced by photosynthetic processes. It is apparent that the xanthophyll cycle is a highly organized dynamic system that can be affected by various photosynthetic processes.

F. The Control of the Xanthophyll Cycle by Light

No significant effects of light on epoxidation have been observed. Studies of the light dependency of epoxidation under conditions where NADPH was generated photosynthetically [35] showed that epoxidation was saturated at such

[1] The xanthophyll cycle occurs in grana as well as in stroma membranes [31]. A light-induced violaxanthin decrease has also been reported in chloroplast envelopes [11], but if so, it appears to be independent of the cyclic xanthophyll reactions in thylakoid membranes.

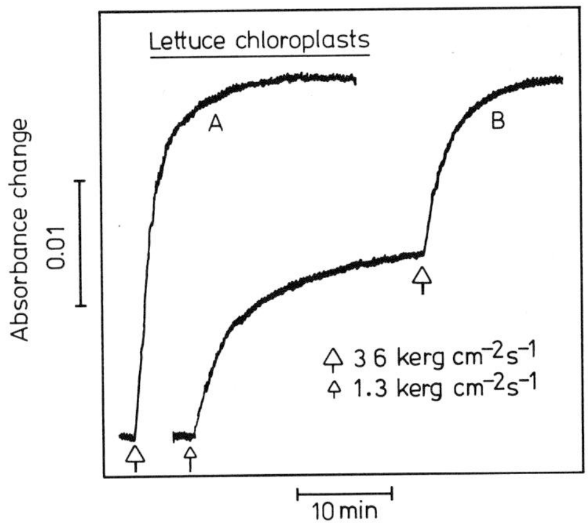

Fig. 8A and B. De-epoxidation in isolated chloroplasts (A), under saturating light intensity and (B), during consecutive phases of increasing light intensities. From Fig. 3 of Ref. [26]

low light intensities that no NADPH production was observed. Thus, even at low NADPH levels, NADPH is not rate-limiting for epoxidation. Since the in vivo level of NADPH in chloroplasts is known to be rather high even in the dark [9], the NADPH requirement for epoxidation in whole cells can be met without light. Also, the oxygen requirement should be saturated by atmospheric oxygen in the dark if its penetration into the leaf tissue is not inhibited by some secondary effect such as closure of the stomata [4].

In isolated chloroplasts, light had at least two effects on de-epoxidation, namely, it not only established the loculus pH required for de-epoxidase activity but affected the fraction of violaxanthin that could be de-epoxidized [26, 27]. This is illustrated in Figure 8, in which it is shown that the amount of violaxanthin available for de-epoxidation was increased when light intensity was increased. Siefermann and Yamamoto [27] suggested that the variable violaxanthin availability might reflect conformational changes of the thylakoid membrane that result in variable exposure of violaxanthin to the de-epoxidase. Figure 9 shows that de-epoxidase activity and violaxanthin availability were both saturated at low light intensities; it further shows that activity and availability reacted to light differently with the former reaching saturation at higher light intensity than the latter. This suggests that violaxanthin availability is not influenced by thylakoid pH but by some other photosynthetic factor. Evidence exists that violaxanthin availability might be related to the redox state of some electron carrier between the two photosystems [27].

In whole cells, the effect of light on the cycle reactions cannot be explained simply by light-induced changes of violaxanthin availability or de-epoxidase activity. While these parameters were affected by low light intensities, the light effects in whole cells were most significant under strong illumination. Figure 10a shows that in *Lemna gibba* at light intensities below 150 kerg/cm^2 sec (white light) there was a transient appearance of zeaxanthin during the first 30 min followed by no further cycle activity. At higher light intensities a steady-state level of zeaxanthin remained in the membrane indicating steady-state turnover of the cycle. Figure 10b shows that the zeaxanthin level increased linearly with

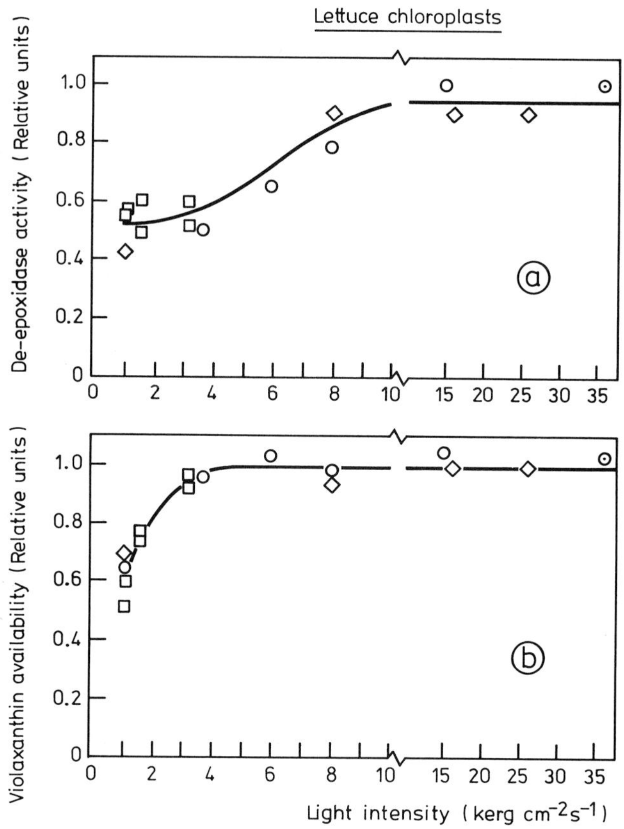

Fig. 9a and b. Effect of light intensity on (a), de-epoxidase activity and (b), violaxanthin availability in isolated chloroplasts. From Fig. 2 of Ref. [26]

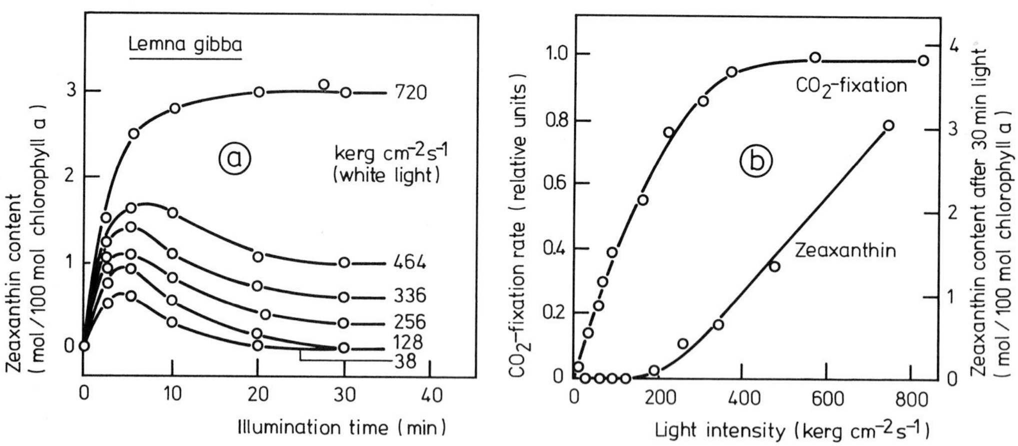

Fig. 10a and b. Effect of light intensity on xanthophyll changes and CO_2-fixation in *L. gibba* clones. (a), kinetics of zeaxanthin changes under various light intensities. Violaxanthin changes (not shown) proceeded antiparallel. (b), effect of light intensity on rate of CO_2-fixation and on steady-state level of zeaxanthin after 30 min of illumination. From Figs. 10 and 18 of Ref. [24]

light intensity while CO_2-fixation was saturated. It appears that the xanthophyll cycle was active when photosynthetic electron transport was more efficient than CO_2 fixation, as should be the case during the first minutes of illumination and at high light intensities [24, 25]. Since a pathway of electron flow from the photosynthetic electron transport chain to the reductant of de-epoxidation has been established [28], it seems reasonable to assume that those conditions induced electron flow to the reductant. Thus, light should also affect de-epoxidation by controlling the level of its reductant, an effect that has not been seen in isolated chloroplast systems presumably because in these systems the reductant has been added in excess.

To summerize the effects of light on the xanthophyll cycle, one may draw the following picture: Under low light, violaxanthin becomes available via some conformational change in the thylakoid membrane; also, the de-epoxidase is activated through photosynthetic acidification of the thylakoid loculus. The transient phenomena (Fig. 10a) show that the cycle is potentially active at low light intensities, but no xanthophyll changes are observed under steady-state conditions. Cycle turnover continues under steady-state conditions only if light intensity is high enough to supply the reducing agent for de-epoxidation. Thus light, through its various effects on the de-epoxidation side of the xanthophyll cycle, exerts effective control on the violaxanthin–zeaxanthin ratio in chloroplast membranes.

G. Hypothesis on the Function of the Xanthophyll Cycle

The occurrence of the xanthophyll cycle within the plant kingdom summarized in Figure 11 shows that it takes place in most photosynthetic organisms. In algae two modifications of the cycle have been observed, one in which violaxanthin is converted to zeaxanthin upon illumination, and another in which diadinoxanthin, a monoepoxide compound with an alkine bond in 7,8-position, is de-epoxidized [32]. After the xanthophyll cycle had been developed in algae, it was not lost during phylogenetic evolution to higher plants. This suggests some physiologic significance of the cycle, but its function is still unknown. Various proposals have been made: none has gained wide support and in some cases the evidence appears to be contradictory.

The ability of carotenoids to function as protective agents against photooxidation is well known (see Ref. [13] for a review) and has been suggested as a possible role for carotenoid epoxides. According to Krinsky [12], in lyophylized cells of *Euglena gracilis* epoxidation of diatoxanthin to diadinoxanthin, then thought to be zeaxanthin and antheraxanthin, respectively, was light-dependent. He postulated that some light-activated oxygen compound, which otherwise would damage the photosynthetic apparatus, was consumed through epoxidation. Hager [4] has questioned Krinsky's results and suggested that they could be artifacts resulting from differential photooxidative pigment destruction. In contrast, epoxidation of zeaxanthin in higher plants is a dark process that

No cyclic carotenoid changes	Diadinoxanthin 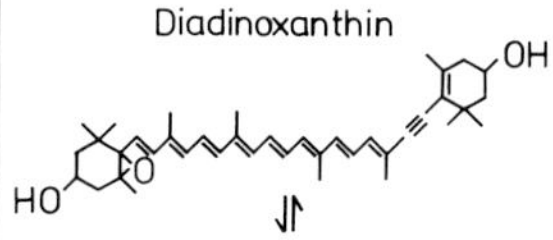 ⇅ Diatoxanthin 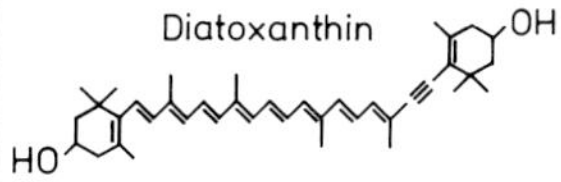	Violaxanthin 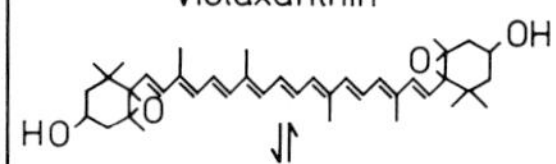 ⇅ Antheraxanthin 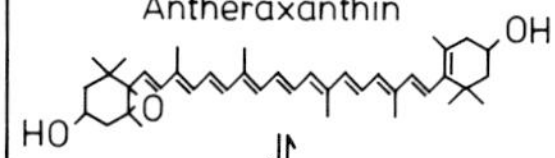 ⇅ Zeaxanthin 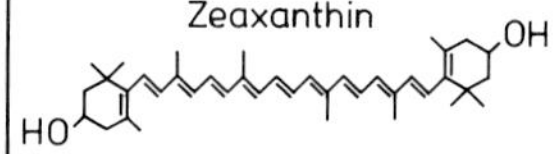
Photobacteria Cyanophyceae Cryptophyceae Rhodophyceae	Euglenophyceae Dinophyceae Chrysophyceae Bacillariophyceae Xanthophyceae Chloromonadophyceae	Chlorophyceae Phaeophyceae Higher plants

Fig. 11. Distribution of the xanthophyll cycle in the plant kingdom. From Fig. 3 of Ref. [6]

does not require light-activated oxygen. Consequently, the xanthophyll cycle does not appear to be involved in protection of the photosynthetic apparatus against light-activated lethal compounds.

Sapozhnikov [17] has proposed that the xanthophyll cycle functions as a pathway for photosynthetic oxygen evolution. However, the mechanisms for de-epoxidation and for epoxidation that were suggested to support this view can be questioned (see Sections C and D); also, photosynthetic activity of whole cells does not require turnover of the xanthophyll cycle (Fig. 10b).

According to Hager [6] the cycle might improve the starting conditions for photosynthetic electron transport at the beginning of illumination by decreasing the NADPH level in the thylakoid partition. Such a function of the cycle seems rather inefficient since NADPH consumption by the cycle is small compared to photosynthetic NADPH production [2].

Speculations about the function of the xanthophyll cycle have to contend with the fact that its turnover is slow as compared with photosynthetic reactions. The cycle has little effect on the levels of its cosubstrates but is effectively controlled by the level of its reductant. It is also affected by photosynthetic proton pumping and the state of the chloroplast membrane. Thus, the violaxanthin–zeaxanthin ratio of the membrane responds to changes of various photosynthetic activities, or in other words, the pigment ratio 'monitors' those changes.

[2] Initial rates of de-epoxidation in whole leaves are in the range of 2–3 μmol violaxanthin/mg chlorophyll × h (spinach leaves, calculated from Ref. [4]); epoxidation, with rates near 0.3 μmol zeaxanthin/mg chlorophyll × h [4], is slower and appears to be rate-limiting for cycle turnover. If the reductant of de-epoxidation is regenerated by NADPH [6], initial rates of NADPH oxidation of 4–6 μmol/mg chlorophyll × h can be expected, and steady-state rates are less. In contrast, photosynthetic NADP reduction is approximately two orders of magnitude higher.

Siefermann and Yamamoto [30] have therefore suggested that the xanthophyll cycle might be part of a regulatory system for photosynthesis that funtions by influencing membrane properties.

Acknowledgements. The author thanks Drs. A. Hager and H.Y. Yamamoto for critically reading the manuscript. Her work has been supported in part by the *Deutsche Forschungsgemeinschaft.*

References

1. Berzborn, R.: Z. Naturforsch. **24b**, 436–446 (1969)
2. Hager, A.: Ber. Dtsch. Bot. Ges. **79**, 94–107 (1966)
3. Hager, A.: Planta **74**, 148–172 (1967)
4. Hager, A.: Planta **76**, 138–148 (1967)
5. Hager, A.: Planta **89**, 224–243 (1969)
6. Hager, A.: Ber. Dtsch. Bot. Ges. **88**, 27–44 (1975)
7. Hager, A., Meyer-Bertenrath, T.: Planta **69**, 198–217 (1966)
8. Hager, A., Perz, H.: Planta **93**, 314–322 (1970)
9. Heber, U., Krause, G.H.: In: Proc. 2nd Internat. Congr. Photosynth. Res. Forti, G., Avron, M., Melandri, A. (eds.). The Hague: W. Junk, Vol. 2, 1972, pp. 1023–1033
10. Heldt, H.W., Werdan, K., Milovancev, M., Geller, G.: Biochim. Biophys. Acta **314**, 224–241 (1973)
11. Jeffrey, S.W., Douce, R., Benson, A.A.: Proc. Natl. Acad. Sci. U.S. **71**, 807–810 (1974)
12. Krinsky, N.J.: In: Biochemistry of Chloroplasts. Goodwin, T.W. (ed.). London: Academic Press, 1966, Vol. 1, pp. 423–430
13. Krinsky, N.J.: In: Photophysiology: Current Topics. Giese, A.C. (ed.). New York and London: Academic Press, 1968, Vol. 3, pp. 123–195
14. Lee, K.H., Yamamoto, H.Y.: Photochem. Photobiol. **7**, 101–107 (1968)
15. Perz, H.F.: Dissertation, Univ. München (1970)
16. Sapozhnikov, D.I.: In: Progress Photosynth. Res. Metzner, H. (ed.). Tübingen: H. Laupp jr., 1969, Vol. 2, pp. 694–700
17. Sapozhnikov, D.I.: J. Pure Appl. Chem. **35**, 47–61 (1973)
18. Sapozhnikov, D.I., Kolotova, L.R., Giller, Y.G.: Dokl. Acad. Nauk USSR **171**, 740–741 (1966)
19. Sapozhnikov, D.I., Krasovskaya, T.A., Mayeskaya, A.N.: Dokl. Akad. Nauk USSR **113**, 465–467 (1957)
20. Sapozhnikov, D.I., Maslova, T.G., Bazhanova, N.V., Popova, O.F.: Biofizika **10**, 349–351 (1965)
21. Sapozhnikov, D.I., Ivantsova, L.V., Maslova, T.G., Popova, I.A., Korolyova, O.Ya.: Dokl. Akad. Nauk USSR **189**, 1135–1137 (1969)
22. Sapozhnikov, D.I., Mayeskaya, A.N., Krasovskaya, T.A., Prialgauskaite, L.L., Torchina, V.S.: Biokhimiya **24**, 39–41 (1959)
23. Sapozhnikov, D.I., Eidelman, Z.M., Bazhanova, N.V., Maslova, T.G., Popova, O.F., Shirjayeva, G.A.: Botan. Zh. **49**, 1463–1465 (1964)
24. Siefermann, D.: Dissertation, Univ. Tübingen (1971)
25. Siefermann, D.: In: Proc. 2nd Internat Congr. Photosynth. Res. Forti, G., Avron, M., Melandri, A. (eds.). The Hague: W. Junk, 1972, Vol. 1, pp. 629–635
26. Siefermann, D., Yamamoto, H.Y.: Biochim. Biophys. Acta **357**, 144–150 (1974)
27. Siefermann, D., Yamamoto, H.Y.: Biochim. Biophys. Acta **387**, 149–158 (1975)
28. Siefermann, D., Yamamoto, H.Y.: Proc. 3rd Internat. Congr. Photosynth. Res. Avron, M. (ed.). Amsterdam: Elsevier, 1975, Vol. 3, pp. 1991–1998
29. Siefermann, D., Yamamoto, H.Y.: Biochem. Biophys. Res. Commun. **62**, 456–461 (1975)
30. Siefermann, D., Yamamoto, H.Y.: Arch. Biochem. Biophys. **171**, 70–77 (1975)
31. Siefermann, D., Yamamoto, H.Y.: Plant Physiol. **57**, 939–940 (1976)

32. Stransky, H., Hager, A.: Arch. Mikrobiol. **73**, 315–323 (1970)
33. Takeguchi, C.A., Yamamoto, H.Y.: Biochim. Biophys. Acta **153**, 459–465 (1968)
34. Weedon, B.C.L.: In: Carotenoids. Isler, O. (ed.). Basel und Stuttgart: Birkhäuser Verlag, 1971, pp. 29–59
35. Westerhoff, N.: Ber. Dtsch. Bot. Ges. **87**, 545–551 (1974)
36. Witt, H.T.: Quart. Rev. Biophys. **4**, 365–477 (1971)
37. Yamamoto, H.Y., Chichester, C.O.: Biochim. Biophys. Acta **109**, 303–305 (1965)
38. Yamamoto, H.Y., Kamite, L.: Biochim. Biophys. Acta **267**, 538–543 (1972)
39. Yamamoto, H.Y., Takeguchi, C.A.: In: Proc. 2nd Internat. Congr. Photosynth. Res. Forti, G., Avron, M., Melandri, A. (eds.). The Hague: W. Junk, 1972, Vol. 1, pp. 622–627
40. Yamamoto, H.Y., Chang, J.L., Aihara, M.S.: Biochim. Biophys. Acta **141**, 342–347 (1967)
41. Yamamoto, H.Y., Chenchin, E.E., Yamada, D.K.: Proc. 3rd Internat. Congr. Photosynth. Res. Avron, M. (ed.). Amsterdam: Elsevier, 1975, Vol. 3, pp. 1999–2006
42. Yamamoto, H.Y., Kamite, L., Wang, Y.-Y.: Plant Physiol. **49**, 224–228 (1972)
43. Yamamoto, H.Y., Nakayama, T.O.M., Chichester, C.O.: Arch. Biochem. Biophys. **97**, 168–173 (1962)

CHAPTER 13

Regulation of Prenylquinone Synthesis in Higher Plants

H.K. LICHTENTHALER

A. Introduction

Benzo- and naphthoquinone derivatives with a lipophilic isoprenoid side chain (prenylquinones) are ubiquitous in nature [28, 49, 51, 52, 66]. All living organisms contain at least one lipoquinone. These prenylquinones belong to the group of prenyllipids that consists of the pure isoprenoid compounds (sterols, carotenoids) and of the mixed terpenoid lipids with an isoprenoid side chain bound to a nonterpenoid nucleus (chlorophylls, prenylquinones). The lipoquinone composition in heterotrophic or photoautotrophic bacteria varies greatly. Algae and higher plants exhibit, in turn, a uniform prenylquinone pattern [28].

There are two localization sites for lipoquinones within the cell: mitochondria and plastids. The plant mitochondria in general contain only one prenylquinone, the ubiquinone, a 2-methyl-5,6-dimethoxy-1,4-benzoquinone with 9 or 10 iso-

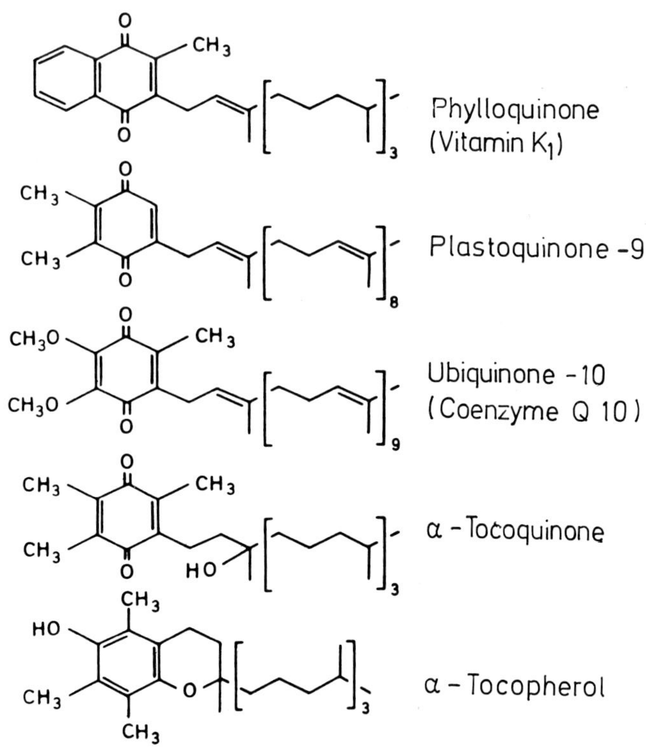

Fig. 1. Chemical structure of the main prenylquinones from higher plants and algae

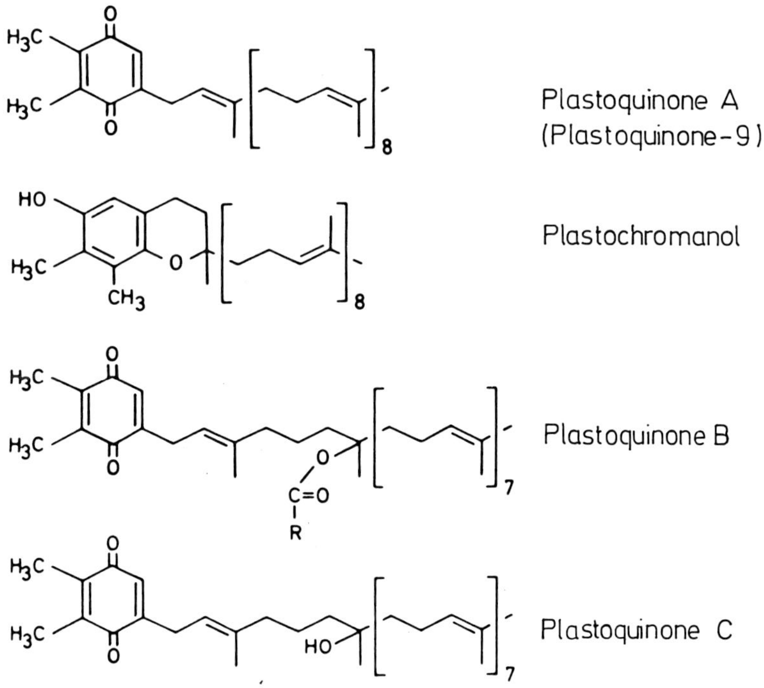

Fig. 2. Chemical structure of plastoquinone-9 derivatives, which may appear in older plant tissues in minor amounts. The exact association of the one OH group with a specific isoprene unit (shown here with the second), has not yet been established for the individual components of the plastoquinone B and C series [62]

prenoid units in the side chain (UQ-9 or UQ-10). Chloroplasts of algae and higher plants always contain several prenylquinones simultaneously [16, 28, 32]. The major components of the plastid quinones are: (1) *plastoquinone-9* (PQ-9), which always occurs together with its reduced form plastohydroquinone-9 (PQ-9 $\cdot H_2$); (2) *α-tocoquinone,* which only occasionally can also be present as α-tocohydroquinone; (3) *α-tocopherol* (vitamin E or α-tocochromanol), a chromanol that can arise from α-tocohydroquinone by ring closure and water removal and that delivers α-tocoquinone upon oxidation; (4) the naphthoquinone derivative *phylloquinone* (vitamin K_1). The chemical structure of the major prenylquinones is shown in Figure 1.

Besides these major quinone components, some additional lipoquinones occasionally occur in minor concentrations; they represent derivatives, precursors, or decomposition products of the main prenylquinones. Thus, besides plastoquinone-9 one can find, especially in chloroplasts from older plant tissues, the plastoquinone C group [16, 62], with one OH group in the prenyl side chain, and the plastoquinone B group [16, 62], which comprises the fatty acid esters of the plastoquinone C group (Fig. 2). Plastochromanol, the cyclic form of plastohydroquinone-9, has been found after column chromatography of lipid extracts from several plants on alumina columns [17]. In our own investigations using a fast adsorption or partition chromatography on silica gel plates (Fig. 3),

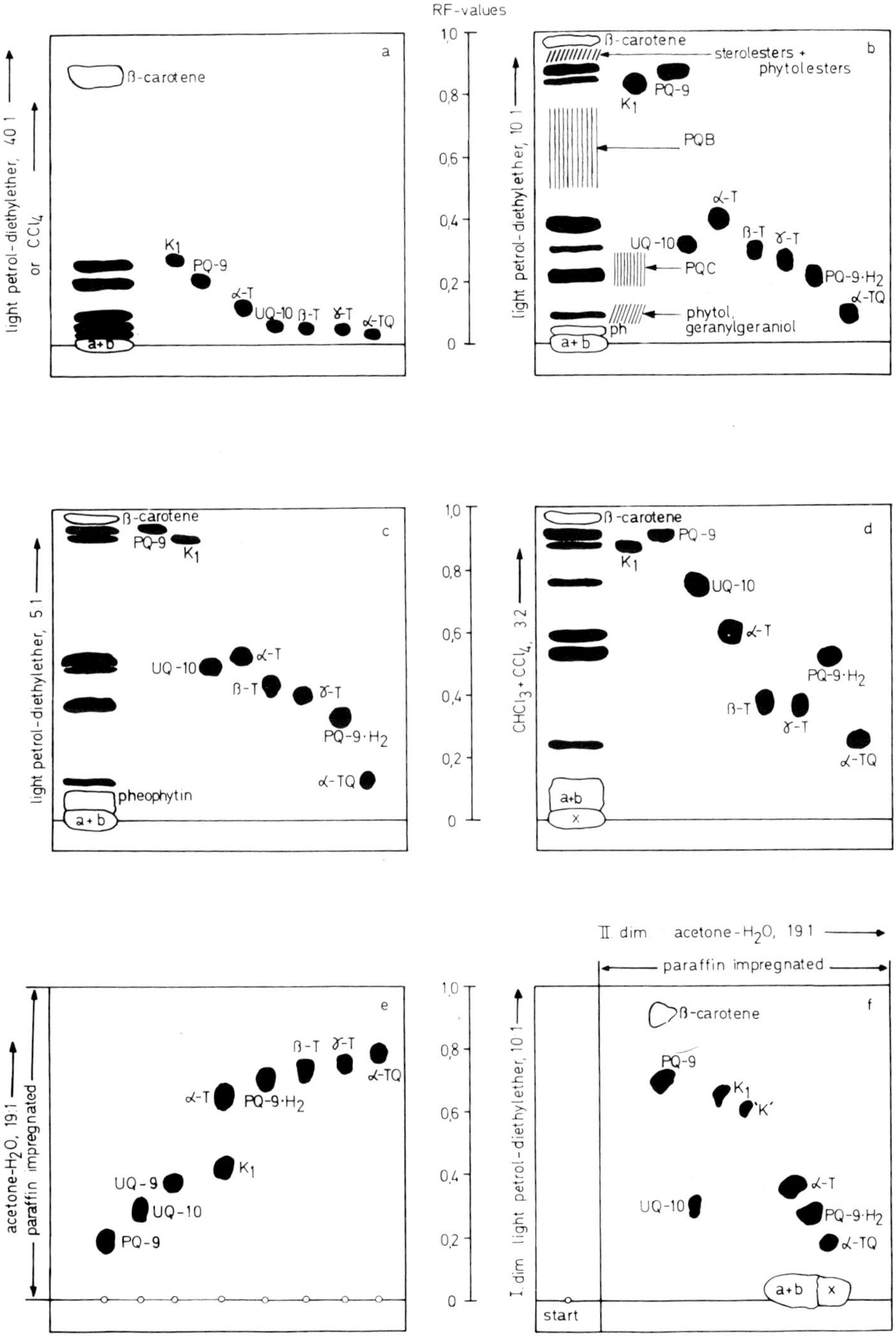

Fig. 3a–f. Adsorption and partition chromatography of prenylquinones on silica gel plates. *PQ-9*, *PQ-9·H_2*: plastoquinone-9, plastohydroquinone-9; *K_1*: vitamin K_1 (phylloquinone); *UQ-9*, *UQ-10*: ubiquinone-9 or -10; α-, β-, γ-T: α-, β-, γ-tocopherol; α-*TQ*: α-tocoquinone; *PQB*, *PQC*: plastoquinone B and C series

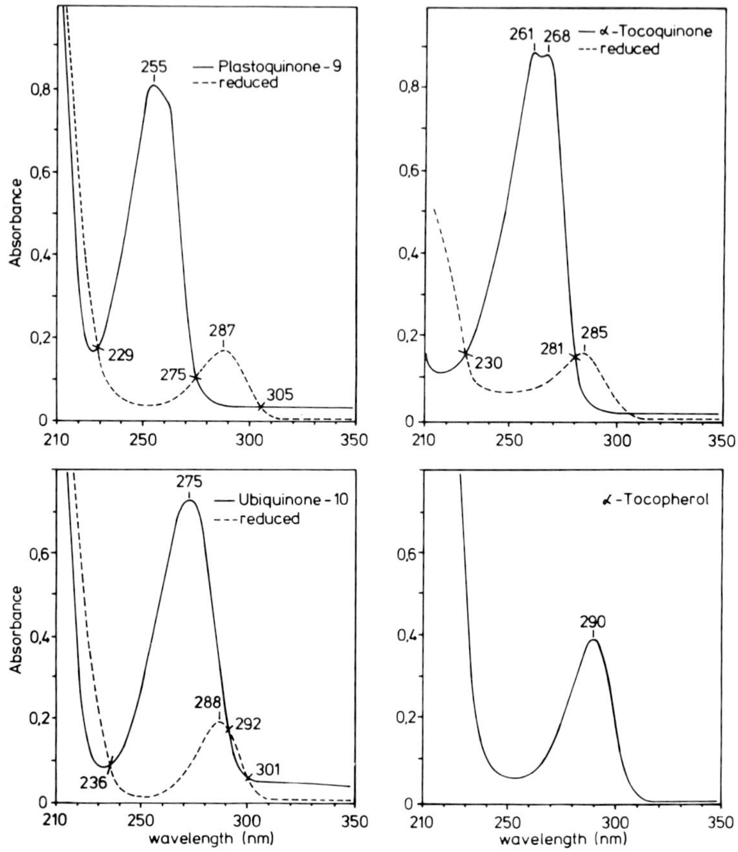

Fig. 4. Absorption spectrum in UV light of the main prenylquinones

it could not be detected and may thus be an artifact obtained on the column. We also found that chromatography of pure plastoquinone-9 on basic alumina columns results in the formation of at least 5 different compounds, which exhibit UV-absorption spectra like those of plastoquinone-9 and R_f values like those of the PQB and PQC series.

In etiolated tissue one finds a second lipophilic naphthoquinone that has chromatographic properties like those of either desmethylvitamin K_1 or a vitamin K_1 with a more unsaturated prenyl side chain. This lipoquinone seems to be a precursor of K_1: its pool rapidly decreases during the light-induced formation of the photosynthetic apparatus [37]. Besides the trimethyl-benzoquinone derivative α-tocopherol, one may find the dimethyl derivatives β- and γ-tocopherol with the corresponding β- and γ- and also δ-tocoquinones [16, 49, 66].

The prenylquinones possess typical absorption spectra in UV light that can be used for their identification and quantitation (Fig. 4). A new and effective method for prenylquinone determination is high-pressure liquid chromatography (Fig. 5).

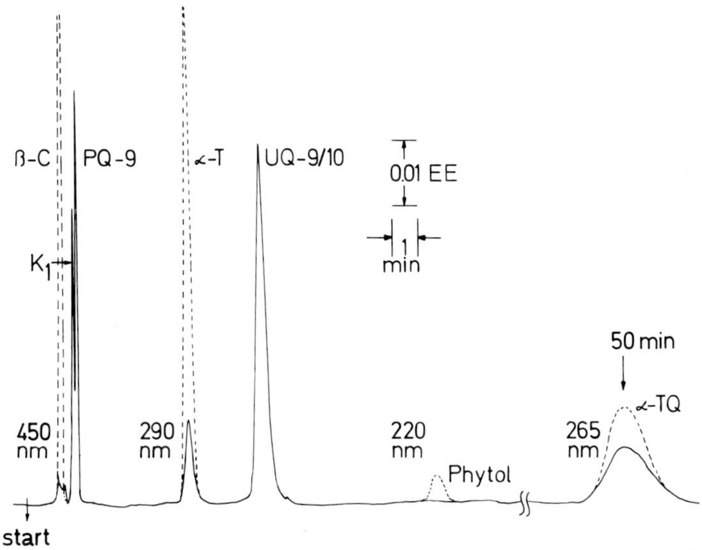

Fig. 5. High-performance liquid chromatography of β-carotene and prenylquinones on LiChrosorb Si 60 5 μ with 1.25% dioxan in hexane as mobile phase

B. Function of Prenylquinones

The lipophilic prenylquinones are potential electron carriers and are thus essential constituents of the biomembranes in mitochondria and chloroplasts. Ubiquinones catalyze electron transport in the respiratory chain [49]. Among the lipoquinones of plastids, plastoquinone-9, the terminal electron acceptor of the photosynthetic light reaction II, plays an important role in the photosynthetic electron transport chain [63]. α-Tocoquinone and phylloquinone are additional lipophilic electron carriers of the photosynthetic biomembranes. They are accumulated parallel to chlorophyll during light-induced thylakoid formation [30]; their functional site, however, is not yet known. Their concentration in the photosynthetic membrane is similar to that of other electron carriers such as P700 or cytochromes (Table 1).

Table 1. Concentrations of prenylquinones and other components of the photosynthetic electron transport chain in pigment system I particles from spinach chloroplasts

Substance	Mol	Reference
Chlorophyll a	150	
Vitamin K_1	2	[30]
Plastoquinone-9	2	[30]
P700	1	[1]
Cytochrome b_6	1	[11]
Cytochrome f	0.5	[1]
α-Tocoquinone	0.5	[30]

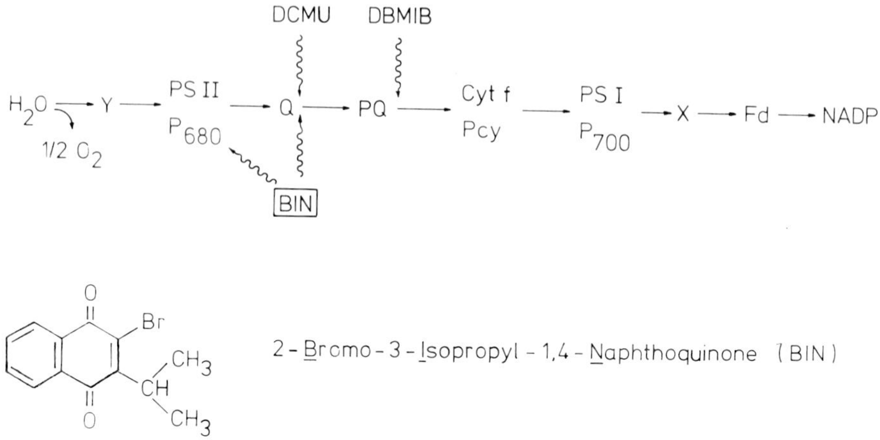

Fig. 6. Photosynthetic electron transport chain with indication of site of inhibition of several inhibitors. (Lichtenthaler and Pfister, 1976, unpublished)

From recent observations with halogenated naphthoquinones one may conclude that the functional site of phylloquinone K_1 is near the reaction center of photosystem II. Thus 2-bromo-, 3-isopropyl-1,4-naphthoquinone (BIN) blocks photosynthetic oxygen evolution and electron transport specifically in photosystem II, while photosystem I reactions are unaffected (Fig. 6). The inhibition site of the bromonaphthoquinone is at an earlier position in the electron transport chain than that of a bromobenzoquinone (DBMIB), which is known to block the oxidation of the reduced plastoquinone-9. In analogy to this, one can assume that the bromonaphthoquinone blocks at the functional site of vitamin K_1. BIN not only inhibits photosynthetic electron transport but, in addition to this, quenches chlorophyll fluorescence in photosystem II. We also found that pure vitamin K_1 exerts similar quenching properties on chloroplasts. Thus a double function as quencher and electron carrier near the reaction center of photosystem II can be postulated for the endogenous vitamin K_1.

Another function for chloroplast prenylquinones is their role as lipid-antioxidants. Plastoquinone-9 and plastohydroquinone-9 form a lipophilic redox system of chloroplasts, as do α-tocoquinone and α-tocopherol. Both can protect the photosynthetic membrane lipids against photooxidative degradation.

C. Biosynthesis of Prenylquinones

The biosynthesis of the benzo- and naphthoquinone residues of the prenylquinones proceeds via the aromatic products of the shikimic acid pathway, while the side chains derive from the prenyl metabolism. The biosynthetic routes have been described in detail in the review article by Threlfall and Whistance [62]. The main precursor for the benzoquinone ring is homogentisic acid, which

Fig. 7. Scheme of the biosynthesis route leading to prenyllipids

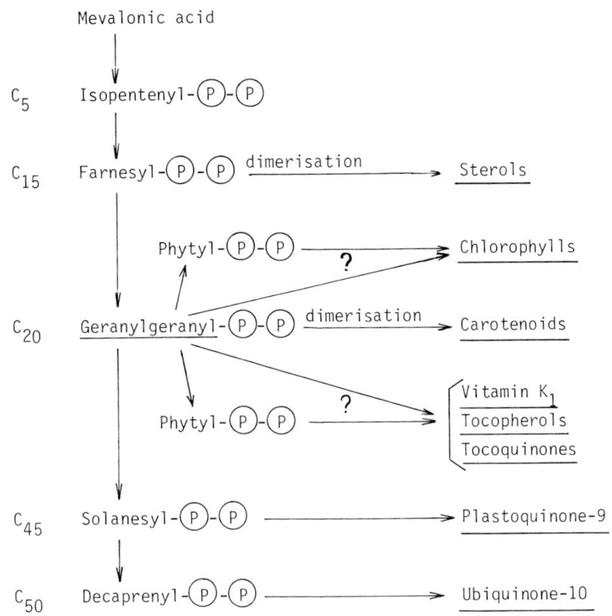

can be prenylated with a C_{45} chain to yield finally the plastoquinones or with a C_{20} prenyl chain to give tocopherols and tocoquinones [61]. The naphthoquinone residue, apparently formed by condensation of succinate with shikimic acid, is bound to a C_{20} prenyl chain to give eventually vitamin K_1. Certain aspects of lipoquinone and chromanol biosynthesis are reported in this book by Goodwin (see Chapter 3).

The detection of tocotrienols [52, 62, 67, 69], of a second lipophilic naphthoquinone [37], as well as the phototransformation of geranylgeraniol-protochlorophyll to geranylgeraniol-chlorophyll a (see Liljenberg, Chapter 14), and the appearance of geranylgeraniol-chlorophyll a in amitrole-treated plants [54], may indicate that in the case of the mixed prenyllipids the unsaturated C_{20} prenyl chain (geranylgeranyl-P-P or free geranylgeraniol) is first bound to the nonisoprenoid nucleus (porphyrin, benzoquinone, or naphthoquinone ring) before its saturation to the phytyl or hydroxyphytyl chain (Fig. 7). Further investigation may thus reveal that the formation of phytyl-Ⓟ-Ⓟ or of free phytol is not required for chlorophyll, tocopherol, and vitamin K_1 synthesis.

D. Site of Synthesis of Prenylquinones and Chromanols in the Plant Cell

At present it appears that there are two or three sites for the biosynthesis of terpenoids in the green plant cell: (1) an intrachloroplastidic site that would be responsible for the synthesis of carotenoids and the prenyl side chains of

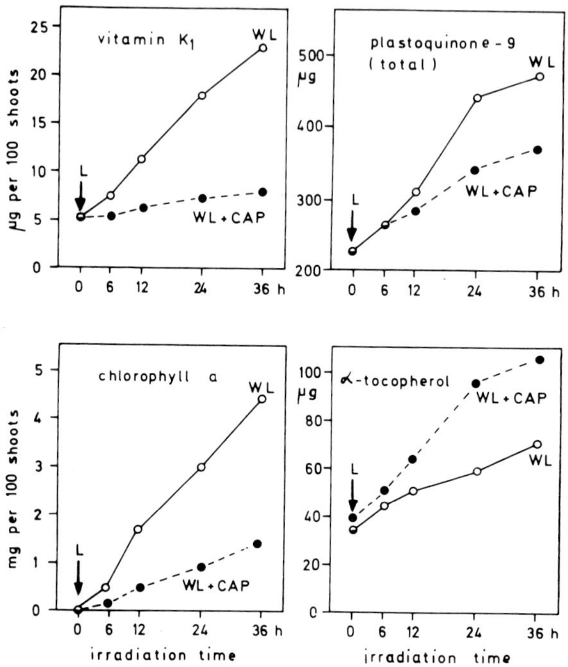

Fig. 8. Inhibition of white light (WL)-induced formation of prenylquinones and chlorophylls in 7-day old *Hordeum* seedlings by chloramphenicol applied via the roots 12 h before illumination (0.1 mg/ml nutrient solution)

chlorophylls, plastoquinones, tocopherols, tocoquinones, and phylloquinone; and (2) an extrachloroplastidic site where sterols (cytoplasm?) and the prenyl chains of ubiquinones (mitochondria?) are formed. Thus in greening seedlings, a considerable amount of radioactivity from $^{14}CO_2$ was incorporated into the chloroplast components β-carotene, vitamin K_1, plastoquinone-9, α-tocopherol, and α-tocoquinone, while only a little was incorporated into ubiquinone and sterols. 2-^{14}C-Mevalonic acid, in turn, was well incorporated in sterols and ubiquinone but only poorly into the chloroplast prenyllipids [62]. The incorporation of mevalonic acid into plastoquinone-9 by isolated chloroplasts [68] is further evidence for the assumption that chloroplasts possess their own biosynthetic system for prenyl formation. The incorporation of $^{14}CO_2$ into β-carotene and plastoquinone-9 by an isolated chloroplast fraction has now been reported [10]. Another group assumes that the prenyl chain of plastoquinone-9 is synthesized in the cytoplasm and transported to the chloroplasts [24]. These data, however, need confirmation and could be explained in a different way.

Whether the chloroplast is capable of synthesizing the different aromatic precursors of the benzoquinone and naphthoquinone rings (e.g., shikimic acid, phenylpyruvic acid, or homogentisic acid) has not yet been established. It seems clear that at least the final steps in lipoquinone biosynthesis should proceed in chloroplasts (for plastoquinone-9, α-tocopherol, α-tocoquinone, and phyllo-

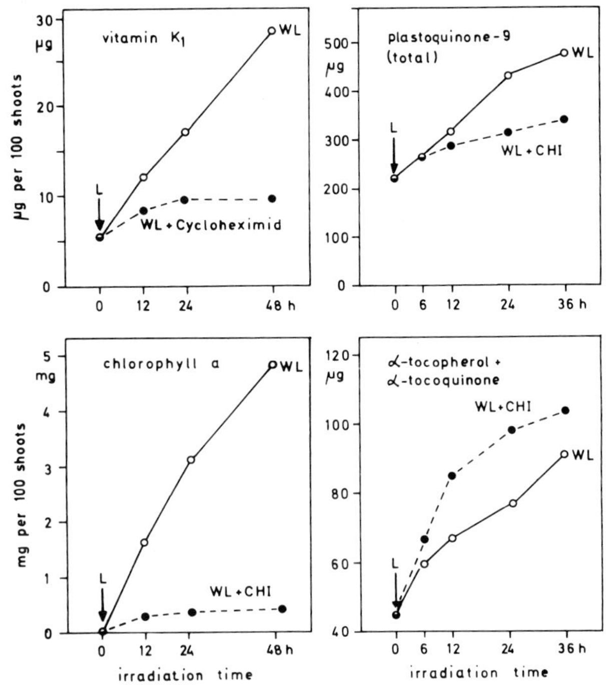

Fig. 9. Inhibition of white light (WL)-induced accumulation of prenylquinones and chlorophylls in *Hordeum* seedlings by cycloheximide applied via the roots 12 h prior to onset of illumination (0.2 mg/ml nutrient solution)

quinone) and in mitochondria (for ubiquinone). From feeding experiments with D,L-tyrosin (3-^{14}C), it was concluded that α-tocopherol is synthesized outside the chloroplast [55]. Another paper assumes that tocopherols may be synthesized outside the chloroplast and transported into plastids before or after the final methylation step [50].

In any case, chloroplasts, which are thought to be former endosymbiotic prokaryotic organisms, are not autonomous with regard to their lipoquinone synthesis. This can be seen from studies with inhibitors of protein synthesis. Chloramphenicol, which is known to bind specifically to chloroplast ribosomes, blocks the formation of vitamin K_1, chlorophylls, carotenoids, and prenylbenzoquinones (Fig. 8), as does cycloheximide, which is known to inhibit cytoplasmatic protein synthesis (Fig. 9). This shows that a fully functional metabolism in both compartments—chloroplasts and cytoplasm—must proceed to yield prenylquinone, chlorophyll, and carotenoid formation in chloroplasts.

The biosynthesis of the benzoquinone residue of ubiquinone proceeds via phenylalanin, p-hydroxy-cinamic acid and p-hydroxy-benzoic acid, which then is bound to a nona- or decaprenyl side chain [62]. It is of interest in this respect that a mitochondrial preparation of *Vicia faba* seeds is capable of synthesizing nonaprenylpyrophosphate from isopentenylpyrophosphate, which then is transferred to p-hydroxy benzoate, the distal precursor of the p-benzoquinone ring of ubiquinone [61]. Whether mitochondria can form p-hydroxybenzoate or whether this compound must be delivered from the cytoplasm, is not yet known.

Future investigations on the enzyme equipment of chloroplasts and mitochondria will elucidate which steps in prenylquinone biosynthesis are carried out by the two cell organelles and which are performed by the cytoplasm. The isolation of 3-hydroxy-3-methyl glutaryl-coenzyme A (HMG-CoA) reductase from both chloroplasts and microsomal fractions is a first step in this direction [2, 13].

E. Regulation of Light-induced Prenylquinone Synthesis by Phytochrome P_{fr}

The lipophilic prenylquinones are formed in all differentiation and developmental stages of plastids [29]. Leucoplasts and amyloplasts have a low lipoquinone content with α-tocopherol as the main component. Etioplasts of dark-grown seedlings and chromoplasts, e.g., from fruits or petals, possess a good level of lipoquinones, predominantly α-tocopherol (+α-tocoquinone) followed by plastoquinone-9 with small amounts of phylloquinone [29]. Chloroplasts, in turn, contain higher amounts of plastoquinone-9 than of α-tocopherol and in particular a higher proportion of vitamin K_1 than etioplasts. In young chloroplasts the prenylquinones, making up 3% to 5% of the total thylakoid lipids, are quantitatively localized together with chlorophylls, carotenoids, phospho- and galactolipids in the photochemically active thylakoids [40, 44].

Light induces in etioplasts, parallel to the formation of chlorophylls and thylakoids, an enhanced and specific synthesis of prenylquinones and carotenoids. This changes the prenylquinone and carotenoid pattern of etioplasts to that of chloroplasts within 20 to 30 h of illumination [30, 38]. The following changes of lipoquinone metabolism can be observed:

1. Enhanced synthesis and accumulation of vitamin K_1 and plastoquinone-9, α-tocoquinone and to a lower degree also α-tocopherol. The promotion of K_1 formation is greater than that of plastoquinone-9.
2. Among the benzoquinone derivatives, formation of more total plastoquinone-9 (oxidized and reduced form) than α-tocopherol (+α-tocoquinone).
3. As compared to darkness, the oxidized form, plastoquinone-9, is enriched to a greater degree than the reduced form, plastohydroquinone-9.
4. Accumulation of a higher percentage of α-tocoquinone (oxidized form) and less α-tocopherol (reduced form) than in the dark.
5. The pool size of the second liponaphthoquinone vitamin 'K', which is accumulated in etioplasts in addition to K_1, decreases to zero with the concomitant formation of vitamin K_1.

It has been shown that these white light-induced changes can be initiated by illumination with continuous far-red light in *Raphanus* and barley seedlings

Table 2. Promotion of prenylbenzoquinone synthesis in etiolated *Raphanus* seedlings treated with 3 short pulses (5 min) of red and/or far-red light at different times after sowing (hours are indicated)

3 Pulses at 42, 48, 60 h	Red	Red/far-red	Far-red	Dark
Chlorophyll a	22.5	16.7	5.3	0
Protochlorophyllide	trace	trace	1.8	3.6
Plastoquinone-9	36	24	22	23
Plastohydroquinone-9	276	240	223	145
Total plastoquinone-9	312	264	245	168
α-Tocopherol	170	160	153	106
α-Tocoquinone	6	4	5	4
Total α-tocopherol	176	164	158	110

Plants were analyzed 24 h after 1st pulse (μg/400 cotyledons, mean values from 6 separate determinations, standard deviation ±3%)

[37, 38]. These results indicate that active phytochrome P_{fr}, at least in its excited state (high irradiance response), initiates and regulates light-induced prenylquinone synthesis related to thylakoid formation.

An induction effect by short red light pulses and its subsequent reversion by short-term far-red pulses are taken as criteria for phytochrome action in its ground state. When dark-grown, etiolated *Raphanus* seedlings are treated with red light pulses at certain intervals, the formation of prenylbenzoquinone is greatly enhanced as compared to that in the dark-grown controls (Table 2). The synthesis of total plastoquinone-9 is promoted to a greater extent than that of total α-tocopherol, as is known from white light [30]. Far-red light pulses given subsequently to red light pulses reverse the red light-induced promotion effects to about the level of those observed in the far-red controls. This is to be expected, since far-red also induces a low quantity of physiologically active phytochrome P_{fr}. Thus the criteria for phytochrome action (induction and reversion) are fulfilled. Particularly in the case of plastoquinone-9, it is seen that far-red pulses subsequent to red ones do not result in a reversion to the levels observed in the far-red controls. This effect is explained by the conclusion that the P_{fr}-induced promotion of plastoquinone-9 accumulation is a very fast phytochrome effect, which after a 5-min red pulse, can no longer be fully reversed by subsequent far-red light pulses.

The formation of protochlorophyllide is also stimulated by the red and far-red light treatment as can be seen from the accumulation of chlorophyll a (or chlorophyllide), to which protochlorophyllide is photoreduced by the light pulses (Table 2).

Similar red light pulse-induced promotion effects can be seen with the naphthoquinone vitamin K_1 (Table 3). In this case far-red pulses given subsequently to red light pulses cause a reversion to the levels of the far-red controls. This

Table 3. Vitamin K_1 and 'K' content (μg in 400 cotyledons) of etiolated *Raphanus* seedlings treated with 2 or 3 short pulses (5 min) of red and/or far-red light at different times after sowing

	2 Pulses at 42 h and 48 h	3 Pulses at 42, 48, and 60 h	2 Pulses at 66 and 72 h
Vitamin K_1			
Red	5.4	7.0	7.0
Red/far-red	3.6	4.4	5.1
Far-red	3.1	4.6	5.2
Dark	2.3	2.6	3.7
Vitamin 'K'			
Red	+	2.2	1.9
Red/far-red	+	2.1	1.8
Far-red	+	2.0	2.5
Dark	+	1.5	2.6

Plants were extracted in each case 24 h after 1st pulse (mean values of 5 separate determinations, standard deviation ±4%)

is to be expected, since the promotion of K_1 synthesis seems to be a slow phytochrome reaction. Even in continuous white light its accumulation is enhanced only after 5 h of illumination [9]. The pool size of the second lipophilic naphthoquinone, here termed 'K' (desmethyl-vitamin K_1? or K_1 with a more unsaturated side chain?), which is accumulated in continuous darkness, decreases following exposure to red light pulses in the older seedlings, as known from continuous white light. Subsequent far-red pulses cause a reversion to the higher 'K' level of the far-red control. The decrease of the 'K' pool size seems to be again a fast phytochrome reaction.

The data described here thus show that active phytochrome P_{fr}, also in its ground state, induces the light-triggered specific prenylquinone synthesis that is connected to thylakoid formation. Together with this promotion of prenylquinone synthesis, phytochrome also enhances the prenyl chain accumulation (C_{20} in α-tocopherol and vitamin K_1, C_{45} in plastoquinone) by an increased metabolite flow via the biosynthetic pathways (Fig. 7), but it has little or no effect on the pattern of prenyl chain formation in darkness [35]. The strong enhancement of C_{20} prenyl chain accumulation, which is correlated to chlorophyll synthesis, is not a phytochrome effect but is induced by white light.

In contrast to the prenylquinones of plastids, the formation of which is greatly enhanced by active phytochrome, the level of the mitochondrial ubiquinone is not directly under light control. Illumination of etiolated seedlings with continuous white light did not, or only to a limited degree, increase the ubiquinone concentration in the first 24 h, during which time plastoquinone formation is strongly promoted [19]. With red and far-red light pulses we did not observe a change in the ubiquinone level, although prenylquinone accumulation in plastids is stimulated.

F. Regulation of Lipoquinone Synthesis by Blue and Red Light

Blue light has specific effects on growth and metabolism of plants that are not observed with red light. Greening of etiolated plants in blue light proceeds more quickly than in red light of about equal light quanta intensity [21, 26]. Blue light particularly increases the accumulation of plastoquinone-9, phylloquinone, α-tocopherol, and α-tocoquinone as compared to red light [21, 26]. Whether there are also differences in the concentration of ubiquinone has not yet been investigated. Chlorophyll and carotenoid levels are changed too but to a considerably lesser degree than those of the chloroplast prenylquinones. In any case, blue light results in a completely different prenyl lipid composition of the photosynthetic apparatus than does red light. It is of particular interest that the differences found between blue- and red-light-grown plants are similar to those between sun and shade leaves or between plants grown at high or low light intensities [12, 29, 33, 70].

It is well known that sun-exposed leaves exhibit significantly higher values for the chlorophyll a/b ratio, moderately higher values for the ratio of chlorophylls to carotenoids (a + b/x + c), and lower values for the ratio, xanthophylls to carotene (x/c) [57]. This was found also for the plants grown in blue light (Table 4). In addition to this the ratios of chlorophyll to the individual chloroplast prenylquinones are much lower in sun-exposed leaves or in plants grown at high light intensity [29, 33, 70]. This again can be seen in plants grown in blue as compared to those kept in red light (Table 4). The higher prenylquinone content on a chlorophyll basis in blue light is correlated with the higher photosynthetic activity of the isolated chloroplasts (Table 5). This difference in Hill-activity was also found in chloroplasts from *Sinapis* plants grown at high and low light intensities [70]. Since plastoquinone-9, phylloquinone, and α-tocoquinone are potential photosynthetic electron carriers, the enhancement of their accumulation together with an increased photosynthetic activity means that the photosynthetic units in blue light and at high light intensities are built up in a different way with more photosynthetic reaction centers on a chlorophyll basis than in red light.

The differences in prenyllipid composition and photosynthetic activity are correlated with differences in chloroplast ultrastructure. Thus chloroplasts from

Table 4. Differences in the weight ratios of prenyllipids in 8- to 9-day-old barley plants greened in blue or red light [21, 26]

Ratio	Blue light		Red light	
	24 h	48 h	24 h	48 h
a + b/x + c	5.9	6.2	4.1	5.2
a/b	3.9	3.8	3.5	3.2
x/c	2.4	2.1	3.7	2.9
a/PQ-9ox	17	19	32	38
a/K_1	145	168	204	266

Table 5. Differences in Hill-activity of chloroplast from 8- to 12-day-old *Hordeum* seedlings grown in blue or red light (μmol O_2/mg chlorophyll · h). (After Buschmann, 1976, unpublished)

	Dichlorophenol-indophenol		Methyl-viologen	
	Blue	Red	Blue	Red
24 h Light	111	80	67	54
48 h Light	83	58	67	49
96 h Light	52	27	21	15

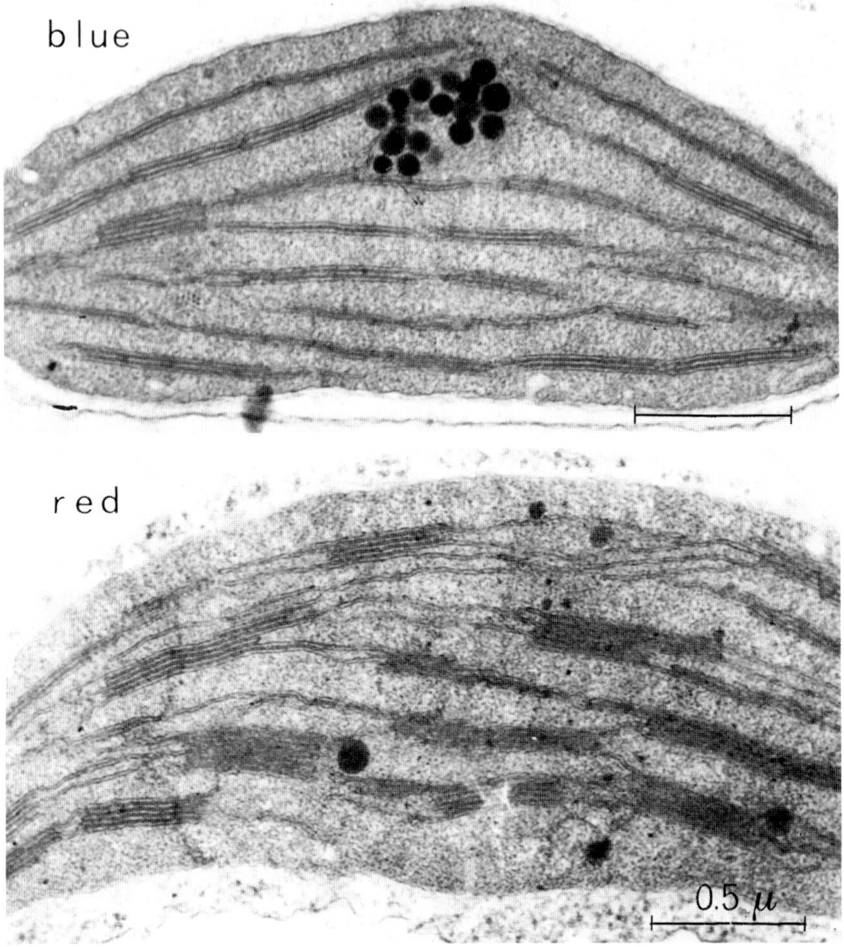

Fig. 10. Chloroplasts from 7-day-old barley seedlings grown in blue or red continuous light (illumination time 24 h; glutaraldehyde + OsO_4; 20,000 × 2). (Photo D. Meier, 1976)

plants grown in blue light possess after a 24-h illumination period fewer grana thylakoids and more plastoglobuli than red-light chloroplasts (Fig. 10). The latter are narrower and longer than blue-light chloroplasts. These differences are maintained and can still be seen after 72 h of illumination. Chloroplasts from sun leaves also have fewer grana and lower stacks than chloroplasts from shade leaves [12, 56]. Since, in principle, blue light has the same effect on the development of a functional photosynthetic apparatus as white light of high intensity, it is concluded that the strong light effects on the composition of the chloroplasts as found in nature are initiated by blue light. In any case these data show that prenylquinone synthesis in chloroplasts is regulated together with the size of the photosynthetic unit and its activity.

G. Influence of Phytohormones on Lipoquinone Metabolism and Photosynthetic Activity

Phytohormones are well-known regulators of plant growth and cell metabolism. Abcissic acid inhibits light-induced chlorophyll-, lipoquinone- and thylakoid formation [36]. Gibberellic acid (GA_3) may increase the germination rate of *Raphanus* seedlings, but blocks the formation of prenylquinones, chlorophylls, and carotenoids [60] at a later stage of development. Indoleacetic acid (IAA) and kinetin enhance the light-induced lipoquinone accumulation in chloroplasts [59, 60], which is initiated by active phytochrome, and thus act synergistically with light. The influence of phytohormones on the level of ubiquinone has not yet been assayed.

The formation of plastoquinone-9 is strongly promoted by the phytohormones IAA and kinetin, while that of α-tocopherol ($+\alpha$-tocoquinone) is suppressed (Fig. 11). This means that under IAA and kinetin influence the common benzoquinone ring precursor, homogentisic acid, is prenylated to a greater extent with a C_{45} prenyl chain and to a lesser extent with a C_{20} chain than in the controls (Fig. 12). The accumulation of the naphthoquinone vitamin K_1 is increased too by IAA and kinetin (Fig. 13). The enrichment of chlorophylls is only minimally enhanced; the formation of carotenoids is suppressed. The higher K_1 and plastoquinone-9 levels are correlated with higher photosynthetic activity of the isolated chloroplasts (Hill reaction) for different developmental stages. IAA and kinetin also increase the concentration of the reaction center chlorophyll P700 (Table 6).

This enhanced synthesis of several components of the photosynthetic electron transport chain together with the increased Hill activity indicates that IAA and kinetin increase the number of photosynthetic units on a chlorophyll basis

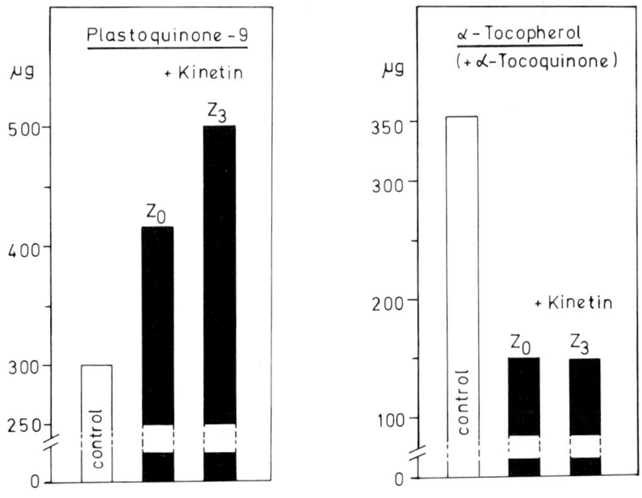

Fig. 11. Promotion of plastoquinone-9 and suppression of α-tocopherol accumulation in 8-day-old *Raphanus* seedlings by kinetin (2 ppm) applied to the nutritive solution either at time of sowing (Z_0) or to 3-day-old plants (Z_3). μg/100 plants [60]

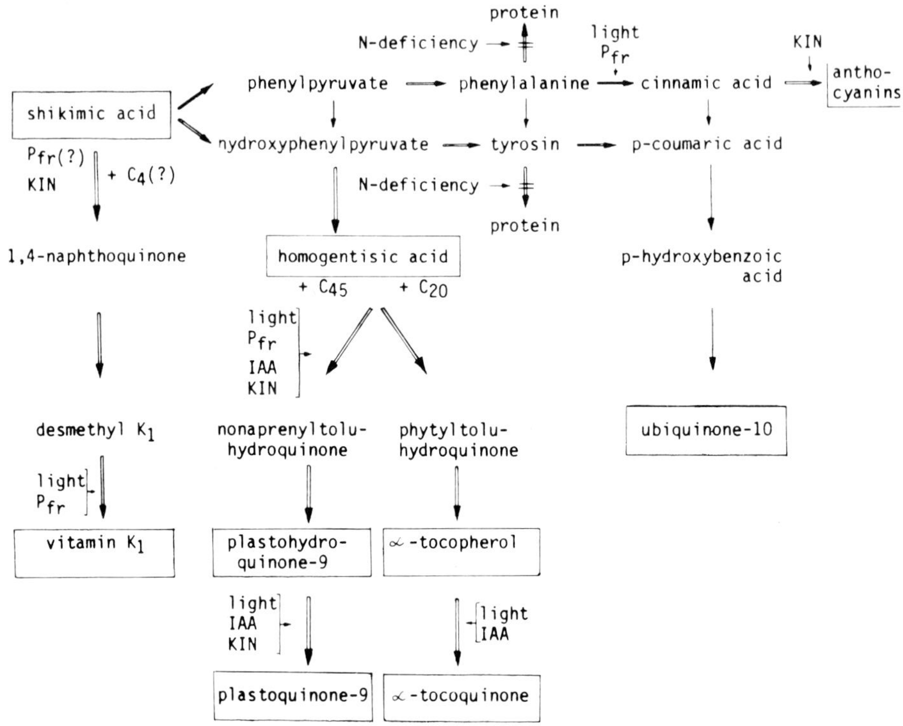

Fig. 12. Scheme of prenylquinone biosynthesis [62]. Changes of metabolite flow, induced by light, phytochrome P_{fr}, and phytohormones, are indicated

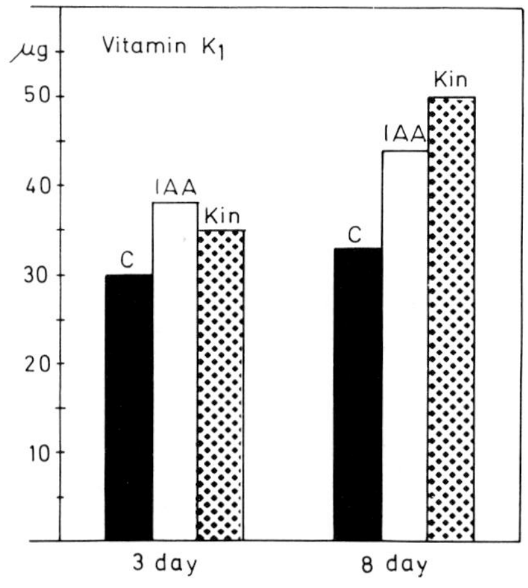

Fig. 13. Stimulation of vitamin K_1 formation in *Raphanus* seedlings by IAA and kinetin. Both phytohormones were applied to nutritive solution (2 ppm each) on sowing. µg/100 plants (values from Straub-Uhrig, 1976, unpublished)

[15]. The data show that phytohormones control the size and variability of the photosynthetic unit, as has been shown for blue light and high light intensity. From the data available at present it is clear that the lipoquinone content, the prenyllipid composition, and the photosynthetic activity of chloroplasts depend very much on the growth conditions under which the plant is kept. This interrelationship is summarized in Table 7.

Table 6. % Promotion of Hill-activity, P700, plastoquinone-9, and K_1 contents, with reference to chlorophyll a, in 8-day-old *Raphanus* seedlings by phytohormones IAA (2 ppm) and kinetin (2 ppm)

Raphanus	IAA	Kinetin
Methylviologen reduction	+12	+ 36
Methylviologen reduction with NH_4Cl	+16	+ 53
DCPIP reduction	+10	+ 35
DCPIP reduction with NH_4Cl	+32	+115
P700	+17	+ 40
Vitamin K_1	+ 2	+ 23
Plastoquinone-9	+52	− 21
α-Tocopherol	−38	− 65
Carotenoids	−32	− 27

Table 7. Differences in prenyllipid composition and photosynthetic activity under different growth conditions

Growth condition I	Growth condition II
Shade leaf	Sun leaf
Weak light	Strong light
Red light	Blue light
Control	+IAA
Control	+kinetin

Growth condition II:
higher values for a/b and a+b/x+c
lower values for x/c and a/PQ-9
a/K_1, a/TQ, a/T
higher Hill-activity
higher photosynthetic capacity
fewer grana thylakoids per chloroplast

H. Formation of Excess Prenylquinones

After completion of the greening phase of leaves, which is identical with the multiplication phase of thylakoids, the chlorophyll and carotenoid contents reach a maximum level that may be more or less constant throughout the vegetation period [33]. At that leaf stage, the prenylquinones of chloroplasts are almost quantitatively localized together with the green and yellow pigments in the photochemically active thylakoids [32, 48]. The phylloquinone, α-tocoquinone and plastoquinone-9 (oxidized form) levels in green leaves also remain then more or less the same with increasing leaf age. The reduced benzoquinone forms, plastohydroquinone-9 and α-tocopherol, however, accumulate continuously after the end of the greening phase, which increases the total plastoquinone-9 and α-tocopherol levels to high values (Fig. 14). This formation of excess chloroplast prenylquinones [31], which are deposited in the osmiophilic plastoglobuli of the plastid stroma [45, 48], is found particularly at high light intensities (e.g., sun leaves) and is much lower in weak light (shade leaves) (Fig. 14). Sun-exposed leaves or plants grown at high light intensities possess in all stages, before and after production of excess chloroplast prenylquinones, a higher prenylquinone level than shade leaves or plants grown under weak light conditions [29, 33, 70].

Lipoquinone formation and protein synthesis compete for the common aromatic precursors of the shikimic acid pathway (Fig. 12). The strong light-dependent lipoquinone accumulation can start only when the protein-rich thylakoids [44] in chloroplasts have been formed. Then more intermediates of the shikimic acid pathway are available for the formation of excess chloroplast prenylquinones. The fact that this lipoquinone accumulation proceeds mainly under strong light conditions can be explained by a higher rate of synthesis of aromatic

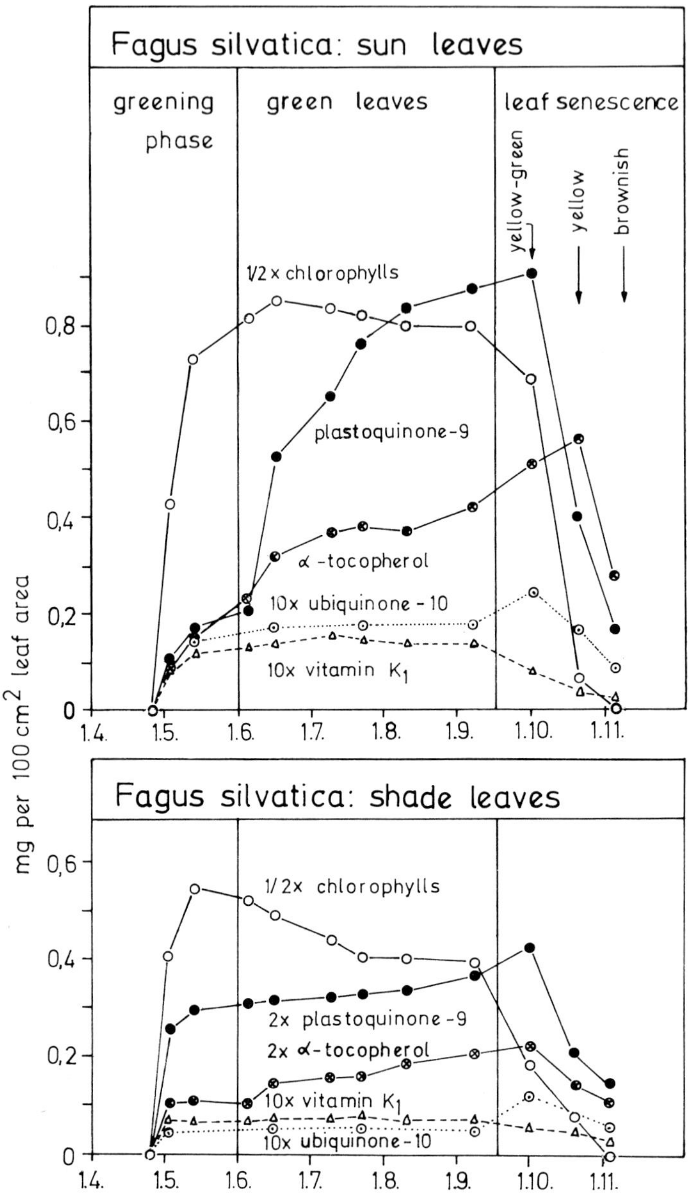

Fig. 14. Increase of the plastoquinone-9 and α-tocopherol levels in sun and shade leaves of *Fagus silvatica*

precursors and an increased production of photosynthetically formed ATP and NADPH. On the other hand, the accumulation of the reduced benzoquinones seems to be related to thylakoid turnover. During natural breakdown of thylakoids, the levels of chlorophylls, carotenoids, and phylloquinone continuously decrease while plastoquinone-9 and α-tocopherol continue to accumulate [33]

(Fig. 14) up to the yellow-green stage of leaf senescence. From this, one would conclude that the turnover of thylakoids and their prenyllipids proceeds in sun leaves and under high light intensities at higher rates than in shade leaves or at low light intensities.

In contrast to the isoprenoid chloroplast benzoquinones, the level of mitochondrial ubiquinone remains fairly constant after the end of leaf development throughout the green vegetation period. Its synthesis seems to be directly regulated with mitochondrial activity, and it is apparently never formed in excess amounts. Sun-exposed leaves have in all development stages higher ubiquinone-10 levels than the shade leaves. The significant increase in the ubiquinone level both in sun and shade leaves during leaf senescence and chlorophyll breakdown (Fig. 14) must be seen in relation to higher mitochondrial activity and respiration in the leaf cells at a stage where photosynthesis and photosynthetic energy production (ATP, NADPH) cease.

I. Stimulation of Prenylquinone Synthesis by Supplying Additional Aromatic Compounds

Protein and prenylquinone synthesis compete for the aromatic products of the shikimic acid pathway (Fig. 12). During the synthesis of thylakoids, which consist of 50% protein [44], the intermediates of the shikimic acid pathway are used for prenylquinone synthesis and for the formation of aromatic amino acids. The formation of excess chloroplast lipoquinones, which are deposited in the plastoglobuli, thus starts only when thylakoid synthesis is completed. It is to be expected that an additional supply of metabolites of the shikimic acid pathway should result in increased lipoquinone formation. This is in fact the case in etiolated barley seedlings in which the accumulation of prenylbenzoquinones and vitamin K_1 is stimulated by shikimic acid, L-tyrosin, and L-phenylalanin to a greater degree than that of chlorophylls or carotenoids [34]. Our present preliminary data show that similar promotion effects can also be obtained with green barley seedlings.

In algae (*Scenedesmus*) shikimic acid, L-tyrosin, and L-phenylalanin induce an enhanced multiplication rate that results in promotion of all thylakoid prenyllipids, prenylquinones, chlorophylls, and carotenoids. Application of D,L-phenylalanin has little effect on cell multiplication, reduces carotenoid and chlorophyll formation, and induces, with reference to chlorophyll a, a specific promotion of phylloquinone, plastoquinone-9, α-tocoquinone, and α-tocopherol accumulation (Table 8).

Nitrogen deficiency gradually reduces and eventually blocks protein synthesis in the cell. The synthesis of chlorophylls and carotenoids proceeds only at a low rate. The formation of chloroplast prenylquinones, in turn, is little influenced [65]. With reference to chlorophyll, nitrogen deficiency results in a specific promotion of prenylquinone synthesis and accumulation. This together with the increased synthesis of flavonoids [27] shows that the formation of

Table 8. Influence of L- and D,L-phenylalanin on growth and on pigment and prenylquinone accumulation in *Scenedesmus* cultures

	Control	+L-Phenylalanin 10^{-4} M	+D,L-Phenylalanin 10^{-4} M
A. Data/suspension culture:			
Cell number	3.4×10^8	4.6×10^8	3.7×10^8
Dry weight	34 mg	36 mg	31 mg
Total lipids	9 mg	13 mg	10 mg
Protein content (after Lowry)	14 mg	17 mg	15 mg
Chlorophylls	390 μg	652 μg	160 μg
Carotenoids	44 μg	85 μg	31 μg
Prenylbenzoquinones	26 μg	29 μg	34 μg
Vitamin K_1	0.08 μg	0.19 μg	0.18 μg
B. Prenylquinones/1 mg chlorophyll a:			
Total plastoquinone-9	46 μg	26 μg	102 μg
α-Tocopherol	38 μg	28 μg	178 μg
α-Tocoquinone	2.1 μg	3.2 μg	3.8 μg
Vitamin K_1	0.26 μg	0.38 μg	1.5 μg

Algae were collected and analyzed 5 days after application of phenylalanin. Initial cell number was in all cases 0.8×10^8. (After Hanigk, 1976, unpublished)

intermediates of the shikimic acid pathway is not blocked in nitrogen deficiency. The aromatic precursors are then directed not via protein synthesis but into other biosynthetic pathways such as lipoquinones and flavonoids (Fig. 12).

J. Turnover of Prenylquinones

A continuous lipoquinone, chlorophyll, and thylakoid breakdown, which is correlated with the loss of photosynthetic activity, occurs in green leaves of *Hordeum* seedlings. This degradation of the photosynthetic apparatus is analytically visualized only in an artificial day–night rhythm with a prolonged dark phase, when light-induced, new formation of lipoquinones, chlorophylls, and thylakoids is excluded [41, 42]. Re-illumination, in turn, after a dark period, results in new synthesis and accumulation of all thylakoid prenyllipids (lipoquinones, chlorophylls, carotenoids), and in the re-formation of photosynthetic reaction centers and thylakoids as seen from the strong increase in photosynthetic activity [42]. A faster destruction rate in the darkness (plastoquinone-9 > vitamin K_1 > chlorophylls) corresponds to a faster re-accumulation rate in the light. The decline of the photosynthetic activity of the intact leaf (oxygen evolution, variable fluorescence) in continuous darkness and its reappearance upon re-illumination proceed faster than the changes in prenyllipid composition. The half-life, which can be estimated to a first approximation from these investigations, is for plastoquinone-9 and α-tocoquinone ca. 2.5 days, for vitamin K_1, ca. 4 days, chlorophylls, ca. 6 days, and for photosynthetic activity, ca. 1.7 days [42].

In labeling experiments of *Chlorella* suspensions under photosynthetic conditions, with a constant pigment and lipoquinone content, the highest degree of labeling was observed, after a 2-h $^{14}CO_2$ exposure, in α- and β-carotene (19% each), in vitamin K_1 (12%), and in plastohydroquinone-9 (5%). The highest mole-specific radioactivity was found in these compounds and in the sterol fraction. This provides evidence that incorporation of ^{14}C into prenyl chains and prenyllipids from photosynthetically fixed $^{14}CO_2$ takes place very quickly. From the labeling kinetics, one can calculate half-lives for the different prenylquinones, carotenoids, and chlorophylls in *Chlorella* to be in the range of 15 to 60 min. This is an astonishingly high turnover rate [20].

Metabolism in algae is certainly faster than in higher plants. In $^{14}CO_2$ tracer experiments in fully developed green leaves of several higher plants, the shortest half-life among the carotenoids was found for β-carotene, i.e., 6 to 7 h [58]. From this one may expect that the turnover of prenylquinones and other prenyllipids in higher plants is faster than the values calculated above from the experiments with prolonged dark phases. The data available thus far indicate that prenylquinones and other prenyllipids undergo a relatively fast turnover, even at a constant lipoquinone and pigment level per leaf. This should be considered in all investigations concerning lipoquinone metabolism.

K. Regulation of Prenylquinone Synthesis by Herbicides

Herbicides may influence plant growth and metabolism in many ways. The majority of herbicides are inhibitors of photosynthesis (block of electron transport in pigment system II), while the remainder interferes mainly with the function of phytohormones. It is to be expected and has been shown in some cases that herbicides have more than one effect on plants. At present there is still discussion on what are primary and secondary herbicide effects. Concerning the prenyllipid and chloroplast prenylquinone metabolism, herbicides can be grouped in two classes:

1. Those that do promote or minimally influence chlorophyll and carotenoid formation but reduce the accumulation of prenylquinones and

2. Those that block chlorophyll and carotenoid synthesis (formation of albino plants) but do not or only minimally effect the formation of chloroplast prenylquinones.

Methabenzthiazuron (MBT) and bentazone belong to the first group. Both inhibit photosynthetic electron transport, which is thought to be their primary effect [18, 53]. Both herbicides, as compared to controls, induce a shade type adaptation (growth condition I in Table 7), which results in changed prenyllipid composition with lower values for the ratio a/b and higher values for the ratio x/c and chlorophyll a to prenylquinones (Table 9). This situation is achieved

Table 9. Change of prenyllipid relations (weight ratios) induced by the herbicide MBT in 13-day-old barley seedlings and by bentazone in 8-day-old *Raphanus* seedlings

	a/b	x/c	a+b/ x+c	a/PQ-9	a/K_1	a/α-T	a/α-TQ
Hordeum:							
Control	3.5	1.6	6.2	10	237	30	238
+MBT	2.7	3.1	6.1	17	347	35	361
Raphanus:							
Control	3.3	2.6	6.8	14	110	20	98
+bentazone	2.4	4.7	6.7	22	160	78	150

MBT (3 ppm) was applied during germination and bentazone, (10^{-4} M) via the roots to the 3-day-old seedlings [25, 39]

with MBT by enhanced formation of chlorophylls and carotenoids coupled with a reduced accumulation rate for chloroplast prenylquinones [25]. Bentazone, in turn, decreases the accumulation of all prenyllipids, but reduces the formation of prenylquinones to a greater degree than that of chlorophylls and carotenoids [39].

The second herbicide group comprises the two pyridazinon derivates San 6706 and San 9785 as well as dichlormate and pyrichlor, all of which are inhibitors of the photosynthetic photosystem II. To this group also belong amitrole, which does not inhibit photosynthesis, and difunone (EMD-IT-5914). All of these herbicides block carotenoid and chlorophyll synthesis simultaneously [3, 4, 14, 22, 43, 54, 64]. It has been shown that carotenoid synthesis in all cases is inhibited at the desaturation step of the first C_{40} isoprenoid chains, which accumulates phytoene, phytofluene, ζ-carotene, or lycopene. Inhibition of chlorophyll synthesis is known for difunone, which blocks the formation of the enzyme porphobilino-genase [22] and for amitrole, which accumulates the presumable chlorophyll a precursors geranylgeraniol-chlorophyll a and dihydrogeranyl-geraniol-chlorophyll a [54]. With San 6706 we found in barley and *Raphanus* seedlings a directly correlated inhibition of chlorophyll and carotenoid synthesis but much less influence on prenylquinone accumulation (Table 10). Similar results have been obtained with nitrogen deficiency. From these

Table 10. % Inhibition of prenyllipid synthesis by the herbicide San 6706 in 8-day-old *Raphanus* seedlings and in 7-day-old *Hordeum* seedlings

	Raphanus		*Hordeum*	
	1 μM	10 μM	1 μM	10 μM
Chlorophylls	69±4	98±4	39±3	91±6
Carotenoids	70±5	92±3	35±6	93±5
Plastoquinone-9	46±3	65±4	11±3	61±5
α-Tocopherol (+α-Tocoquinone)	17±3	19±3	5±3	56±4

The herbicide was present in the nutrient solution from sowing [43]

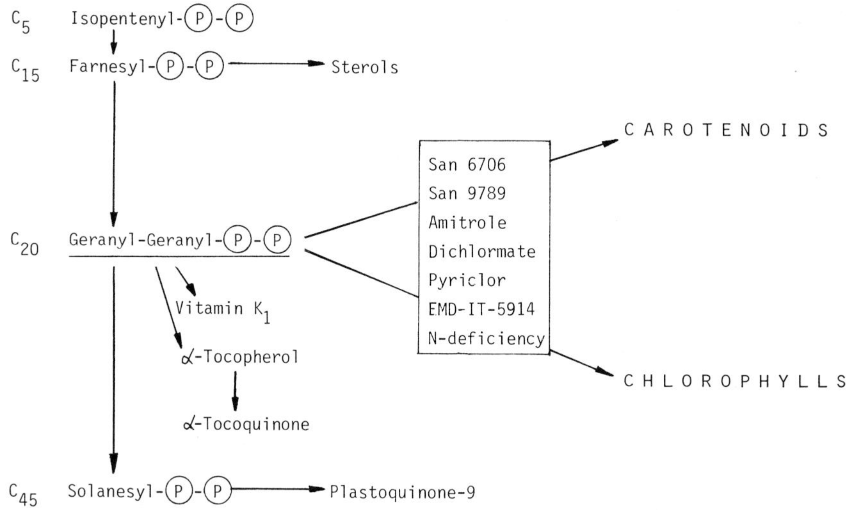

Fig. 15. Scheme for the formation of prenyllipids with inhibition site of some herbicides indicated

results it is clear that the inhibition site for San 6706, N-deficiency, and probably also for the other herbicides, does not lie in an early step of prenyl chain metabolism, because the prenyl chains of the prenylquinones are still being formed (Fig. 15).

Some authors postulated that the primary action site of these herbicides is blockage of carotenoid synthesis and that the chlorophylls are secondarily bleached out [14]. This is unlikely for many reasons, the main argument being as follows: It has been shown that the chloroplast ribosomes are missing in amitrole- and dichlormate-treated plants [5–8]. San 6706 reduces the number of chloroplast ribosomes [4]. Thus the central inhibition mechanism for the action of the preceding herbicides seems to be their interference with the formation of chloroplast ribosomes. This would result in a simultaneous inhibition of synthesis of enzymes for both carotenoid and chlorophyll metabolism. That lipoquinone formation is much less affected would mean that the enzymes for prenylquinone synthesis either possess a longer half-life or are not formed on chloroplast ribosomes.

In any case the present herbicide data show that the formation of chlorophylls and carotenoids is regulated in a correlated manner more or less independently from prenylquinone metabolism.

L. Formation of Prenylquinones in Plant Tissue Cultures

Analysis of several differently pigmented faint green, yellow-green, or green sterile callus cultures of *Pimpinella, Nicotiana, Petunia, Ginkgo,* and *Silybum* shows that plant tissue cultures, like intact plants, possess the biochemical capacity to synthesize the different plastidic prenyllipids such as chlorophylls,

Table 11. Prenyllipid content and weight ratios of green cotyledons, a tissue culture, and a cell suspension culture of *Silybum marianum* (μg/10 g fresh weight). (After Lichtenthaler and Uhrig-Straub, 1976, unpublished)

	Green cotyledons	Green callus culture	Greenish cell suspension
Chlorophyll a	4,152	111	16
Chlorophyll b	1,591	39	5.5
Chlorophyll a+b	5,743	150	21.5
Xanthophylls (x)	735	33	7
β-carotene	219	6	1.8
Carotenoids (x+c)	954	39	8.8
Plastoquinone-9 (PQ-9)	87	32	6
Plastohydroquinone-9	260	21	10
Total PQ-9	347	53	16
α-Tocopherol (α-T)	357	274	58
a/b	2.6	2.9	2.9
x/c	3.4	5.5	3.9
a+b/x+c	6.0	3.8	2.4
a/PQ-9 Total	12.0	2.1	1.0
a/α-T	116	0.4	0.27

carotenoids, and the lipophilic benzo- and naphthoquinones [46, 47]. The mitochondrial ubiquinone is also present [23].

The pigment and prenylquinone contents of callus cultures (dry weight or fresh weight basis) are, however, much lower than those of green leaf tissue [46, 47]. Cell suspension cultures contain all prenylquinones, but their levels on a fresh weight basis are even lower than those of tissue cultures (Table 11). By using chlorophyll as a reference, cell suspension and plant tissue cultures exhibit a much higher prenylquinone content than green leaves, while the level of carotenoids is about equal. With regard to their prenyllipid and lipoquinone accumulation, plant tissue cultures and cell suspension cultures behave in a manner similar to that of green leaf tissue during latent nitrogen deficiency. Also in the latter case the synthesis rate of chlorophylls and carotenoids is diminished as compared to that in control plants, while accumulation of the prenylquinones is less affected, which results in a promotion of prenylquinone formation with reference to chlorophyll [65].

On the other hand, the prenyllipid composition of plant tissue cultures is similar to that of leaves from aurea mutants. This has been shown in detail for *Nicotiana* tissue cultures, which like aurea leaf exhibit low ratios of chlorophyll a to plastoquinone-9 and K_1. This indicates that plant tissue cultures, like aurea plants or plants grown in blue light or at high light intensities, contain smaller photosynthetic units with more reaction centers and less antenna chlorophyll than chloroplasts from fully green leaves. That plant tissue cultures are photosynthetically active has been shown by measuring their variable fluorescence and oxygen evolution in the intact tissue [46, 47] and by isolation of photochemically active chloroplasts.

Some plant tissue cultures do not form chlorophylls. A yellowish chlorophyll-free *Hordeum* tissue culture contains only trace amounts of carotenoids, a high level of prenylquinones [46], and thus behaves like root or other chlorophyll-free tissue of intact plants [29]. Application of IAA to a *Pimpinella* tissue culture increases the formation of plastoquinone-9 [47] in a way similar to that in intact *Raphanus* plants [59]. Additional supply of tissue cultures with aromatic precursors of the shikimic acid pathway specifically promotes, as in intact plants, the synthesis of chloroplast prenylquinones but has little influence on the level of chlorophylls and carotenoids [46]. From the data available so far it appears that the same regulatory principles that work in intact green plants also control prenylquinone synthesis and accumulation in plant tissue cultures.

M. Conclusions

The data reported here show that light-induced prenylquinone formation of chloroplasts is initiated by active phytochrome P_{fr} and is regulated by environmental factors such as red and blue light, low and high light intensities, and by the phytohormones IAA and kinetin. This light-triggered lipoquinone synthesis is correlated with differences in composition, structure, and photochemical activity of the photosynthetic apparatus. The level of vitamin K_1 and the oxidized benzoquinone derivatives plastoquinone-9 and α-tocoquinone are, with reference to chlorophyll, in some way representative of the number of photosynthetic units and of the photosynthetic capacity of chloroplasts. The formation of excess prenylquinones in fully developed leaves is controlled in a different way, by light intensity and substrate availability (aromatic precursors).

By a change of growth conditions (internal and external factors) and the application of herbicides, it has been shown that chlorophyll and carotenoid formation are regulated in a correlated manner together with thylakoid formation and in some way independent of chloroplast prenylquinone synthesis. The accumulation of mitochondrial ubiquinone seems to be regulated in relation to mitochondrial activity and, in some way, by other mechanisms, apart from the synthesis of prenylquinones in plastids.

Future investigations with endogenous ^{14}C-metabolites will increase our knowledge of biosynthesis and turnover of prenylquinones in the compartments plastids and mitochondria and will give further insight into the regulation mechanisms of plant cell metabolism.

Acknowledgements. The work described here was sponsored by a grant from the *Deutsche Forschungsgemeinschaft*. We thank Prof. Dr. H. Musso, Department of Organic Chemistry, University of Karlsruhe, for the synthesis of bromonaphthoquinone, and Mrs U. Prenzel for skillful technical assistance.

References

1. Anderson, J.M., Fork, D.C., Amesz, J.: P 700 and cytochrome f in particles obtained by digitonin fragmentation of spinach chloroplasts. Biochem. Biophys. Res. Commun. **23**, 874–879 (1966)
2. Bach, T.J., Lichtenthaler, H.K., Rétey, J.: The occurrence of 3-hydroxy-3-methylglutaryl-coenzyme A (HMG-CoA) reductase in higher plants. In: Abstracts of the Symposium on Lipids and Lipid Polymers in Higher Plants. Botanical Institute, Univ. of Karlsruhe, 1976, p. 96
3. Bartels, P.G., McCullough, C.: A new inhibitor of carotenoid synthesis in higher plants 4-chloro-5-(dimethylamino)2-α,α,α(trifluoro-m-tolyl)-3(2H)-pyridazinone (Sandoz 6706). Biochem. Biophys. Res. Commun. **48**, 16–22 (1972)
4. Bartels, P.G., Hyde, A.: Chloroplast development in 4-chloro-5-(dimethylamino)-2-(α,α,α-trifluoro-m-tolyl)-3(2H)-pyridazinone(Sandoz 6706)-treated wheat seedlings: a pigment, ultrastructural and ultracentrifugational study. Plant Physiol. **45**, 807–810 (1970a)
5. Bartels, P.G., Hyde, A.: Buoyant density studies of chloroplast and nuclear deoxyribonucleic acid from control and 3-amino-1,2,4-triazole-treated wheat seedlings, Triticum vulgare. Plant Physiol. **46**, 825–830 (1970b)
6. Bartels, P.G., Matsuda, K., Siegel, A., Weier, T.E.: Chloroplastic ribosome formation: inhibition by 3-amino-1,2,4-triazole. Plant Physiol. **42**, 736–741 (1967)
7. Bartels, P.G., Pegelow, E.J.: The action of sirmate (3,4-dichlorobenzylmethyl-carbamate) in chloroplast ribosomes of *Triticum vulgare L.* seedlings. J. Cell Biol. **37**, 1–6 (1968)
8. Bartels, P.G., Weier, T.E.: The effect of 3-amino-1,2,4-triazole on the ultrastructure of plastids of *Triticum vulgare* seedlings. Am. J. Botany **56**, 1–7 (1969)
9. Becker, H.: Untersuchungen über den Pigment- und Lipochinonstoffwechsel bei der Entwicklung funktionstüchtiger Chloroplasten. Dissertation, Univ. of Münster, 1971
10. Bickel, H., Schultz, G.: Biosynthesis of plastoquinone and β-carotene in isolated chloroplasts. Phytochemistry **15**, 1253–1255 (1976)
11. Boardman, N.K., Anderson, J.M.: Fractionation of the photochemical system of photosynthesis. II Cytochrome and carotenoid contents of particles isolated from spinach chloroplasts. Biochim. Biophys. Acta **143**, 187–203 (1967)
12. Boardman, N.K., Björkman, O., Anderson, J.M., Goodschild, D.J., Thorne, S.W.: Photosynthetic adaptation of higher plants to light intensity: relationship between chloroplast structure, composition of the photosystems and photosynthetic rates. Proc. 3rd Internat. Congress on Photosynthesis. Avron, M. (ed.). Amsterdam: Elsevier, 1974, pp. 1809–1827
13. Brooker, J.D., Russel, D.W.: Subcellular localization of 3-hydroxy-3-methyl-glutaryl coenzyme A reductase in *Pisum sativum* seedlings. Arch. Biochem. Biophys. **167**, 730–737 (1975)
14. Burns, E.R., Buchanan, G.A., Carter, M.C.: Inhibition of carotenoid synthesis as a mechanism of action of amitrole, dichlormate and pyrichlor. Plant Physiol. **47**, 144–148 (1971)
15. Buschmann, C., Lichtenthaler, H.K.: Influence of IAA and kinetin on composition and photosynthetic activity of chloroplasts. In: Plant Growth Regulation. Pilet, P.E. (ed.). Berlin-Heidelberg-New York: Springer, 1977
16. Crane, F.L., Henninger, M.D., Wood, P.M., Barr, R. In: Biochemistry of Chloroplasts. Goodwin, T.W. (ed.). London: Academic Press, 1966, Vol. 1, pp. 133–151
17. Dunphy, P.J., Whittle, K.J., Pennock, J.F.: Plastochromanol. In: Biochemistry of Chloroplasts. Goodwin, T.W. (ed.). London: Academic Press, 1966, Vol. 1, pp. 165–171
18. Fedtke, C.: Changed physiology in wheat plants treated with the herbicide methabenzthiazuron. Naturwissenschaften **61**, 272 (1974b)
19. Goodwin, T.W.: Regulation of terpenoid synthesis in higher plants. In: Biosynthetic Pathways in Higher Plants. Pridham, J.B., Swain, T. (eds.). London: Academic Press, 1965, pp. 57–71
20. Grumbach, K.H.: Untersuchungen über die Inkorporation von $^{14}CO_2$ in die photosynthetischen Pigmente und Lipochinone von *Chlorella pyrenoidosa*. Dissertation. Univ. of Karlsruhe, 1976
21. Grumbach, K.H., Lichtenthaler, H.K.: Kinetic of lipoquinone and pigment synthesis during red light-induced thylakoid formation in etiolated barley seedlings. Z. Naturforsch. **30c**, 337–341 (1975)
22. Hampp, R., Sankhla, N., Huber, W.: Effect of EMD-IT-5914 on chlorophyll synthesis in leaves of *Pennisetum typhoides* seedlings. Physiol. Plantarum **33**, 53–57 (1975)

23. Ikeda, T., Matsumoto, T., Noguchi, M.: Formation of ubiquinone by tobacco plant cells in suspension culture. Phytochemistry **15**, 568–569 (1976)
24. Janiszowska, W., Michalski, W., Kasprzyk, Z.: Polyprenylquinones and α-tocopherol in *Calendula officinalis*. Phytochemistry **15**, 125–127 (1976)
25. Kleudgen, H.K.: The influence of the herbicide methabenzthiazuron on the accumulation of chloroplast prenyllipids in *Hordeum* seedlings. In preparation, 1977
26. Kleudgen, H.K., Lichtenthaler, H.K.: Die Synthese von Lipochinonen und isoprenoiden Pigmenten bei der Ergrünung von Etioplasten im Blaulicht. Z. Naturforsch. **29c**, 142–146 (1974)
27. Krause, J., Reznik, R.: Der Einfluß der Phosphat- und Nitratversorgung auf den Phenylpropanstoffwechsel in Buchweizenblättern (*Fagopyrum esculentum Moench*). Z. Pflanzenphysiol. **68**, 134–143 (1972)
28. Lichtenthaler, H.K.: Verbreitung und relative Konzentration der lipophilen Plastidenchinone in grünen Pflanzen. Planta **81**, 140–152 (1968a)
29. Lichtenthaler, H.K.: Die Verbreitung der lipophilen Plastidenchinone in nichtgrünen Pflanzengeweben. Z. Pflanzenphysiol. **59**, 195–210 (1968b)
30. Lichtenthaler, H.K.: Light-stimulated synthesis of plastid quinones and pigments in etiolated barley seedlings. Biochim. Biophys. Acta **184**, 164–172 (1969a)
31. Lichtenthaler, H.K.: Die Bildung überschüssiger Plastidenchinone in Blättern von *Ficus elastica Roxb*. Z. Naturforsch. **24b**, 1461–1466 (1969b)
32. Lichtenthaler, H.K.: Localization and functional concentration of lipoquinones in chloroplasts. Progress in Photosynthesis Res. Vol. I, 304–314 (1969c)
33. Lichtenthaler, H.K.: Die unterschiedliche Synthese der lipophilen Plastidenchinone in Sonnen- und Schattenblättern von *Fagus silvatica L*. Z. Naturforsch. **26b**, 832–842 (1971)
34. Lichtenthaler, H.K.: Regulation der Lipochinonsynthese in Chloroplasten. Ber. Dtsch. Bot. Ges. **86**, 313–329 (1973)
35. Lichtenthaler, H.K.: Regulation of prenyl chain synthesis in etiolated *Hordeum* seedlings by far-red and white light. Physiol. Plantarum **33**, 241–244 (1975)
36. Lichtenthaler, H.K., Becker, K.: Inhibition of the light induced vitamin K_1 and pigment synthesis by abscisic acid. Phytochemistry **9**, 2109–2113 (1970)
37. Lichtenthaler, H.K., Becker, K.: Changes of the plastidquinone and carotenoid metabolism associated with the formation of functioning chloroplasts in continuous far-red and white light. Proc. IInd Internat. Congress on Photosynthesis Research. The Hague: Dr. V. Junk N.V. Publ., 1972, Vol. 3, pp. 2451–2459
38. Lichtenthaler, H.K., Becker, K.: Kinetic of lipoquinone synthesis in etiolated *Raphanus* seedlings in continuous far-red and white light. Z. Pflanzenphysiol. **75**, 296–302 (1975)
39. Lichtenthaler, H.K., Burkard, G., Prenzel, U.: Effect of the photosynthesis inhibitor bentazone on pigment and prenylquinone accumulation in *Raphanus* seedlings. Weed Science, in preparation, 1977
40. Lichtenthaler, H.K., Calvin, M.: Quinone and pigment composition of chloroplasts and quantasome aggregates from *Spinacia oleracea*. Biochim. Biophys. Acta. **79**, 30–40 (1964)
41. Lichtenthaler, H.K., Grumbach, K.H.: Kinetic of lipoquinone and pigment synthesis in green *Hordeum* seedlings during an artificial day-night rhythm with a prolonged dark phase. Z. Naturforsch. **29c**, 532–540 (1974a)
42. Lichtenthaler, H.K., Grumbach, K.H.: Observations on the turnover of thylakoids and their prenyl lipids in *Hordeum vulgare L*. Proc. 3rd Internat. Congr. on Photosynthesis. Avron, M. (ed.), Amsterdam: Elsevier, 1974b, Vol. III, pp. 2007–2015
43. Lichtenthaler, H.K., Kleudgen, H.K.: Effect of the herbicide San 6706 on biosynthesis of photosynthetic pigments and prenylquinones in *Raphanus* and *Hordeum* seedlings. Z. Naturforsch. **32c**, 236–240 (1977)
44. Lichtenthaler, H.K., Park, R.B.: Chemical composition of chloroplast lamellae from spinach. Nature (London) **198**, 1070–1072 (1963)
45. Lichtenthaler, H.K., Sprey, B.: Über die osmiophilen globulären Lipideinschlüsse der Chloroplasten. Z. Naturforsch. **21b**, 690–697 (1966)
46. Lichtenthaler, H.K., Straub, V.: Die Bildung von Lipochinonen in Gewebekulturen. Planta Med. **Suppl.**, 198–213 (1975)
47. Lichtenthaler, H.K., Straub, V., Grumbach, K.H.: Unequal formation of prenyllipids in a plant tissue culture and in leaves of *Nicotiana tabacum L*. Plant Sci. Lett. **4**, 61–65 (1975)

48. Lichtenthaler, H.K., Weinert, H.: Die Beziehungen zwischen Lipochinonsynthese und Plastoglobulibildung in den Chloroplasten von *Ficus elastica Roxb.* Z. Naturforsch. **25b**, 619–623 (1970)
49. Morton, R.A.: Ubiquinones, plastoquinones and vitamin K. Biol. Rev. **46**, 47–96 (1971)
50. Newton, R.P., Pennock, J.F.: The intercellular distribution of tocopherols in plants. Phytochemistry **10**, 2323–2328 (1971)
51. Pennock, J.F.: Occurrence of vitamin K and related quinones. Vitamins and Hormones **24**, 307–329 (1966)
52. Pennock, J.F.: Chemistry of isoprenoid quinones. In: Terpenoids in Plants. Pridham, J.B. (ed.). London: Academic Press, 1967, pp. 129–146
53. Pfister, K., Buschmann, C., Lichtenthaler, H.K.: Inhibition of the photosynthetic electron transport by bentazone. Proc. 3rd Internat. Congr. on Photosynthesis. Avron, M. (ed.). Amsterdam: Elsevier, 1974, Vol. I, pp. 675–681
54. Rüdiger, W., Benz, J., Lempert, U., Schoch, S., Steffens, D.: Hemmung der Phytolakkumulation mit Herbiciden. Geranylgeraniol- und Dihydrogeranyl-geraniolhaltiges Chlorophyll aus Weizenkeimlingen. Z. Pflanzenphysiol. **80**, 131–143 (1976)
55. Schultz, G., Huchzermeyer, Y., Reupke, B., Bickel, H.: On the intracellular site of biosynthesis of α-tocopherol in *Hordeum vulgare*. Phytochemistry **15**, 1383–1386 (1976)
56. Skene, D.S.: Chloroplast structure in mature apple leaves grown under different levels of illumination and their response to changed illumination. Proc. Royal Soc. (London) Ser. B. **186**, 75–78 (1974)
57. Seybold, A., Egle, K.: Lichtfeld und Blattfarbstoffe I. Planta **26**, 491–515 (1937)
58. Stransky, H.: Studies on carotenoid metabolism in higher plants. In: Abstracts of the Symposium on Lipids and Lipid Polymers in Higher Plants. Karlsruhe: Botanical Institute, Univ. of Karlsruhe, 1976, p. 89
59. Straub, V., Lichtenthaler, H.K.: Die Wirkung von β-Indolylessigsäure auf die Bildung der Chloroplastenpigmente, Plastidenchinone und Anthocyane in *Raphanus*-Keimlingen. Z. Pflanzenphysiol. **70**, 34–45 (1973a)
60. Straub, V., Lichtenthaler, H.K.: Die Wirkung von Gibberellinsäure A_3 und Kinetin auf die Bildung der Photosynthesepigmente, Lipochinone und Anthocyane in *Raphanus*keimlingen. Z. Pflanzenphysiol. **70**, 308–321 (1973b)
61. Thomas, G., Threlfall, D.R.: Synthesis of Polyprenyl-toluquinols from Homogentisate and Polyprenyl-pyrophosphates in Particulate Fractions of *Euglena* and Sugar Beet. Biochem. J. **142**, 437–440 (1974)
62. Threlfall, D.R., Whistance, G.R.: Biosynthesis of isoprenoid quinones and chromanols. In: Aspects of Terpenoid Chemistry and Biochemistry. Goodwin, T.W. (ed.). London: Academic Press, 1971, pp. 357–359
63. Trebst, A.: Energy conservation in photosynthetic electron transport of chloroplasts. Ann. Rev. Plant Physiol. **25**, 423–458 (1974)
64. Urbach, D., Suchanka, M., Urbach, W.: Effect of substituted pyridazinon herbicides and of difunone (EMD-IT 5914) on carotenoid synthesis in green algae. Z. Naturforsch. **31c**, 652 (1976)
65. Verbeek, L., Lichtenthaler, H.K.: Der Einfluß von Stickstoffmangel auf die Lipochinon- und Isoprenoidsynthese der Chloroplasten von *Hordeum vulgare L.* Z. Pflanzenphysiol. **70**, 245–258 (1973)
66. Wallwork, J.C., Crane, F.L.: The nature, distribution, function and biosynthesis of prenyl phytoquinones and related compounds. In: Progress in Phytochemistry. Reinhold, L., Liwschitz, Y. (eds.). London: Interscience Publ., 1970, Vol. 2, pp. 267–341
67. Wellburn, A.R.: Studies on the biosynthesis of the tocopherols in higher plants. Phytochemistry **9**, 743–748 (1970)
68. Wellburn, A.R., Hemming, F.W.: The subcellular distribution and biosynthesis of castaprenols and plastoquinone in leaves of *Aesculus hippocastanum*. Biochem. J. **104**, 173–177 (1967)
69. Whittle, K.J., Andley, B.G., Pennock, J.F.: The incorporation of ^{14}C methionine into chromanols and quinones by *Hevea brasiliensis* latex. Biochem. J. **103**, 21c–22c (1967)
70. Wild, A., Ruhle, W., Grahl, H.: The effect of light intensity during growth of *Sinapis alba* on the electron-transport and the noncyclic photophosphorylation. In: Environmental and Biological Control of Photosynthesis. Marcelle, R. (ed.). The Hague: Dr. W. Junk Publishers, 1975, pp. 115–121

CHAPTER 14

Chlorophyll Formation: The Phytylation Step

C. LILJENBERG

A. Introduction

Chlorophyll a has been shown to have two different functions in the thylakoid membranes of the chloroplast, (1) as "antenna" chlorophylls and (2) as reaction center chlorophylls. It is reasonable to assume a connection between the structure of the chlorophyll molecule and its different functions. The last step of the biosynthesis of chlorophyll, the addition of the fatty alcohol, phytol, to the propionic acid residue of pyrrole ring IV of the chlorophyllide molecule implies a thorough modification of the physicochemical properties of the molecule. Several of the earlier porphyrin intermediates in the biosynthetic chain leading to chlorophyll a as well as its immediate precursors, protochlorophyllide and chlorophyllide, are rather hydrophilic. The esterification of chlorophyllide with phytol will increase the hydrophobic nature of the molecule without changing the light-absorbing properties—at least not within the visible wavelength region. Because the photosynthetic pigments are bound to membrane systems rich in lipids, the degree of esterification and also the nature of the alcohols involved must be important factors for the photosynthetic function of the chloroplasts. It is obvious that a hydrophobic tail on the chlorophyll makes it possible for the pigment to be associated either to the molecules of the lipid bilayer in the membrane or to the membrane proteins. A characteristic feature is that the building up of these membranes parallels the synthesis of the chlorophyll pigments.

B. Biosynthetic Pathways for the Different Protochlorophyll Forms to Chlorophyll a

The last part of the biosynthetic chain leading to chlorophyll a involves photoreduction of Mg-vinylphaeoporphyrin a_5 (protochlorophyllide) and esterification of the reduced chlorophyllide with phytol (see Scheme a). Evidence for these steps seems to be well documented. According to this pathway, the esterification takes place after the photoreduction to chlorophyllide a has taken place. The alcohol involved in this case is phytol [20, 21, 27, 44]. It is, however, also well documented that nonirradiated plants contain some esterified pigments. Upon irradiation, at least some of these pigments will be photoreduced, thus forming some kind of chlorophyll a (see Schemes b and c). According to Rebeiz and Castelfranco [43], these two pathways for formation of chlorophyll a can be summarized as follows (Scheme b): From a common precursor, protochlorophyllide as well as prenylprotochlorophyllide can be formed. In the first case, the protochlorophyllide is photoreduced to chlorophyllide, followed by the esterification (mainly with phytol). In the second case, the esterification takes place prior to the photoreduction.

The hypothesis for the last steps of chlorophyll a formation is based on kinetic studies of ^{14}C-δ-aminolevulinic acid incorporation in the acid- and ester-forms of the pigment and on the different cofactor requirements for biosynthesis of the two forms [42]. According to a scheme proposed by Smith [49], which is very similar to the second case in the scheme of Rebeiz and Castelfranco [43], the prenylprotochlorophyllide should actually be formed through esterification of already existing protochlorophyllide (Scheme c).

(a)

(b)

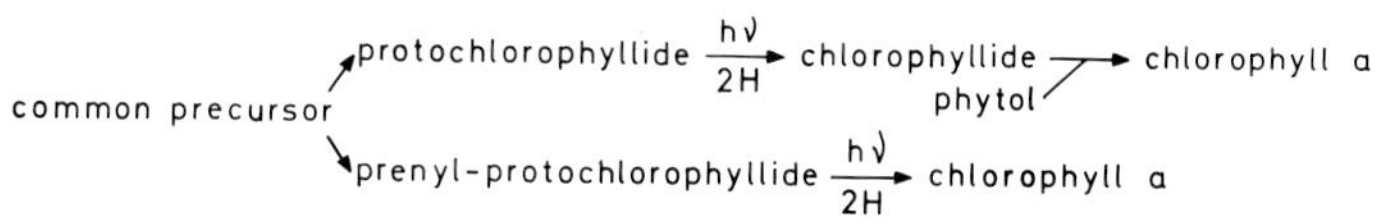

(c)

Rudolph and Bukatsch [45] proposed a specific enzyme, protochlorophyllase, for this reaction. According to Jones and Ellsworth [25], such an enzyme is present in plants, but in a later paper by Ellsworth and Nowak [12], it was shown that it was not possible to catalyze the esterification of protochlorophyllide with a prenol using a protochlorophyllase preparation from dark-grown wheat leaves. The prenols used were phytol and farnesol. This fact thus raised the question of whether the esterifying alcohol of the protochlorophyllide ester could be something other than these two prenols. The pigment has, since the work of Fischer and Oestreicher [15] on porphyrin synthesis, been regarded as the methyl-phytyl-ester of Mg-vinylpheoporphyrin a_5. Efforts have been made to isolate and identify the long-chain alcohol of the ester.

Fischer and Rüdiger [14] could not detect any phytol in the protochlorophyllide ester fraction isolated from dark-grown barley leaves. Nor did Rebeiz and Castelfranco [44] detect any phytol in the protochlorophyllide ester isolated from etiolated cucumber cotyledons. This is contrary to Godnev et al. [17], who found the esterifying alcohol of this pigment from barley to be phytol. The protochlorophyll pigments found in the inner seed coats of pumpkin seeds do not photoconvert to chlorophyll a. Ellsworth [13] showed that these pigments were esterified with farnesol and a series of C_{20}-prenols from geranylgeraniol to the saturated phytanol, including those with two and three double bonds.

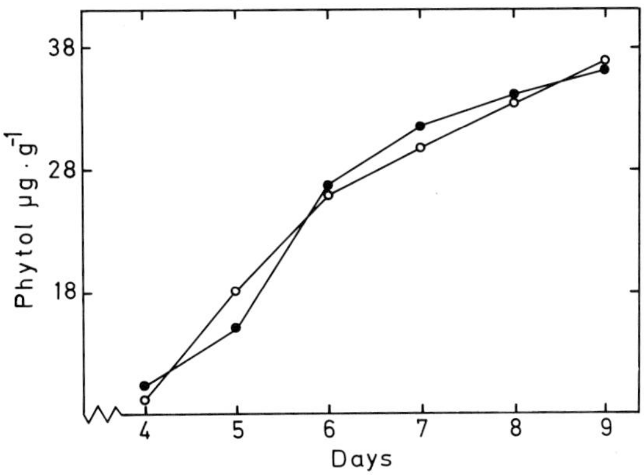

Fig. 1. Phytol content in relation to age in dark-grown leaves of barley. *Filled circles:* free phytol, before alkaline hydrolysis; *open circles:* total phytol after alkaline hydrolysis. *Ordinate:* µg/g fresh weight. (Liljenberg [34])

The amount of phytol before and after alkaline hydrolysis of pigment extracts from dark-grown leaves of different ages was determined by Liljenberg [33]. The total amount of phytol (free and bound by esterification) and the free amount of phytol, respectively, are shown in Figure 1. If esterified protochlorophyllide had contained phytol, then the total amount of the alcohol would have exceeded that of free phytol, which was not the case. Furthermore, this relation was observed at different ages. This indicated that the alcohol of esterified protochlorophyllide from barley was not phytol.

The pigment was isolated from large amounts of dark-grown leaves of barley (2.3 kg fresh weight) and purified. After saponification of the pigment the esterifying alcohol was further purified by thin-layer chromatography and run on a combined gas chromatograph-mass spectrometer [34].

The mass spectrum of the alcohol is shown in Figure 2 together with a spectrum of geranylgeraniol that was run under the same conditions as the alcohol from the protochlorophyllide ester. The two spectra correspond well except in the m/e region 290–296. The discrepancy here is probably due to impurities of a tetrachlorobiphenyl (mol wt 290) that displays a characteristic isotopic cluster of ions at m/e 290, 292, 294, and 296. The comparison between the two mass spectra speaks in favor of an identity of the two substances. No other alcohol was found. Part of the esterified protochlorophyllide in the dark-grown leaves cooled down to 2° C could be photoconverted to chlorophyll a by irradiation with consecutive light flashes. The pigments were measured fluorometrically [34]. The results indicate that at least a small part of the chlorophyll in the greening barley leaves should contain geranylgeraniol. This fact could be the basis for speculations about the role of this pigment in the plastid (see below). Except for this report, Wellburn [61] has also recently shown that newly emerging leaves of horse chestnut (*Aesculus hippocastanum*) contain small amounts of chlorophyll-containing geranylgeraniol.

Except for these reports dealing with higher plants, the bacteriochlorophyll in *Rhodospirillum* species has been shown to have geranylgeraniol instead of phytol as the esterifying alcohol [3, 26].

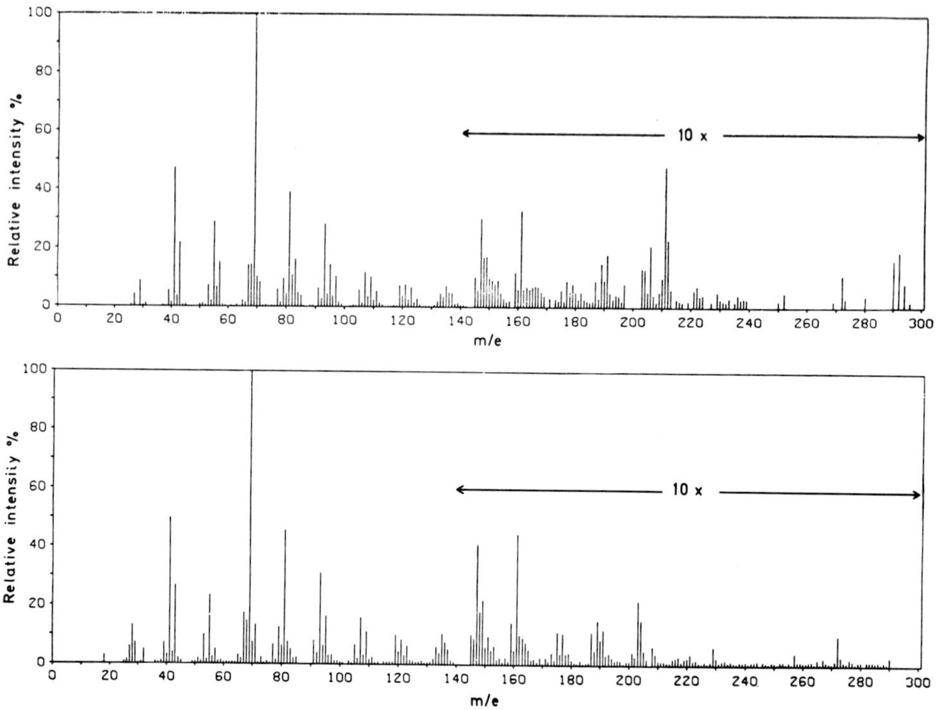

Fig. 2. Mass spectra of the alcohol from the protochlorophyllide ester and of authentic geranylgeraniol. From m/e 140 and higher the relative intensities are multiplied by a factor of 10. (Liljenberg [34])

C. The Occurrence of the Two Forms of Protochlorophyll Pigments in Dark-grown Plants and Their Photoreduction

The proportions of the acid- and ester-forms of the protochlorophyll pigment differ between species [18] and is also age-dependent [48]. In a study of 17 species the percentage of the esterified form (of total protochlorophyll pigments) varied from 10 to 88.

The literature dealing with the phototransformation of protochlorophyllide ester is confusing. Complete phototransformation has been reported by Godnev et al. [18] and by Rudolph and Bukatsch [45], and partial phototransformation, by Rebeiz and Castelfranco [44]. Later Godnev et al. [19] revised their earlier statements concerning the photoreduction of the protochlorophyllide ester. They did not find any conversion to chlorophyll a. In another case, the protochlorophyllide ester was reported to remain unchanged after irradiation [56].

According to Liljenberg [34], using gas chromatographic methods for alcohol analysis, the protochlorophyllide ester of dark-grown barley seedlings having geranylgeraniol as the esterifying alcohol can be partially photoconverted to the corresponding chlorophyll a. This would imply that, besides the main bulk of chlorophyll a having phytol as the main esterifying alcohol, small amounts of chlorophyll a should be esterified with geranylgeraniol.

These minor amounts of chlorophyll a (geranylgeranyl chlorophyllide) may have specific structural roles. Because photosystem I is the first system to start working in photosynthesis after only a few minutes of irradiation, and this at extremely low chlorophyll concentrations, the question can be asked, whether the reaction center molecules first formed are esterified with this alcohol.

D. The Esterification of Chlorophyllide a

Many reports have appeared concerning the enzyme chlorophyllase (EC 3.1.1.14) [1, 2, 4, 5, 23, 24, 28, 37, 38, 40, 41, 46, 51, 52].

Chlorophyllase in situ is probably membrane-bound, and chlorophyll–chlorophyllide conversion has been studied by Terpstra [53, 54] in various chloroplast lamellar fragments.

Most of the preceding authors describe the properties of the solubilized enzyme upon the hydrolysis of chlorophyll a and pheophytin a (in some cases also chlorophyll b and pheophytin b). Because this enzyme catalyzes the hydrolysis of chlorophyll to chlorophyllide and free phytol, it has been regarded to function also at the esterification of chlorophyllide. In the presence of detergents or acetone, chlorophyllase preparations can esterify pheophorbide with phytol to pheophytin a [9, 10].

Also transesterification of methylpheophorbide a and methylchlorophyllide a with phytol to pheophytin a and chlorophyll a, respectively, has been reported [5, 9, 10, 60]

Chlorophyllase preparations from wheat seedlings catalyze hydrolysis of pheophytin a, transphytylation of methylpheophorbide a to pheophytin a, and phytylation of pheophorbide a to pheophytin a [11]. Because activities of these reactions are differently affected by changing the reaction conditions, such as acetone and buffer concentrations, pH, and temperature, Ellsworth [11] suggested that his chlorophyllase preparation contained more than one enzyme. This interpretation has been supported by Ichinose and Sasa [24], who reported similar effects of an enzyme preparation extracted from *Chlorella*.

It has been suggested that chlorophyll a in situ might be synthesized from methylchlorophyllide via a transesterification to chlorophyll a [5]. The presence of small amounts of methylchlorophyllide in expanding wheat leaves and in leaves of passion flower has also been interpreted as evidence that chlorophyllase can work as a transesterification enzyme exchanging methanol for phytol [22].

Granick [21] has suggested that "the mechanism of phytylation is probably via phytol pyrophosphate rather than by the hydrolytic enzyme, chlorophyllase." Also Goodwin [20] has proposed that, in analogy with other terpenoid reactions, phytyl pyrophosphate should be the activated substrate in this reaction. The presence of phytyl pyrophosphate in higher plants has been reported by Watts and Kekwick [58], who found free phytol as well as phytyl pyrophosphate

Table 1. Molar ratios of different forms of phytol to protochlorophyllide from etioplasts and the relative ratio between the two forms of phytol from etioplasts and 8-h irradiated etiochloroplasts (Liljenberg [35])

Ratio	Etiochloroplasts	
	nonirradiated	irradiated
free phytol/protochlorophyllide	1.8	—
phytyl pyrophosphate/protochlorophyllide	1.0	—
free phytol/phytylpyrophosphate	1.8	2.7

in greening leaves of *Phaseolus vulgaris*. These analyses were, however, made on whole leaves; therefore no conclusions can be made from these experiments about the contents of the plastids.

Phytyl pyrophosphate has been isolated from the etiochloroplast of dark-grown barley seedlings that had been irradiated for 8 h [35]. The molar ratios of free phytol/protochlorophyllide, phytyl pyrophosphate/protochlorophyllide, and free phytol/phytyl pyrophosphate are shown in Table 1.

The presence of free phytol with a molar ratio of free phytol/protochlorophyllide in the nonirradiated etioplast of 1.8 might be regarded as a pool of phytol for the esterification of chlorophyllide [33]. The absolute amount of free phytol in the etiochloroplasts taken from seedlings irradiated for 8 h did not differ remarkably from the values for nonirradiated seedlings. The small difference in the content of the two forms of phytol between the nonirradiated and the irradiated seedlings was supposed to reflect, in some way or other, interconnection between the synthesis of the porphyrin moiety of the chlorophylls and the synthesis of the phytol. These two processes do not proceed independently of each other. The synthesis of phytyl pyrophosphate in etioplasts and in etiochloroplasts was supposed to be coupled to the phytylation of the chlorophyll pigments. There are other terpenoid derivatives in the plastid containing so-called phytol chains, such as α-tocopherol, α-tocoquinone and vitamin K_1, but the amount of these substances increases very little or not at all during the first 6–8 h of illumination compared to phytol synthesis during the same period [29, 33].

The fact that phytol is present in the etioplast in an energy-favored pyrophosphate complex and with a molar ratio (phytyl pyrophosphate/protochlorophyllide) of 1.0 supports the idea that this substance is the reactant of the esterification.

A possible connection between the esterification of chlorophyllide a and the change in the structure of the prolamellar body has been reported by Treffry [55]. At 0° C complete photoconversion of protochlorophyllide occurs without any loss of regularity of the prolamellar body. If the temperatures is then increased to 26° C, transformation of the prolamellar body and esterification of chlorophyllide a occur concurrently.

E. Light-induced Changes in the Rate of the Esterification Reaction

Pretreatment with a short light impulse followed by darkness for 2 to 3 h will change the rate of esterification of chlorophyllide a in previously dark-grown seedlings. The same amount of chlorophyllide a will, after such pretreatment, be esterified in almost half the time compared with samples that were not pretreated [32]. The effect of pretreatment was later confirmed by Shlyk [47]. The spectral dependence of this pretreatment effect has been investigated (Fig. 3).

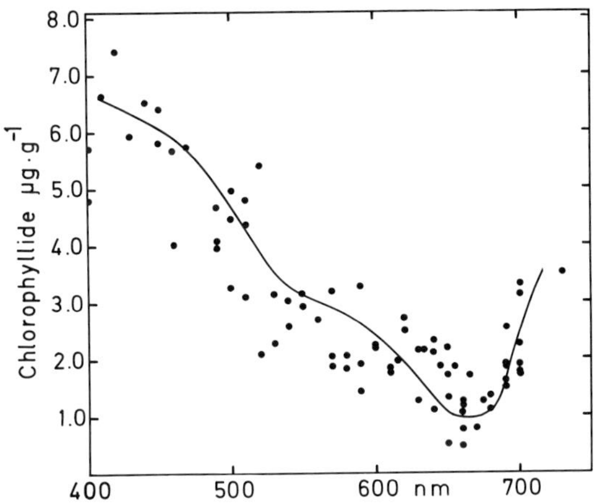

Fig. 3. Effect of preirradiation with monochromatic light on the phytylation of chlorophyllide a. Figure shows the remaining amount of chlorophyllide a in dark-grown wheat leaves given the following light and dark treatments: monochromatic light (12.3 mJ/cm^2)+3 h of darkness+light flashes+20 min of darkness. (Liljenberg [32])

The most effective wavelength region has been shown to be around 660–670 nm. This eliminates the possibility of protochlorophyllide being the light-absorbing pigment at this reaction. It is probable that this effect is caused by light absorbed by phytochrome in analogy with light pretreatment effects on carotenoid synthesis and the lag phase of chlorophyll synthesis [6, 57]. The question then arises on which system the phytochrome might be acting, because pretreatment with light does not lead to a higher content of free phytol [33]. Phytochrome might instead activate or increase the synthesis of the enzyme catalyzing the esterification reaction. These results are in agreement with the observations of Lichtenthaler [30] and Lichtenthaler and Becker [31] on prenyl chain formation in dark-grown seedlings of *Hordeum* and *Raphanus*, respectively. Continuous far-red light enhances the synthesis of phytyl chains as compared to continuous darkness and, significantly, to synthesis of vitamin K_1. No formation of free phytol has been reported. Stimulation by continuous white light is much greater than that induced by far-red light. With white light chlorophyllide (via protochlorophyllide) is formed and esterification starts. It has been proposed that chlorophyllide formation determines the rate of enhanced phytyl chain production.

F. Biosynthesis of Esterifying Prenols

The generally accepted pathway for the formation of geranylgeraniol and phytol is condensation of isopentenyl pyrophosphate by enzyme systems, i.e., in sequenses of five carbon additions. In this way the prenols would be synthesized in the etiochloroplast as alcohol pyrophosphates, which is the most accepted view for geranylgeranyl pyrophosphate formation. From results of in vivo labeling studies, Costes [7] and Wellburn et al. [59] have suggested that the immediate precursor of phytol is geranylgeraniol. A less likely model for the building up of the prenyl chain of the chlorophyll molecule would be through a sequential addition of five carbon units to the porphyrin nucleus in analogy with prenyl chain additions to benzoquinone derivatives in animal tissues as postulated by Martius [36].

Another possible way of synthesis proposed by Wellburn [61] would be that complete molecules of geranylgeraniol are first coupled to chlorophyllide a, probably as pyrophosphate and then hydrogenated to phytol. According to Wellburn this would facilitate the stereospecific reduction of geranylgeraniol to phytol and explain the tight coupling between phytol formation and chlorophyll synthesis. This pathway is supported by the recently reported short lifetime intermediate in the phytylation of chlorophyllide a in vivo [39] showing chromatographic characteristics similar to those of geranylgeranyl chlorophyllide and an absorption spectrum very similar to that of chlorophyll a.

It is obvious, however, that esterification of chlorophyllide a with phytyl pyrophosphate or with free phytol is a pathway that cannot be ruled out, particularly because free phytol and phytyl pyrophosphate have been found in etiochloroplasts [35].

G. The Pools of Free Phytol and Phytol Bound as Acyl Esters During Greening

It has been possible with gas chromatographic technique to show that free phytol exists in dark-grown leaves [33, 58]. This means that a pool of free phytol is available for esterification of the chlorophyllide formed from the protochlorophyllide accumulated in the dark.

This pool of free phytol found in dark-grown plants has been shown to change during irradiation. The first effect of light is an immediate decrease, which is interpreted as reflecting the esterification of chlorophyllide. The decrease is later followed by an increase during continuous irradiation. The concentration, however, never exceeds that in dark-grown leaves [33].

After 4 h of irradiation there is more esterified phytol present than could be accounted for from the measured amount of chlorophyll pigments. This is due to the presence of acceptors other than chlorophyll to which this amount of phytol produced in excess is bound [35]. These have been shown to be saturated fatty acids or acids with only one double bond. Dark-grown seedlings

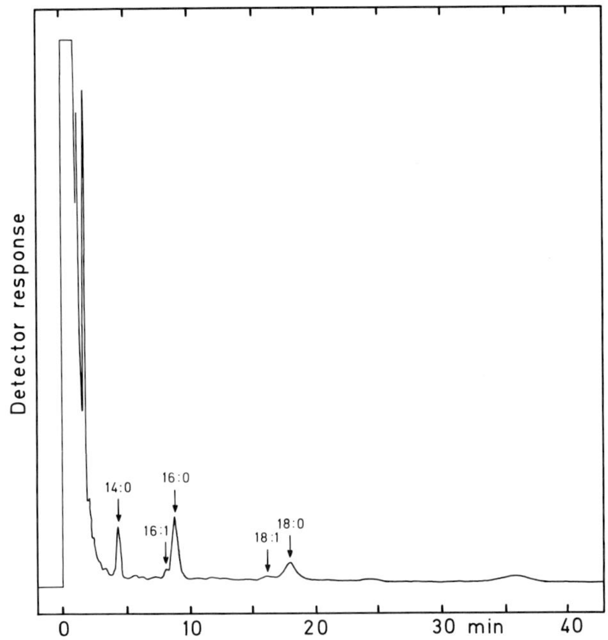

Fig. 4. Gas chromatogram of methyl esters of the acids esterified to phytol in greening barley seedlings. Dark-grown seedlings had been irradiated for 8 h prior to isolation and determination. (Liljenberg [35])

Table 2. Fatty acid content of the phytyl esters

Fatty acid		mole percentage (% of total identified acids)
Myristic	14:0	18.3
Palmitic	16:0	42.1
Palmitoleic	16:1	10.1
Stearic	18:0	21.7
Oleic	18:1	7.8

Phytyl esters were isolated from dark-grown seedlings irradiated for 8 h (Liljenberg [35])

after irradiation were thus shown to contain acyl esters of phytol (Fig. 4, Table 2).

Except for greening leaves of higher plants there are only a few reports on the presence of fatty acid esters of phytol in plant material. Steffen et al. [50] have reported that phytol is esterified to fatty acids in seeds of different species. Because the phytol content decreased during germination in the dark, these authors have questioned whether the free phytol pool of the dark-grown leaves really is the result from synthesis in situ. Anyhow, the existing free phytol together with the phytyl pyrophosphate is located in the etioplasts.

Yellow leaves of *Acer* platanoides have been shown to contain phytyllinole-

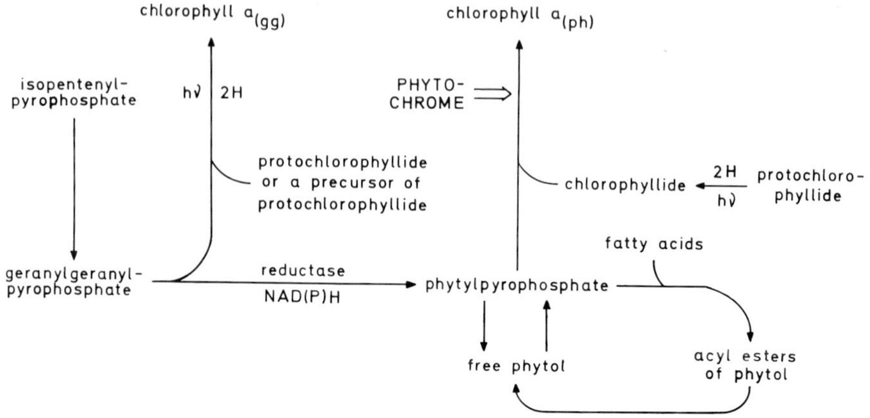

Fig. 5. Chlorophyll $a_{(gg)}$ and chlorophyll $a_{(ph)}$ represents chlorophyll a with geranylgeraniol and phytol, respectively, as esterifying alcohol

nate [8]. There are also observations that mosses contain acyl esters of phytol [16].

In all reported cases the fatty acids probably function as temporary acceptors for phytol during the biosynthesis and during the breaking down of chlorophyll in the chloroplast.

Figure 5 summarizes most of the reactions known, together with possible but not fully proven pathways, involved in phytylation of the chlorophyll formation.

References

1. Ardao, C., Vennesland, B.: Chlorophyllase activity of spinach chloroplastin. Plant Physiol. **35**, 368–371 (1960)
2. Bacon, M.F., Holden, M.: Chlorophyllase of sugar-beet leaves. Phytochemistry **9**, 115–125 (1970)
3. Brockmann, H., Knobloch, G., Schweer, I., Trowitzsch, W.: Die Alkoholkomponente des Bacteriochlorophyll a auf Rhodospirillum rubrum. Arch. Mikrobiol. **90**, 161–164 (1973)
4. Böger, P.: Chlorophyllase of *Chlorella vulgaris Beijerinck*. Phytochemistry **4**, 435–443 (1965)
5. Chiba, Y., Aiga, I., Iclemori, M., Satoh, Y., Matsushita, K., Sasa, T.: Studies on chlorophyllase of *Chlorella protothecoides* I. Enzymatic phytylation of methyl chlorophyllide. Plant Cell Physiol. (Tokyo) **8**, 623–635 (1967)
6. Cohen, R., Goodwin, T.W.: The effect of red and far red light on carotenoid synthesis by etiolated maize seedlings. Phytochemistry **1**, 67–72 (1962)
7. Costes, C.: Biosynthèse du phytol des chlorophylles et du squelette tétraterpénique des caroténoides dans les feuilles vertes. Phytochemistry **5**, 311–324 (1966)
8. Csupor, L.: Das Phytol in vergilbten Blättern. Planta Med. **19**, 37–41 (1970)
9. Ellsworth, R.K.: Studies on chlorophyllase I. Hydrolytic and esterification activities of chlorophyllase from wheat seedlings. Photosynthetica **5**, 226–232 (1971)
10. Ellsworth, R.K.: Studies on chlorophyllase. III. Differences in gel chromatographic behavior of the hydrolytic and esterification activities obtained from wheat seedlings. Photosynthetica **6**, 276–281 (1972a)

11. Ellsworth, R.K.: Studies on chlorophyllase II. The effects of changes in reaction conditions on hydrolytic and esterification activities observed in preparations from wheat seedlings. Photosynthetica **6**, 32–40 (1972b)
12. Ellsworth, R.K., Nowak, C.A.: The inability of crude homogenates of etiolated wheat seedlings containing protochlorophyllase to convert ^{14}C-protochlorophyllide to ^{14}C-protochlorophyll. Photosynthetica **7**, 246–251 (1973)
13. Ellsworth, R.K., Nowak, C.A.: A gas chromatographic–mass spectrometric analysis of the esterifying alcohols of pumpkin seed protochlorophylls. Anal. Biochem. **57**, 534–546 (1974)
14. Fischer, G.G., Rüdiger, W.: Bildung und Vorkommen von Phytol, II Über den Phytolgehalt der Protochlorophylle und des im Dunkeln gebildeten Chlorophylls. Liebigs Ann. Chem. **627**, 35–46 (1959)
15. Fischer, H., Oestreicher, A.: Über Protochlorophyll und Vinylporphyrine. Ein Beitrag zur Kenntnis der Oxo-Reaktion. Hoppe-Seylers Z. Physiol. Chem. **262**, 243–269 (1939/40)
16. Gellerman, J.L., Anderson, W.H., Schlenk, H.: Synthesis and analysis of phytyl and phytenoyl wax esters. Lipids **10**, 656–661 (1975)
17. Godnev, T.N., Kaler, V.L., Rotfarb, R.M.: On the question of the presence of phytol as a component of the protochlorophyll of etiolated leaves. Dokl. Akad. Nauk SSSR **140** (6), 1445–1447 (1961)
18. Godnev, T.N., Rotfarb, R.M., Akulovich, N.K.: On the nature of certain reactions in the biosynthesis of chlorophyll and factors influencing them. Photochem. Photobiol. **2**, 119–128 (1963)
19. Godnev, T.N., Galaktionov, S.G., Raskin, V.I.: Steric conditions of the reactions of hydrogenation at positions 7 and 8 of pyrrole ring IV of protochlorophyll pigments. Dokl. Akad. Nauk USSR **181**, 237–240 (1968)
20. Goodwin, T.W.: Biosynthesis by chloroplasts. In: Structure and Function of Chloroplasts. Gibbs, M. (ed.). Berlin-Heidelberg-New York: Springer, 1971, pp. 215–276
21. Granick, S.: The heme and chlorophyll biosynthetic chain. In: Biochemistry of Chloroplasts. Goodwin, T.W. (ed.). London and New York: Academic Press, 1967, Vol. 2, pp. 373–410
22. Hines, G.D., Ellsworth, R.K.: Methylchlorophyllide a as a probable intermediate in the chlorophyll a pathway. Plant Physiol. **44**, 1742–1744 (1969)
23. Holden, M.: The breakdown of chlorophyll by chlorophyllase. Biochem. J. **78**, 359–364 (1961)
24. Ichinose, N., Sasa, T.: Studies on chlorophyllase of *Chlorella protothecoides* III. Purification and catalytic properties. Plant Cell Physiol. (Tokyo) **14**, 1157–1166 (1973)
25. Jones, C.B., Ellsworth, R.K.: Evidence for the presence of protochlorophyllase in etiolated wheat seedlings. Plant Physiol. **44**, 1478–1480 (1969)
26. Katz, J.J., Strain, H.H., Harkness, A.L., Studier, M.H., Svec, W.A., Janson, T.R., Cope, B.T.: Esterifying alcohols in the chlorophylls of purple photosynthetic bacteria. A new chlorophyll, bacteriochlorophyll$_{(gg)}$, all-transgeranylgeranylbacteriochlorophyllide a. J. Am. Chem. Soc. **94**, 7938–7939 (1972)
27. Kirk, J.T.O.: Biochemical aspects of chloroplast development. Ann. Rev. Plant Physiol. **21**, 11–42 (1970)
28. Klein, A.O., Vishniak, W.: Activity and partial purification of chlorophyllase in aqueous systems. J. Botan. Chem. **236**, 2544–2547 (1961)
29. Lichtenthaler, H.K.: Light-stimulated synthesis of plastid quinones and pigments in etiolated barley seedlings. Biochim. Biophys. Acta **184**, 164–172 (1969)
30. Lichtenthaler, H.K.: Regulation of prenyl chain synthesis in etiolated Hordeum seedlings by far-red and white light. Physiol. Plantarum **33**, 241–244 (1975a)
31. Lichtenthaler, H.K., Becker, K.: The influence of continuous far-red and white light on prenyl chain synthesis in plastids of Raphanus seedlings. Planta **122**, 255–258 (1975b)
32. Liljenberg, C.: The effect of light on the phytolization of chlorophyllide a and the spectral dependence of the process. Physiol. Plantarum **19**, 848–853 (1966)
33. Liljenberg, C.: Phytol changes in dark grown leaves of barley after varying light treatments. Physiol. Plantarum **25**, 358–362 (1971)
34. Liljenberg, C.: Characterization and properties of a protochlorophyllide ester in leaves of dark grown barley with geranylgeraniol as esterifying alcohol. Physiol. Plantarum **32**, 208–213 (1974)
35. Liljenberg, C.: Chlorophyll formation in dark grown plants. A study of the esterification of chlorophyllide. PhD thesis Göteborg Univ., Göteborg, Sweden (1975)

36. Martius, C.: Metabolism of vitamin E based on tissue analysis. Vitam. Horm. **20**, 457–462 (1962)
37. McFeeters, R.F., Chichester, C.O., Whitaker, J.R.: Purification and properties of chlorophyllase from *Ailanthus altissima* (tree-of-heaven). Plant Physiol. **47**, 609–618 (1971)
38. McFeeters, R.F.: Substrate specificity of chlorophyllase. Plant Physiol. **55**, 377–381 (1975)
39. Ogawa, T., Bovey, F., Shibata, K.: An intermediate in the phytylation of chlorophyllide a in vivo. Plant Cell Physiol. (Tokyo) **16**, 199–207 (1975)
40. Ogura, N., Takamiya, A.: Studies on chlorophyllase of tea leaves. Botan. Mag. (Tokyo) **79**, 588–594 (1966)
41. Ogura, N.: Studies on chlorophyllase in tea leaves III. Properties of soluble and insoluble chlorophyllases. Plant Cell Physiol. (Tokyo) **13**, 971–979 (1972)
42. Rebeiz, C.A., Yaghi, M., Abou-Haidar, M., Castelfranco, P.A.: Protochlorophyll biosynthesis in cucumber (*Cucumis sativus*, L.) cotyledons. Plant Physiol. **46**, 57–63 (1970)
43. Rebeiz, C.A., Castelfranco, P.A.: Protochlorophyll biosynthesis in a cell-free system from higher plants. Plant Physiol. **47**, 24–32 (1971)
44. Rebeiz, C.A., Castelfranco, P.A.: Protochlorophyll and chlorophyll biosynthesis in cell-free systems from higher plants. In: Ann. Rev. Plant Physiol. Briggs, W.R., Green, P.B., Jones, R.L. (eds.). Palo Alto, California: Annual Reviews Inc., 1973, pp. 129–171
45. Rudolph, E., Bukatsch, F.: Die Protochlorophyll(id)-Umwandlung und ihre Beziehungen zur Photooxydation der Ascorbinsäure bei etiolierten Keimpflanzen. Planta **69**, 124–134 (1966)
46. Shimizu, S., Tamaki, E.: Chlorophyllase of tobacco plants I. Preparation and properties of water soluble enzyme. Botan. Mag. (Tokyo) **75**, 462–467 (1962)
47. Shlyk, A.A., Rudoi, A.B., Vezitzkii, A.Y.: Immediate appearance and accumulation of chlorophyll b after a short illumination of etiolated maize seedlings. Photosynthetica **4**, 68–77 (1970)
48. Sironval, C., Michel-Wolwertz, M.R., Madsen, A.: On the nature and possible functions of the 673- and 684-mμ forms in vivo of chlorophyll. Biochim. Biophys. Acta. **94**, 344–354 (1965)
49. Smith, J.H.C.: Protochlorophyll transformations. In: Comparative Biochemistry of Photoreactive Systems. Allen, M.B. (ed.). New York and London: Academic Press, 1960, pp. 257–277
50. Steffens, D., Blos, I., Schoch, S., Rüdiger, W.: Lichtabhängigkeit der Phytolakkumulation. Ein Beitrag zur Frage der Chlorophyll-Biosynthese. Planta **130**, 151–158 (1976)
51. Sudyina, O.H., Dovbysh, E.F.: On problem of chlorophyllase isolation. Ukr. Botan. Zh. **29**, 92–95 (1972a)
52. Sudyina, O.H., Babenko Ye, H., Fomishyna, R.M.: Kinetic of chlorophyllase reaction with different concentration of enzyme. Ukr. Botan. Zh. **29**, 419–424 (1972b)
53. Terpstra, W.: Properties of chloroplasts and chloroplast fragments, as deduced from internal chlorophyll→chlorophyllide conversion. Z. Pflanzenphysiol. **71**, 129–143 (1974)
54. Terpstra, W., Goedheer, J.C.: Chlorophyllase and lamellar structure in *Phaeodactylum tricornutum* I. Chlorophyll–chlorophyllide conversion within the lamellae. Z. Pflanzenphysiol. **75**, 118–130 (1975)
55. Treffry, T.: Phytylation of chlorophyllide and prolamellar-body transformation in etiolated peas. Planta **91**, 279–284 (1970)
56. Virgin, H.I.: Pigment transformations in leaves of wheat after irradiation. Physiol. Plantarum **13**, 155–164 (1960)
57. Virgin, H.I.: Action spectrum for the elimination of the lag phase in chlorophyll formation in previously dark grown leaves of wheat. Physiol. Plantarum **14**, 439–452 (1961)
58. Watts, R.B., Kekwick, R.G.O.: Factors affecting the formation of phytol and its incorporation into chlorophyll by homogenates of the leaves of the french bean *Phaseolus vulgaris*. Arch. Biochem. Biophys. **160**, 469–475 (1974)
59. Wellburn, A.R., Stone, K.J., Hemming, F.W.: The stereochemistry of phytol biosynthesis in the leaves of *Ficus elasticus* and *Aesculus hippocastanum*. Biochem. J. **100**, 23c–25c (1966)
60. Wellburn, A.R.: Studies on the esterification of chlorophyllides. Phytochemistry **9**, 2311–2313 (1970)
61. Wellburn, A.R.: Evidence for chlorophyll esterified with geranylgeraniol in newly greened leaves. Biochem. Physiol. Pflanzen **169**, 265–271 (1976)

CHAPTER 15

Biosynthesis and Degradation of Lipid Polymers

P.E. KOLATTUKUDY

A. Introduction

Biopolyesters derived from hydroxyfatty acids are unique to plants. Cutin, the structural component of plant cuticle, is essentially a polyester derived from hydroxy- and epoxyfatty acids [37, 47]. Suberin, the structural polymer of the cuticle of underground parts of plants, and which is also deposited in wound periderm of plants, also contains substantial amounts of esterified hydroxyfatty acids and dicarboxylic acids, as well as smaller quantities of the more oxygenated monomers similar to those found in cutin [4, 19, 20, 27, 54]. With the advent of modern analytical techniques and biochemical techniques, substantial progress has been made in our understanding of the biochemistry of these unique biopolymers. In this chapter, I shall review the synthesis and degradation of these polymers; other aspects have been discussed in a recent publication [31].

B. Biosynthesis of Cutin

Early suggestions that cutin is formed by autooxidative processes [21, 50] gave way to the speculation that certain ill-defined "fatty acid oxidases" and lipoxidases are involved in cutin biosynthesis [3, 17]. The latter hypothesis is not based on sound biochemical reasons, but was based on certain observations on wounded leaves assuming that cutin resynthesis occurred during wound healing. However, with the recent findings that the wound healing process involves suberization rather than resynthesis of cutin, it appears clear that such observations are not relevant to cutin biosynthesis. Systematic biosynthetic

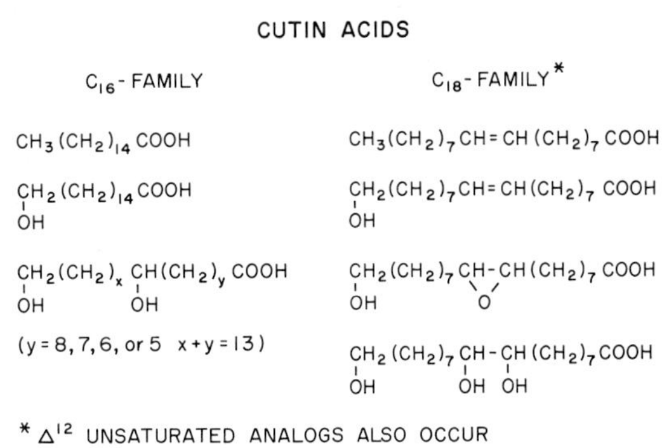

Fig. 1. Structure of the major components of cutin

studies on cutin began with the discovery that exogenous labeled precursors were readily incorporated into cutin in rapidly expanding *Vicia faba* leaves [23, 24]. Subsequently, a variety of other rapidly expanding tissues, as well as cell-free preparations from such tissues, have been used to elucidate the biochemical pathway involved in the biosynthesis of cutin. On the basis of their structure and biosynthesis, the monomer acids of cutin can be classified into two families, a C_{16} family and a C_{18} family; the major components of both families are shown in Figure 1.

I. Biosynthesis of the C_{16} Family of Cutin Acids

1. In vivo Studies

Palmitic acid, 16-hydroxypalmitic acid, and 10,16-dihydroxypalmitic acid and/or its positional isomers comprise the C_{16}-family of cutin acids in most plants. Rapidly expanding leaves of *V. faba* incorporated [1-^{14}C]palmitic acid quite readily into an insoluble fraction that was suspected to be cutin [36]. Exhaustive hydrogenolysis of the labeled residue with $LiAlH_4$ resulted in the release of all of the ^{14}C into products soluble in organic solvents. The bulk of the label was found to be in hexadecane-1,7,16-triol, the product expected to be formed by the reduction of 10,16-dihydroxypalmitic acid, the major component of cutin in this plant. The only other labeled components were hexadecane-1,16-diol and hexadecanol derived from 16-hydroxypalmitic acid and palmitic acid, respectively; these acids are the other two significant components of *V. faba* cutin. Similar results have been obtained also with other plant tissues, such as grape berry and apple fruit skin slices, as well as leaf disks of apple and *S. odoris*. As ω-hydroxyacids are common constituents of cutin, and all dihydroxy C_{16} acids contain the terminal hydroxyl group, it was suspected that ω-hydroxylation and subsequent in-chain hydroxylation were involved in the biosynthesis of the C_{16}-family of cutin acids. However, the time-course of incorporation of [1-^{14}C]palmitic acid into the cutin components of *V. faba* leaf disks failed to reveal the expected precursor–product relationships, because at all times the major radioactive component of cutin was 10,16-dihydroxypalmitic acid, and the distribution of ^{14}C among the three cutin components remained relatively constant [36]. Therefore, an alternate approach was taken to test the possibility that ω-hydroxypalmitic acid is the immediate precursor of dihydroxypalmitic acid. Synthetic 16-hydroxy[G-^{3}H]palmitic acid and biosynthetically labeled 16-hydroxypalmitic acid derived from [1-^{14}C]palmitic acid, when incubated with *V. faba* leaf disks, were incorporated into 10,16-dihydroxypalmitic acid without involving degradation and resynthesis. This direct conversion of 16-hydroxypalmitic acid to 10,16-dihydroxy acid could involve either dehydrogenation followed by hydration or a direct hydroxylation. Since neither [10-^{14}C]palmitoleic acid nor [10-^{14}C]palmitelaidic acid was converted into 10,16-dihydroxypalmitic acid in *V. faba* or apple, the former possibility appeared unlikely [36]. However, there is a possibility that the exogenous unsaturated acids were not available to the same subcellular sites and metabolic pathways as the exogenous saturated acids. In an attempt to eliminate this possibility, [10-^{14}C-9,10-^{3}H]palmitic acid was used as a precursor of the dihydroxy acid. From the $^3H:^{14}C$ isotopic ratios of 16-hydroxypalmitic acid and 10,16-dihydroxypalmitic acid derived from the doubly labeled palmitic acid, it was found that only one of the four H atoms of C-9 and C-10 was lost during the in-chain hydroxylation process. These results strongly suggested that a Δ^9 intermediate was not involved in the introduction of the in-chain hydroxyl group. Consistent with this conclusion was the observation that the conversion of the 16-hydroxy acid to the dihydroxy acid required molecular oxygen and was inhibited by chelators such as phenanthroline, and that this inhibition was reversed by Fe^{+2}. This conclusion, that a mixed function oxidase directly introduces the in-chain hydroxyl group, is also consistent with the observation that positional isomers, in which the in-chain hydroxyl group is at 10, 9, 8, or 7, occur in cutin from many plants [31, 37]. Such experimental results suggested a biochemical pathway that involves ω-hydroxylation followed by a direct hydroxylation at C-10 or at another in-chain carbon.

2. *Enzymatic Synthesis of the C_{16} Family of Cutin Acids*

a) ω-Hydroxylation. Since the in vivo studies strongly suggested a reasonable pathway for the biosynthesis of the C_{16} family of cutin acids, we set out to search for the enzymes that should be involved in the proposed pathway. ω-Hydroxylation, the expected first committed step in cutin biosynthesis, has not been demonstrated in higher plants. Since epidermis appears to be the site of synthesis of cutin [31], epidermal extracts of *V. faba* leaves were used to search for an ω-hydroxylating enzyme. In numerous attempts with labeled palmitic acid, hydroxyfatty acids were detected; but radio gas-liquid chromatographic analysis and analyses of chemical degradation products derived from the hydroxyacids showed that α- and β-hydroxy acids constituted the bulk of the hydroxylated products formed in such cell-free preparations. Recently, ω-hydroxylation has been demonstrated with a microsomal preparation obtained from the shoots of germinating *V. faba* seeds using radio gas-liquid chromatographic and high-pressure liquid chromatographic resolution of the positional isomers of the hydroxy acids [58]. With such a radio-gas chromatographic assay, it was found that the ω-hydroxylase showed a pH optimum near 8.0 and required O_2 and NADPH (but not NADH). This ω-hydroxylase was inhibited by CO, chelators such as phenanthroline, and cytochrome C, but we have been unable to demonstrate reversal of the inhibition caused by CO with light. Therefore, it appears that, if a cytochrome P_{450}-type enzyme is involved in this reaction, this cytochrome must have a higher affinity for CO than those heretofore observed in other hydroxylases.

Since cutin monomers did not accumulate in tissues that were known to synthesize cutin, it was concluded that the monomers are transferred to the polymer as soon as they are generated [36]. As incorporation of the hydroxy acids into cutin is known to require activation [7, 10], it appears possible that palmitoyl-CoA or a similar thioester is the true substrate for ω-hydroxylation. However, in the present case no requirement for ATP and CoA could be shown, neither did palmitoyl-CoA serve as a more efficient substrate than palmitic

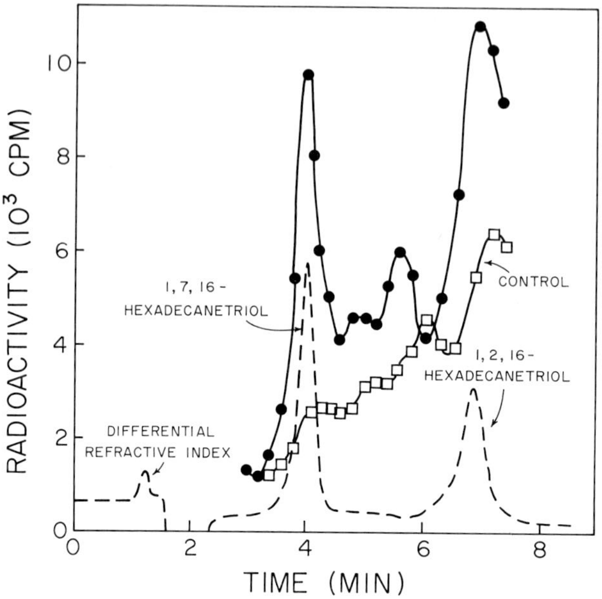

Fig. 2. High-pressure liquid chromatogram demonstrating the in-chain hydroxylation of ω-hydroxy[G-^{3}H]palmitic acid in a microsomal preparation from *Vicia faba* embryonic shoots. Incubation mixture consisted of NADPH, glucose-6-phosphate, and glucose-6-phosphate dehydrogenase, in addition to the substrate, in 0.1 M Tris-HCl buffer containing 0.25 M sucrose, 1 mM dithioerythritol, and 1 mM $MgCl_2$. The lipid extracted after 2 h incubation at 30° C was subjected to hydrogenolysis, and the isolated labeled C_{16}-triol fraction was mixed with the C_{16}-triols indicated and injected into the liquid chromatograph equipped with a 0.25 in. × 1 ft. μBondapak/C_{18} column, and a 25% H_2O/methanol solvent system was used

acid. It is possible that the involvement of a thioester could not be demonstrated because the particulate preparation catalyzed rapid hydrolysis of exogenous palmitoyl-CoA, and because of competing reactions such as β-oxidation [58].

b) In-Chain Hydroxylation. The experimental evidence obtained with tissue slices indicated that a direct hydroxylation is involved in the introduction of the hydroxyl group at C-10 of 16-hydroxypalmitic acid in *V. faba*. Therefore, 16-hydroxy[G-^{3}H]palmitic acid was used as a substrate with cell-free extracts prepared from the excised epidermis of *V. faba* leaves. These extracts catalyzed α- and β-oxidation of the ω-hydroxy acids, in addition to C-10 hydroxylation. As it was difficult to resolve the mixture of positional isomers formed in such reaction mixtures, analysis of the CrO_3 products had to be used as an assay method for this hydroxylase [61]. With such an assay it was found that this hydroxylase, which showed a pH optimum at 7.3, required O_2 and NADPH (but not NADH) as cofactors, as expected of a mixed function oxidase. In addition to these cofactors, ATP and CoA were also required, suggesting that activation of the carboxyl group, eight methylenes away from the hydroxylation site, is required. Thus, 16-hydroxypalmitoyl-CoA probably undergoes hydroxylation at C-10, and the dihydroxyacyl group could then be transferred directly to the hydroxyl group of the cutin primer as discussed in a succeeding section.

Since CrO_3 oxidation is extremely tedious and not reproducibly quantitative, we resorted to a high-pressure liquid chromatographic technique to separate the 10-hydroxylation product from α- and β-oxidation products [59]. Using such a direct measurement, it was demonstrated that C-10 hydroxylation of 16-hydroxy[G-^{3}H]palmitic acid is catalyzed by a microsomal preparation from the shoots of germinating *V. faba* seeds (Fig. 2). With this assay it was found that O_2 and NADPH (but not NADH) are the required cofactors. In this system no requirement for ATP and CoA could be demonstrated. In any case, with this newly developed assay it is feasible to examine this key enzyme involved in cutin biosynthesis.

The presence of small amounts of dicarboxylic acids in cutin preparations, and the recent identification of 16-oxo-9-hydroxy- and 16-oxo-10-hydroxypalmitic acids as significant components of cutin from embryonic *V. faba,* suggest that the terminal hydroxyl group can undergo further oxidation [25, 26]. In tissue slices of embryonic shoots of *V. faba,* 10,16-dihydroxy[G-^{3}H]palmitic acid was shown to be oxidized to the 16-oxo derivative. Enzymatic conversion of 16-hydroxypalmitic acid to hexadecane-1,16-dioic acid was also demonstrated in cell-free preparations from the epidermis of *V. faba* leaves [32]. This ω-hydroxy acid dehydrogenase, located in the 100,000 *g* supernatant, exhibit a pH optimum near 8.0. Unlike alcohol dehydrogenase and the ω-hydroxy acid dehydrogenase from animal liver, the epidermal enzyme showed a strong preference for NADP. The apparent K_m values for 16-hydroxypalmitic acid and NADP were 1.3×10^{-5} M and 3.6×10^{-4} M, respectively. That a 16-oxo intermediate is involved in the conversion of the ω-hydroxyacid to the dicarboxylic acid was strongly suggested by the observations that the 16-oxo intermediate could be trapped as the 2,4-dinitrophenylhydrazone, and that the cell-free preparation also catalyzed the oxidation of synthetic 16-oxo-palmitic acid to hexadecanedioic acid. Thiol reagents such as N-ethylmaleimide and *p*-chloromercuribenzoate inhibited the ω-hydroxy acid dehydrogenase.

The results discussed above strongly support the pathway shown in Figure 3 for the biosynthesis of the C_{16} family of cutin monomers. Even though CoA esters are postulated to be substrates for all of the reactions involved in this pathway, conclusive evidence is not available concerning the nature of the substrates. However, the observations, that in the enzyme preparations from *V. faba* the C-10 hydroxylase and the transacylase involved in the formation of the

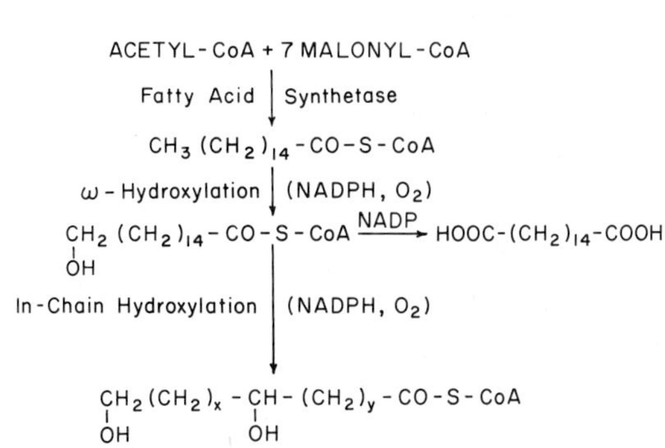

Fig. 3. A highly probable biosynthetic pathway for the C_{16} family of cutin acids

polymer from monomer acids require activation of the carboxyl groups, support the above hypothesis. Furthermore, such a hypothesis would provide an adequate explanation for the finding that only trace quantities of hydroxy acids are found in tissues that rapidly synthesize cutin.

II. Biosynthesis of the C_{18} Family of Cutin Acids

1. *In vivo Experiments*

In fairly slowly growing plant tissues with well-developed cuticles, such as apple fruits, the C_{18} family of acids constitutes a substantial proportion of the cutin monomers [14, 47, 60]. With rapidly growing (young) apple fruit, it was shown that exogenous precursors were incorporated into the C_{18} family of cutin acids only in the skin slices, but not in the internal tissue [39]. Labeled acetate was incorporated into all cutin acids in apple skin slices, whereas palmitic acid was incorporated only into the C_{16} family of cutin acids [39]. Exogenous [1-^{14}C]stearic acid was not incorporated into the in-chain hydroxylated C_{18} acids, suggesting that a direct hydroxylation similar to that involved in the biosynthesis of the C_{16} family of cutin acids is not involved in the biosynthesis of the C_{18} family of acids. However, exogenous [1-^{14}C]oleic acid was readily converted into 18-hydroxyoleic acid, 18-hydroxy-9,10-epoxystearic acid, and 9,10,18-trihydroxystearic acid. Similar results were also obtained with leaves of apple, *S. odoris,* and *R. officinalis.* All of the products were identified by applying radio chromatographic techniques to various derivatives of the metabolic products, as well as to their chemical degradation products. These results strongly suggested that the in-chain hydroxylated C_{18} acids are produced by modification of the Δ^9 double bond of 18-hydroxyoleic acid [38, 39]. As expected from such a hypothesis, exogenous 18-hydroxy[18-^{3}H]oleic acid was converted into 18-hydroxy-9,10-epoxystearic acid and 9,10,18-trihydroxystearic acid of cutin in apple fruit skin and in leaves of apple and *S. odoris* [9]. Similar experiments with synthetic 18-hydroxy-9,10-epoxy[18-^{3}H]stearic acid showed that this epoxy acid was directly converted into 9,10,18-trihydroxystearic acid. These studies with tissue slices very strongly suggested that ω-hydroxylation of oleic acid, followed by epoxidation of the Δ^9 double bond, and subsequent hydration of the epoxide give rise to some of the major components of the C_{18} family of cutin acids.

A reaction sequence similar to that proposed above could give rise to the Δ^{12} unsaturated members of the C_{18} family of cutin acids if linoleic acid serves as the precursor, and the epoxidation and subsequent hydration are specific for the Δ^9 double bond. In fact, in apple skin slices exogenous [1-^{14}C]linoleic acid was incorporated specifically into 18-hydroxylinoleic acid, 18-hydroxy-9,10-epoxyoctadecenoic acid, and 9,10,18-trihydroxyoctadecenoic acid [39]. Similar conversions were also demonstrated with young leaves of *R. officinalis* [8]. The specificity of epoxidation and subsequent hydration was further demonstrated by the observation that exogenous [1-^{14}C]linolenic acid was converted specifically to 18-hydroxylinolenic acid, 18-hydroxy-9,10-epoxyoctadecadienoic acid, and 9,10,18-trihydroxyoctadecadienoic acid. Even though in these cases only the Δ^9 double bond is epoxidized, in some plants such as *R. officinalis* double bonds in other positions can also be epoxidized. Thus, for example, 9,10,12,13,18-pentahydroxyoctadecanoic acid, which has been found in the cutin of *R. officinalis* [5, 8], was suggested to be derived from linoleic acid by ω-hydroxylation, sequential epoxidation, and subsequent epoxide hydration at the two double bonds [37]. Evidence for this hypothesis was provided by the observation that all of the expected intermediates in this pathway were identified in the cutin of *R. officinalis* by gas-liquid chromatography and mass spectrometry [8]. Furthermore, exogenous [1-^{14}C]linoleic acid was converted specifically into 9,10,12,13,18-pentahydroxyoctadecanoic acid, and into all of the intermediates predicted by the preceding hypothesis for the biosynthesis of this pentahydroxy acid. From the foregoing discussion, it is apparent that ω-hydroxylation, epoxidation, and epoxide hydration are the three key reactions involved in the biosynthesis of the C_{18} family of cutin acids. Therefore, we began a search for enzymes capable of catalyzing these reactions.

2. *Enzymatic Synthesis of the C_{18} Family of Cutin Acids*

A cell-free preparation capable of catalyzing ω-hydroxylation of a fatty acid was already discussed in a previous section, and probably such an enzyme is also involved in the hydroxylation of oleic acid.

a) Epoxidation. Since our earlier structural studies showed that 18-hydroxy-9,10-epoxy C_{18} acid is a major component of cutin from grape berries [60], we attempted, in vain, to obtain a cell-free preparation, capable of catalyzing epoxidation, from this tissue. However, young spinach leaves, which also contain 18-hydroxy-9,10-epoxystearic acid as the major component of cutin, provided a 3,000 *g* particulate preparation that catalyzed epoxidation of 18-hydroxy[18-^{3}H]oleic acid [11]. Since the epoxidation product was also hydrated and incorporated into the polymer by this enzyme preparation, both the soluble lipids and the insoluble polymer had to be examined for the enzymatic products. Furthermore, in order to get a reliable measure of epoxidation, both the epoxy acid and the epoxide hydration product had to be measured. Therefore, the reaction products were subjected to acetolysis with acetic acid (to cleave the epoxide), followed by $LiAlH_4$ treatment. Thus, all of the enzymatic epoxidation products were converted to octadecane-1,9,10,18-tetraol, which was identified by radio chromatographic techniques and by chemical degradation. Using such a tedious assay procedure, we have partially characterized this novel biosynthetic epoxidation reaction. As expected of a mixed function oxidase, this epoxidase required NADPH and O_2; but in addition, ATP and CoA were also required, strongly suggesting that activation of the carboxyl group seven methylenes away from the double bond is required for the epoxidation of the olefin. Thus, it appears that 18-hydroxyoleyl-CoA is epoxidized by an extracellular enzyme that is associated with the cuticular matrix, and presumably the hydroxyepoxyacyl moiety would then be transferred to the growing polymer. The epoxidase showed a pH optimum near 9.0, and the K_m and V were 7.5×10^{-5} M and 180 nmol/l-h-mg protein, respectively. Light reversible inhibition by CO suggested the involvement of cytochrome P_{450} in the epoxidation. Chelators inhibited epoxidation and this inhibition was reversed by $FeSO_4$. The epoxidase appears to have a fairly stringent substrate specificity, as 18-hydroxyelaidic acid, 18-acetoxyoleic acid, and oleic acid were not readily epoxidized by this enzyme preparation. Enzyme preparations similar to that described above were also obtained from the skin of young apple fruit and from the epidermis of *S. odoris* leaves.

b) Epoxide Hydration. Even though epoxide hydrases are known to be involved in a variety of catabolic reactions [29, 45, 48, 49], involvement of such an enzyme in biosynthetic processes is not known. However, as the experimental results obtained with tissue slices strongly suggested that epoxide hydration is involved in the biosynthesis of the 9,10,18-trihydroxy-C_{18} acid, we searched for such an enzyme in homogenates of the skin tissue from young apple fruit with synthetic 18-hydroxy-cis-9,10-epoxy[18-^{3}H]stearic acid as the substrate. This epoxy acid was enzymatically converted into *threo*-9,10,18-trihydroxystearic acid, which was identified using radio chromatographic techniques in combination with chemical degradation [12]. That this epoxide hydrase is involved in cutin biosynthesis was suggested by the observation that this enzymatic activity was found almost exclusively ($>90\%$) in the 3,000 *g* particulate fraction that contained the cuticular membranes. This conclusion was further supported by the fact that 3,000 *g* particulate fractions from internal tissue of apple fruit contained less than 6% of the epoxide hydrase activity found in similar preparations from the skin tissue. The epoxide hydrase required no cofactors and showed a pH optimum of 6.5, which is substantially lower than the pH optimum reported for the catabolic epoxide hydrases [48]. The apparent K_m and V for the epoxide hydrase from apple skin was 4×10^{-4} M and 170 µmol/l-h-mg protein, respectively [12]. That the enzymatic activity required essential thiol group(s) was indicated by the observation that *p*-chloromercuribenzoate strongly inhibited the enzyme (80% inhibition at 0.1 mM); iodoacetamide and N-ethylmaleimide were less potent inhibitors. This observation, together with the finding that 1,1,1-trichloropropylene-2,3-oxide was only moderately inhibitory (60% inhibition at 5 mM), suggests that the biosynthetic epoxide hydrase from the plant is quite different from the catabolic epoxide hydrases studied heretofore [45, 49]. In further support of this conclusion is the finding that the epoxide hydrase from apple skin tissue showed fairly stringent substrate specificity, as it did not readily hydrate styrene oxide, methyl-18-hydroxy-9,10-epoxystearate, 18-acetoxy-9,10-epoxystearic acid, methyl-18-acetoxy-9,10-epoxystearate, or 9,10-epoxystearic acid [12]. Particulate preparations capable of hydrating 18-hydroxy-9,10-epoxystearic acid to the trihydroxyacid were also obtained from the leaf epidermis of *S. odoris* and spinach leaves.

All of the preceding experimental results strongly support the pathway shown in Figure 4 for the biosynthesis of the C_{18} family of cutin acids. The substrates for ω-hydroxylation and subsequent epoxidation and epoxide hydration steps are depicted as CoA-esters for the following reasons. The epoxidation catalyzed by the particulate enzyme preparation from spinach leaves required ATP and CoA. As might be expected from the involvement of a thioester, free monomers are barely detectable in soluble lipids. Even though the epoxide hydrase prepara-

Fig. 4. A highly probably biosynthetic pathway for the C_{18} family of cutin acids

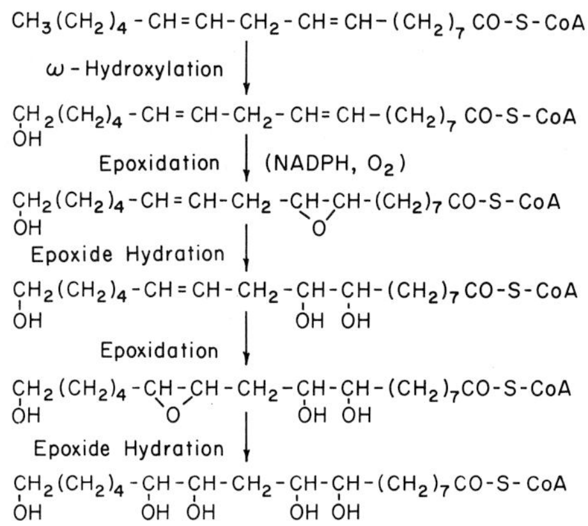

tion from apple skin did not show any requirement for ATP and CoA, a thioester might be involved in vivo. However, the chemical nature of the substrates involved in the pathway shown in Figure 4 is yet to be elucidated; but the various steps indicated appear to be well established.

III. Synthesis of Cutin from Monomers

Because the polymer is deposited outside the cell membrane and the cell wall, it is probable that synthesis of the polymer occurs at an extracellular location, possibly in close association with the preexisting cutin. In fact, in the epidermal homogenates of *V. faba* leaves, incorporation of labeled monomer acids into an insoluble polymer occurred in a 3,000 *g* particulate preparation that contained the cuticular membranes [7]. This incorporation of hydroxy acids into the insoluble material appears to represent cutin biosynthesis, as similar preparations from mesophyl tissue did not catalyze incorporation of hydroxy acids into insoluble polymers. Conclusive evidence that the incorporation of labeled palmitic acid, 16-hydroxypalmitic acid, and 10,16-dihydroxypalmitic acid into the insoluble material, catalyzed by the particulate preparation, did represent cutin biosynthesis, was provided by enzymatic degradation of the insoluble polymer. Thus, cellulase, pectinase, and pronase treatments failed to release the radioactivity from the insoluble material, while a fungal cutinase (discussed in a later section) released the bulk of the radioactivity contained in the polymeric material. Treatment of the insoluble polymer with $LiAlH_4$, $NaBH_4$, alcoholic KOH, and BF_3 in methanol, and analyses of the soluble products released, showed that the carboxyl group of the hydroxy acid was esterified to a hydroxy group of cutin. In support of this conclusion was the observation that methyl esters and fatty alcohols were not significantly incorporated into cutin, but acetates of hydroxy acids were incorporated at 60 to 70% of the rate observed with the parent acid [10].

Incorporation of monomer acids into the polymer required ATP and CoA, strongly suggesting that hydroxyacyl-CoA is the substrate for the esterification reaction. Two pH optima near 7.0 and 8.5 were observed when free acids were used as the substrates; but palmitoyl-CoA incorporation showed a single optimum pH near 8.5, suggesting that activation of the carboxyl group and transacylation occurred optimally at pH 7.0 and 8.5, respectively [7]. The apparent K_m values for palmitic acid, 16-hydroxypalmitic acid, and 10,16-dihydroxypalmitic acid were 2.0×10^{-5}, 6.7×10^{-5}, and 1.1×10^{-4} M, respectively [10]. However, these values may not reflect the true specificity of the transacylase that catalyzes the synthesis of the polymer, as the present apparent K_m values were determined with free acids. It is likely that activation of the monomers is a limiting factor in the enzyme preparation, but in vivo, the hydroxy acids might be synthesized as their CoA esters, and therefore there might not be a requirement for the thiokinase. Substrate specificity of the enzyme preparation did reflect the chain length of the acyl moieties found in *V. faba* cutin, as the C_{16} family of acids was the preferred substrate, although fatty acids from C_{10} to C_{18}, and C_{18} hydroxy acids, could also be incorporated into the polymer by the enzyme preparation. Apparently, other plant tissues also contain similar enzymes, because we have obtained enzyme preparations capable of esterifying acyl moieties to cutin from *V. faba* flower petals and the epidermis of *S. odoris* leaves; however, these preparations have not been examined in detail.

As the particulate enzyme from the epidermal extracts of *V. faba* contained the cuticular membranes, it appeared that the monomer acids were esterified to the free hydroxyl groups of cutin present in the enzyme preparation, and therefore no primer dependence could be demonstrated. However, mild sonication of the particulate preparation provided an enzyme preparation that required an exogenous primer. Purified *V. faba* cutin was the most active primer, although a variety of chemically distinct cutins from several plant species could function as primers [10]. With this enzyme preparation, the hydroxyacyl transferase activity depended on the amount of *V. faba* cutin added up to 3 mg/ml, whereas other possible acyl acceptors, such as cellulose, glycerol, hexadecanol, cholesterol, and polyethylene glycol, were ineffective. The primer-dependent incorporation of monomer acids into cutin showed the same cofactor requirements and substrate specificity as the particulate preparation. With the primer-dependent system, it could be demonstrated that the hydroxyl groups of the primer were the acyl acceptors, as already suggested by the results obtained with the particulate enzyme. Thus, for example, acetylation of *V. faba* cutin decreased its primer efficiency by 70%, whereas methylation of the carboxyl groups of cutin had no effect on its efficiency as a primer. Other chemical treatments of cutin that increased the number of hydroxyl groups or opened the polymer matrix also increased the priming efficiency of cutin. A brief treatment of *V. faba* cutin with a purified fungal cutinase doubled the priming efficiency of the cutin, probably because cutinase hydrolyzed some of the ester bonds, making more hydroxyl groups available for esterification with the monomer acids. Cutin from very young leaves might be expected to be less well cross-linked than cutin from mature leaves, and therefore the former might be expected to have a more open structure. In fact, cutin from very young leaves was substantially more effective as a primer than cutin from mature leaves. Based on the experimental results summarized above, synthesis of the polymer from the monomers may be represented as shown in Figure 5.

Fig. 5. Schematic representation of cutin synthesis from monomers

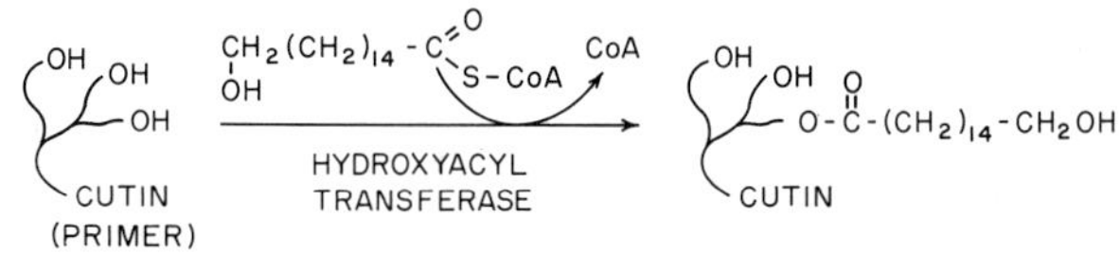

C. Biosynthesis of Suberin

Aliphatic components constitute 25 to 50% of the soluble depolymerization products of suberin [19, 20, 28, 34] and the following discussion will be limited to the aliphatic components of this polymer. The characteristic components of the aliphatic monomers of suberin are ω-hydroxyfatty acids, dicarboxylic acids, very long ($>C_{20}$) acids and alcohols, with very small quantities of in-chain epoxy and hydroxy acids similar to those found in cutin [27].

I. Experiments with Tissue Slices

Since 18-hydroxyoleic acid and the corresponding dicarboxylic acid constitute the major components of the aliphatic monomers of potato suberin [4, 28, 54], [1-^{14}C]oleic acid was incubated with small, rapidly growing potato tubers, and the hydrogenolysate of suberin was examined by radio thin-layer and gas chromatographic techniques. Octadecene-1,18-diol was found to be labeled, showing incorporation of exogenous oleic acid into the major suberin components (unpublished results). However, as it was difficult to obtain uniform samples of rapidly growing intact potato tubers throughout the year, a more convenient alternate experimental system was necessary. Since the composition of the aliphatic components of the suberin deposited in the wound periderm was shown to be identical to that of the natural skin (suberin) of intact potato tubers [33], wound healing potato tuber slices appeared to be a good experimental system with which to study suberin biosynthesis. According to the experimentally determined time-course of suberin deposition in such potato slices [33], the most rapid synthesis of the aliphatic components of suberin should occur about four days after preparation of the slices. In fact, incubation of suberizing potato slices with [1-^{14}C]oleic acid showed that the most rapid incorporation of this exogenous labeled material into insoluble material (suberin) occurred after four days of aging [13]. Hydrogenolysis ($LiAlH_4$) of the insoluble material released the radioactivity into the soluble fraction, which was subjected to radio thin-layer and gas chromatographic analyses. Octadecene-1,18-diol was identified as the major labeled product, and chemical degradation of this diol showed that it was Δ^9 monounsaturated, thus demonstrating conversion of exogenous oleic acid into the major components of suberin. Incorporation into the diol fraction (derived by the $LiAlH_4$ reduction of ω-hydroxyacid and/or dicarboxylic acid) occurred only after two days of aging, and maximal incorporation occurred after 4 days of aging, at which time suberin synthesis is known to occur at maximal rates. Thus, the incorporation of exogenous oleic acid into the insoluble

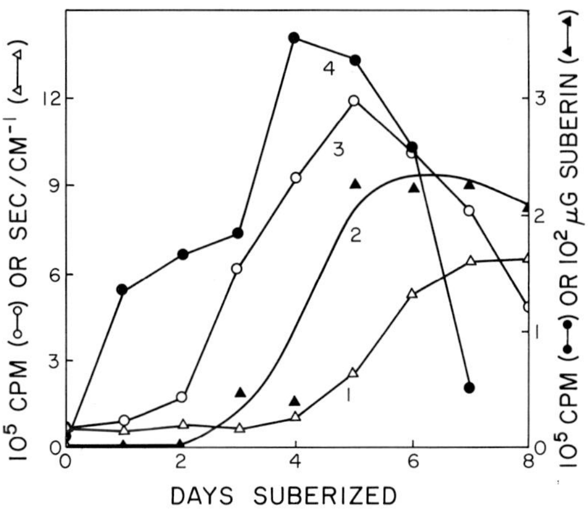

Fig. 6. Changes in incorporation of [1-^{14}C]acetate into suberin (○–○), ω-hydroxyfatty acid dehydrogenase activity (●–●), level of the aliphatic monomers of suberin measured as octadecene-1,18-diol (▲–▲), and resistance of the tissue surface to diffusion of water vapor, (△–△) during aging of potato slices

material does accurately reflect suberin biosynthesis. However, with this substrate, the synthesis of the other characteristic components of suberin, namely very long saturated carbon chains, could not be examined. Therefore, [1-^{14}C]acetate was used as a precursor, and a time-course of incorporation of radioactivity into the insoluble material showed a pattern consistent with the time-course of suberin deposition (Fig. 6). Hydrogenolysis of the insoluble material released ^{14}C into soluble components, mainly fatty alcohols and alkane-α,ω-diols. Radio gas-liquid chromatographic analyses of these two fractions (as their acetates) showed that the alcohols contained substantial quantities of very long-chain ($>C_{20}$) components (Table 1). The major component of the diol fraction was found to be octadecene-1,18-diol (Fig. 7). Thus, acetate incorporation into the insoluble materials did in fact represent incorporation into characteristic suberin components. Even the composition of the soluble lipids reflected the suberization

Table 1. Distribution of radioactivity in the fatty alcohols isolated from the hydrogenolysate of the insoluble residue obtained from suberizing potato disks incubated with [1-^{14}C]acetate

Days aged	Radioactivity (% total)						
	C_{16}	C_{18s+u}[a]	C_{20}	C_{22}	C_{24}	C_{26}	C_{28}
0	ND[b]	ND	ND	ND	ND	ND	ND
1	20	32	18	14	17	ND	ND
2	27	30	14	12	15	3	ND
3	13	12	11	26	24	17	ND
4	11	9	8	27	26	17	4
5	6	6	4	20	29	26	10
6	10	9	6	18	24	22	11
7	14	12	6	19	21	15	12
8	26	14	12	18	11	9	12

Chain length distribution was determined by radio-gas chromatography of the alcohol acetates.
[a] *s*: saturated; *u*: unsaturated. [b] *ND*: not detected

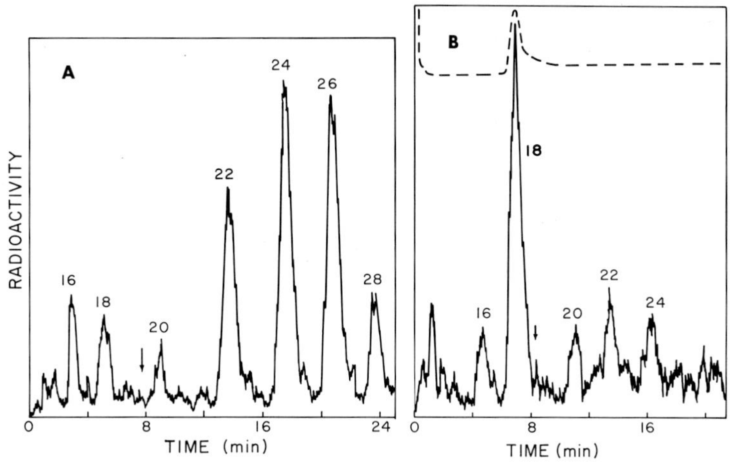

Fig. 7A and B. Radio gas-liquid chromatograms of diol diacetates prepared from the hydrogenolysate of suberin derived from [1-^{14}C]acetate in 5-day aged potato disks. The number on each peak represents chain length

process, in that as suberization proceeded, longer chain ($>C_{20}$) fatty acids appeared mainly in the neutral lipid fraction obtained from the soluble lipids. Thus, the suberizing potato slices offer a convenient system in which to study the biosynthesis of suberization.

II. Experiments with Cell-Free Preparations

The structure of the aliphatic monomers of suberin suggests that chain elongation, ω-hydroxylation, and oxidation of the ω-hydroxyacids to the dicarboxylic acids are the three key reactions involved in the biosynthesis of the major aliphatic monomers of suberin (Fig. 8). Even though the chain elongating enzyme

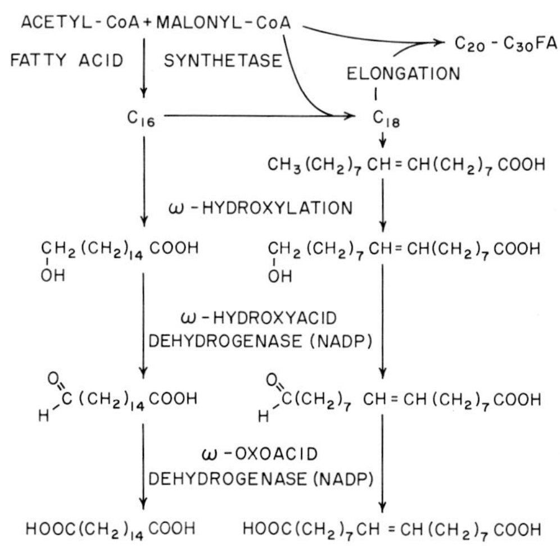

Fig. 8. A probable pathway for the biosynthesis of the major aliphatic monomers of suberin

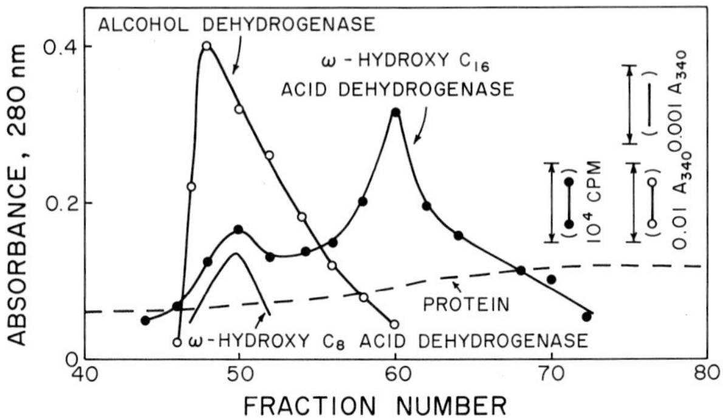

Fig. 9. DEAE-cellulose chromatography of an ω-hydroxyfatty acid dehydrogenase fraction obtained by Sepharose 6-B fractionation [31] of an acetone powder extract prepared from suberizing potato disks [2]

system has not been characterized from suberizing tissues, this process has been studied in other plant tissues [30, 46], and probably a similar system is involved in suberization. From the experimental results obtained with particulate preparations from germinating seeds [46], and with cell-free preparations from the epidermis of *Pisum sativum* [30], it appears that a fatty acyl derivative (probably a thioester) is elongated with malonyl-CoA as the elongating agent and NADPH as the reductant. The nature of the fatty acyl derivative that is elongated remains uncertain except for the first elongation step from C_{16} to C_{18}, which involves ACP [22]. ω-Hydroxylation of fatty acids has not been demonstrated in cell-free preparations from suberizing plant tissues, and the only available example of an ω-hydroxylating enzyme preparation from a higher plant is discussed in a previous section concerning cutin biosynthesis.

The presence of dicarboxylic acids as major components is a hallmark of suberin and, therefore, the ability to convert ω-hydroxyacids to dicarboxylic acids would appear to be a characteristic of suberizing tissues. In fact, cell-free extracts obtained from suberizing potato tuber disks catalyzed the conversion of 16-hydroxy[G-^{3}H]hexadecanoic acid to the corresponding dicarboxylic acid with NADP or NAD as the cofactor, with a slight preference for the former (1). This ω-hydroxy acid dehydrogenase, located mainly in the soluble supernatant fraction, showed a pH optimum near 9.5 and an apparent K_m of 5×10^{-5} M for 16-hydroxypalmitic acid. This enzyme was inhibited by thiol-directed reagents such as *p*-chloromercuribenzoate, N-ethylmaleimide, and iodoacetamide. That 16-oxo-hexadecanoic acid was an intermediate in the conversion of the 16-hydroxy acid to the dicarboxylic acid was suggested by the observation that the same preparation also catalyzed the conversion of 16-oxo[G-^{3}H]hexadecanoic acid to the dicarboxylic acid with NADP as the preferred cofactor. More recently, the ω-hydroxy acid dehydrogenase has been separated from the ω-oxoacid dehydrogenase by a gel filtration technique [2]. Since the ω-hydroxy acid dehydrogenase was not resolved from alcohol dehydrogenase by this fractionation technique, it appeared possible that alcohol dehydrogenase might be responsible for the observed oxidation of the ω-hydroxy acid. However, ω-hydroxy acid dehydrogenase, which, unlike alcohol dehydrogenase, has a specificity for NADP, has been resolved from the latter enzyme by DEAE-cellulose ion exchange chromatography (Fig. 9), and thus, ω-hydroxy acid dehydrogenase is probably an enzyme produced specifically for suberin biosynthesis [2]. The time-course of appearance of the ω-hydroxy acid dehydrogenase activity in suberizing potato slices correlated with the appearance of suberin, supporting the hypothesis that this enzyme is induced for the biosynthesis of suberin (Fig. 8). Preliminary results obtained with cycloheximide and actinomycin D showed that the transcriptional and translational processes responsible for the synthesis of the aliphatic components of suberin occur between 72 and 96 h after preparation of the potato disks.

D. Degradation of Lipid Polymers

I. Degradation by Plants

The extracellular location of the polymers and their important protective role would suggest that they are not subject to turnover during the life of a plant. However, in specialized cases such as during the penetration of the pollen tube into the stigma, cutin hydrolysis probably occurs, and indirect evidence to support this possibility has been presented [40, 44]. With the availability of labeled cutin, it became possible to conveniently test the preceding hypothesis more directly [55]. Extracellular fluid from germinating pollen from nasturtium (*Tropaeolum majus*) catalyzed the release of all classes of monomers from biosynthetically ^{14}C-labeled apple cutin. The pollen cutinase showed a pH optimum of 6.5, which is substantially lower than that found for the fungal cutinase discussed in a succeeding section. The pollen cutinase was insensitive to "active" serine-directed reagents such as diisopropylfluorophosphate, whereas it was very sensitive to thiol-directed reagents such as *p*-chloromercuribenzoate. Thus, the pollen cutinase appears to be an "SH hydrolase" in contrast to the fungal cutinases, which are "serine hydrolases." Maximal amounts of cutinase appear to be released into the germinating medium in a few hours, and this release does not appear to be significantly inhibited either by cycloheximide or by actinomycin D. Therefore, it is tentatively concluded that the release of cutinase does not involve protein synthesis; the enzyme is presumably present in the intine layer of the pollen.

II. Degradation by Microorganisms

1. Isolation and Characterization of Cutinases

Several plant pathogens and saprophytes have been shown to be capable of growth on cutin as the sole source of carbon, and under such conditions these organisms must produce extracellular enzymes capable of degrading the polymer

[15, 16, 18, 43, 51, 56]. Recently, cutinases have been isolated and characterized from the extracellular fluid of pathogens grown on cutin. The first isolation of a homogeneous cutinase was accomplished from the extracellular fluid of *Fusarium solani* f. *pisi* grown on apple cutin [52, 53]. A 6- to 7-fold purification using Sephadex G-100 gel filtration, QAE-Sephadex, and SE-Sephadex chromatography gave two isozymes of cutinase in homogeneous state, as judged by polyacrylamide disc gel electrophoresis and sedimentation equilibrium centrifugation. Both cutinase I and cutinase II gave a molecular weight near 22,000 by Sephadex G-100 gel filtration, sedimentation equilibrium centrifugation, amino acid composition, and sodium dodecyl sulfate (SDS) polyacrylamide electrophoresis. The SDS electrophoresis showed that a small portion of cutinase II contained a proteolytic clip near the middle of the molecule, resulting in two fragments, 10,600 and 9,800 in molecular weight. The amino acid composition of cutinase I was similar to that of cutinase II. Both cutinases also showed similar catalytic properties; with tritiated apple cutin as the substrate, they showed similar substrate concentration dependence, protein concentration dependence, time-course profiles, and pH dependence profiles with an optimum near 10.0 [53]. Both cutinases also catalyzed hydrolysis of methylhexadecanoate, cyclohexylhexadecanoate, and to a much lesser extent, hexadecylhexadecanoate, but not 9-hexadecanoyloxyheptadecane, cholesterol hexadecanoate, hexadecylcinnamate, and α- or β-glucosides of *p*-nitrophenol. Both cutinases hydrolyzed short-chained esters, but not long-chained esters of *p*-nitrophenol, and the highest V was obtained for both cutinases with *p*-nitrophenyl acetate. Both enzymes were also similar in that they showed no metal ion dependence, and they were not affected by thiol-directed reagents such as *p*-chloromercuribenzoate, N-ethylmaleimide, and iodoacetamide; in fact, no free SH group could be detected in these enzymes [42]. The "active" serine-directed reagents such as diisopropylfluorophosphate and paraoxon severely inhibited both cutinases. Treatment with [^{3}H]diisopropylfluorophosphate resulted in covalent attachment of the radioactive moiety to both cutinases. Stoichiometry of this covalent attachment showed that both enzymes contained one "active" serine per molecule, and reaction of this serine residue with the inhibitor completely inactivated them. SDS gel electrophoresis of both of the enzymes, tritiated as indicated above, showed that, in both cases, the protein band at 21,800 molecular weight was radioactive. Only in the case of cutinase II was a small amount of ^{3}H also associated with a minor protein band—one of the two smaller peptides (at 10,600 mol wt). In addition to this difference, it was found that cutinase I hydrolyzed the model substrates mentioned previously at least three times as rapidly as did cutinase II. A structural difference between cutinase I and cutinase II was also indicated by the observation that Ouchterlony double diffusion analysis with rabbit anticutinase I revealed spurs showing that the two enzymes are not immunologically identical [57].

The procedure used to purify the cutinases also resulted in a 34-fold purification of a nonspecific esterase that was electrophoretically homogeneous [52, 53]. Its molecular weight was determined to be about 52,000 by several techniques. SDS electrophoresis indicated that this enzyme contained some proteolytic clips. This enzyme hydrolyzed all of the model substrates that were hydrolyzed by the cutinases and, in addition, it hydrolyzed long-chain esters of *p*-nitrophenol, but it did not catalyze hydrolysis of cutin. This enzyme also contains an "active" serine that is involved in catalysis, as shown by covalent attachment of labeled diisopropylfluorophosphate and the resulting inhibition of the enzymatic activity. The role of this enzyme is not understood, as it catalyzes the hydrolysis of neither cutin nor the oligomers generated from cutin by the cutinases; these oligomers are further hydrolyzed by the two cutinases [53].

Cutin depolymerizing activity has been demonstrated in the extracellular fluid of a variety of other plant pathogens. Recently, from the extracellular fluid of *F. roseum culmorum* grown on cutin, a single cutinase and a nonspecific esterase were isolated utilizing Sephadex G-100, QAE-Sephadex, and SP-Sephadex chromatography [57]. The molecular weight of the cutinase, which was electrophoretically homogenous, was estimated to be 24,300 by SDS polyacrylamide gel electrophoresis. The electrophoretic mobility of this enzyme was between that of cutinases I and II isolated from *F. solani pisi.* The catalytic properties of the enzyme from *F. roseum culmorum,* such as time-course, protein concentration dependence, substrate concentration dependence, pH dependence profile, and substrate specificity, were similar to those observed with the *F. solani* enzymes. With the techniques previously used with the *F. solani* enzyme, it was demonstrated that this enzyme also uses an "active" serine for catalysis. The amino acid composition of the cutinase from *F. roseum culmorum* was similar to that of the two cutinases from *F. solani,* except for the presence of larger numbers of serine, valine, and cysteine residues in the former. Ouchterlony double diffusion studies with a rabbit antibody prepared against cutinase I from *F. solani pisi* demonstrated that the cutinase from *F. roseum culmorum* did not cross-react with anticutinase I, but the enzyme from *F. roseum* was equally sensitive to inhibition by anticutinase I as were the two cutinases from *F. solani pisi.*

2. Characterization of Cutinases as Glycoproteins

Recently cutinase I from *F. solani pisi* was found to be a glycoprotein; it contains about 3.5% carbohydrate [42]. SDS electrophoresis showed that the protein band coincided with the carbohydrate staining band. In the case of cutinase II, both the parent molecule at 22,000 molecular weight and the two faster moving fragments contained carbohydrates. Even though the nature of the carbohydrate has not been defined, the mode of attachment of the carbohydrate to the protein has been studied, and, as a result, novel modes of attachment of carbohydrates to proteins have been revealed. Thus, when cutinase I was treated with NaB^3H_4 in 0.1 M NaOH at 4°C for 216 h, the protein became radioactive, and analysis of the protein hydrolysate with an amino acid analyzer showed that alanine, α-aminobutyric acid, phenylalanine, and tyrosine contained ^3H. Attachment of carbohydrates to serine and threonine of the protein by glycosidic linkages would be expected to generate labeled alanine and α-aminobutyric acid by the above treatment. But the labeling of the two additional amino acids, namely phenylalanine and tyrosine, was unexpected. The most probable

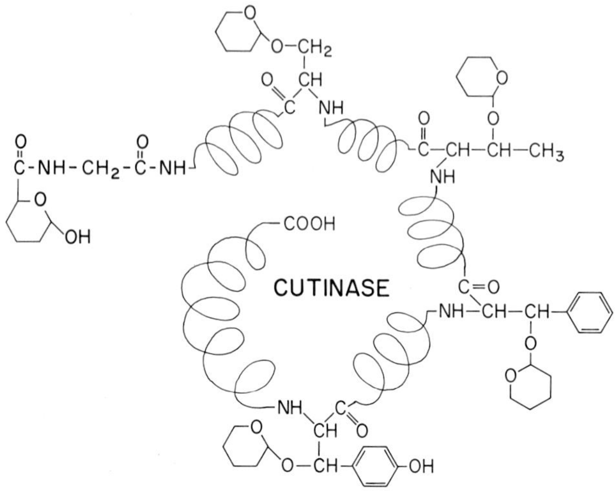

Fig. 10. A schematic representation of the glycoprotein nature of cutinase I isolated from *F. solani pisi*. The pyranoid rings represent carbohydrates

explanation for this labeling is that cutinase contains a carbohydrate attached by glycosidic linkage to β-hydroxyphenylalanine and β-hydroxytyrosine, two amino acids heretofore not found in proteins. This hypothesis would predict that the ^3H in phenylalanine should be entirely in the side chain, and chemical degradation of the labeled phenylalanine showed that 50% of the ^3H was contained in the α-carbon and 50% in the β-carbon. These results together with the finding that the tritiated phenylalanine was a 1:1 mixture of D and L isomers strongly support the concept that a carbohydrate is attached to cutinase by a glycosidic linkage at the hydroxyl group of the β-hydroxyphenylalanine—a novel mode of attachment of a carbohydrate to a protein. Probably a carbohydrate is also similarly attached to the hydroxyl group of β-hydroxytyrosine. Another novel mode of attachment of a carbohydrate to cutinase was strongly suggested by the observation that a carbohydrate residue attached to the protein did not undergo elimination under basic conditions. This carbohydrate residue contained a group reducible by NaB^3H_4, and the tritiated material derived by this reduction was tentatively identified as gulonic acid; it therefore appears that glucuronic acid might be linked to the protein by an amide linkage to either an episilon amino group of a lysine or to the amino group of the N-terminal amino acid. Since the N-group of the enzyme appears to be blocked, the latter possibility deserves serious consideration. On the basis of the experimental evidence thus for obtained, cutinase might be schematically represented as shown in Figure 10.

III. Degradation by Animals

Even though cutin and suberin are components of the normal diet of insects and higher animals, including man, little is known about the fate of these polymeric materials. The observation that polymeric materials present in sewage sludge, upon $LiAlH_4$, treatment, gave rise to monomers expected from cutin and suberin, strongly suggests that these polymers can survive the digestive processes [35]. As it is a common practice to use the plant skin materials generated in the food processing industry in cattle feeds, it is important to determine whether animals can utilize at least a portion of the ingested polymers. In an attempt to answer this question, biosynthetically ^{14}C-labeled apple cutin was fed to rats [6]. Within seven hours, about 10% of the label appeared in respiratory CO_2, and 10 to 15% of the label was found in the liver. The bulk of the label found in the liver was in fatty acids of acyl lipids. All of the tissues contained measurable amounts of ^{14}C. Thus, it appears that at least a portion of the hydroxyfatty acid polymers in the diet can be utilized by the rat, and presumably by other animals. It is not known whether the intestinal microbes play a part in the degradation of the ingested polymers. However, a 30,000 *g* supernatant prepared from the intestinal contents of the rat can catalyze hydrolysis of labeled cutin [6]. Since such a hydrolysis is likely to release the potentially toxic epoxy acids present in cutin, it is important to determine the fate of such acids. An epoxide hydrase was found in the microsomal preparation obtained from the intestinal wall of rabbits, and this activity was highest in the upper part of the small intestine [41]. This microsomal preparation catalyzed hydration of 18-hydroxy-9,10-epoxystearic acid just as rapidly as the commonly used test substrate for epoxide hydrase, namely, styrene oxide. The hydrolysis of the polymer, detoxification of a potentially toxic monomer generated from it, and the catabolism of the monomers, demonstrated by the respiratory CO_2 and by the incorporation of the metabolic products into the usual lipids of the liver, show that animals can utilize at least a portion of the lipid polymers ingested.

E. Conclusions

Even though substantial progress has been made in our understanding of the biochemistry of lipid polymers, virtually nothing is known about many aspects of this problem. Almost totally unexplored are questions concerning the regulation of synthesis of lipid polymers. Very little direct evidence bearing on the function of lipid polymers is available. How the lipid polymers fulfill the commonly assumed role of plant protection is not well understood, and it is quite possible that such information will be useful in devising new approaches to plant protection. With the currently available experimental techniques, it is feasible to seek answers to many of these unanswered questions.

Acknowledgements. This work was supported in part by a grant (BMS74-09351A01) from the National Science Foundation. I thank Linda Brown for assistance in preparing this manuscript and for technical assistance during much of the work summarized in this chapter.

References

1. Agrawal, V.P., Kolattukudy, P.E.: ω-Hydroxyacid dehydrogenase enzyme from suberized potato slices. 29th Annual Northwest Regional Meeting of the Am. Chem. Soc., June 1974, Abstract 202; also Biochemistry of Suberization. ω-Hydroxyacid oxidation in Enzyme Preparations from Suberizing Potato Tuber Disks. Plant Physiol. **59**, 1977, in press
2. Agrawal, V.P., Kolattukudy, P.E.: unpublished results; also see Ref. 31
3. Bredemeijer, G., Heinen, W.: I. Free fatty acid movement during cutin synthesis in injured *Gasteria verrucosa* leaves. Acta Botan. Neerl. **17**, 15–25 (1968)
4. Brieskorn, C.H., Binnemann, P.H.: Carbonsäuren und Alkanole des Cutins und Suberins von *Solanum tuberosum*. Phytochemistry **14**, 1363–1367 (1975)
5. Brieskorn, C.H., Kabelitz, I.: Hydroxyfettsäuren aus dem Curin des Blattes von *Rosmarinus officinalis*. Phytochemistry **10**, 3195–3204 (1971)
6. Brown, A.J., Kolattukudy, P.E.: unpublished results
7. Croteau, R., Kolattukudy, P.E.: Enzymatic synthesis of a hydroxy fatty acid polymer, cutin, by a particulate preparation from *Vicia faba* epidermis. Biochem. Biophys. Res. Commun. **52**, 863–869 (1973)
8. Croteau, R., Kolattukudy, P.E.: Biosynthesis of pentahydroxystearic acid of cutin from linoleic acid in *Rosmarinus officinalis*. Arch. Biochem. Biophys. **162**, 458–470 (1974)
9. Croteau, R., Kolattukudy, P.E.: Direct evidence for the involvement of epoxide intermediates in the biosynthesis of the C_{18} family of cutin acids. Arch. Biochem. Biophys. **162**, 471–480 (1974)
10. Croteau, R., Kolattukudy, P.E.: Biosynthesis of hydroxyfatty acid polymers. Enzymatic synthesis of cutin from monomer acids by cell-free preparations from the epidermis of *Vicia faba* leaves. Biochemistry **13**, 3193–3202 (1974)
11. Croteau, R., Kolattukudy, P.E.: Biosynthesis of hydroxyfatty acid polymers. Enzymatic epoxidation of 18-hydroxyoleic acid to 18-hydroxy-*cis*-9,10-epoxystearic acid by a particulate preparation from spinach (*Spinacia oleracea*). Arch. Biochem. Biophys. **170**, 61–72 (1975)
12. Croteau, R., Kolattukudy, P.E.: Biosynthesis of hydroxyfatty acid polymers: enzymatic hydration of 18-hydroxy-*cis*-9,10-epoxystearic acid to *threo*-9,10,18-trihydroxystearic acid by a particulate preparation from apple (*Malus pumila*). Arch. Biochem. Biophys. **170**, 73–81 (1975)
13. Dean, B.B., Kolattukudy, P.E.: Biochemistry of Suberizations: Incorporation of [1-^{14}C]oleic acid and [1-^{14}C]acetate into the aliphatic components of suberin in potato tuber disks. Plant Physiol. **59**, 1977, in press
14. Englinton, G., Hunneman, D.H.: Gas chromatographic-mass spectrometric studies of long chain hydroxy acids. I. The constituent cutin acids of apple cuticle. Phytochemistry **7**, 313–322 (1968)
15. Hankin, L., Kolattukudy, P.E.: Utilization of cutin by a pseudomonad isolated from soil. Plant Soil **34**, 525–529 (1971)
16. Heinen, W.: Über den enzymatischen Cutin-Abbau I. Mitteilung: Nachweis eines 'Cutinase'-Systems. Acta Botan. Neerl. **9**, 167–190 (1960)
17. Heinen, W., Brand, I.V.D.: Enzymatische Aspekte zur Biosynthese des Blatt-Cutins bei *Gasteria verricuosa*-Blättern nach Verletzung. Z. Naturforsch. **186**, 67–79 (1963)
18. Heinen, W., Linskens, H.F.: Cutinabbau durch Pilzenzyme. Naturwissenschaften **47**, 18 (1960)
19. Holloway, P.J.: The composition of suberin from the corks of *Quercus suber L.* and *Betula pendula Roth.* Chem. Phys. Lipids **9**, 158–170 (1972)
20. Holloway, P.J.: The suberin composition of the cork layers from some *Ribes* species. Chem. Phys. Lipids **9**, 171–179 (1972)
21. Huelin, F.E.: Studies in the natural coatings of apples. IV. The nature of cutin. Australian J. Biol. Sci. **12**, 175–180 (1959)
22. Jaworski, J.G., Goldschmidt, E.E., Stumpf, P.K.: Fat metabolism in higher plants: properties of the palmityl acyl carrier protein: stearyl acyl carrier protein elongation system in maturing safflower seed extracts. Arch. Biochem. Biophys. **163**, 769–776 (1974)
23. Kolattukudy, P.E.: Biosynthesis of a lipid polymer, cutin: the structural component of plant cuticle. Biochem. Biophys. Res. Commun. **41**, 299–305 (1970)
24. Kolattukudy, P.E.: Cutin biosynthesis in *Vicia faba* leaves: effect of age. Plant Physiol. **46**, 759–760 (1970)
25. Kolattukudy, P.E.: Identification of 16-oxo-9-hydroxyhexadecanoic acid, a novel monomer,

as a major component of the biopolymer, cutin. Biochem. Biophys. Res. Commun. **49**, 1040–1046 (1972)

26. Kolattukudy, P.E.: Biosynthesis of a hydroxy fatty acid polymer, cutin. Identification and biosynthesis of 16-oxo-9 or 10-hydroxypalmitic acid, a novel compound in *Vicia faba*. Biochemistry **13**, 1354–1363 (1974)
27. Kolattukudy, P.E.: Biochemistry of cutin, suberin and waxes, the lipid barriers on plants. In: Recent Advances in the Chemistry and Biochemistry of Plant Lipids. Galliard, T., Mercer, E.I. (eds.). New York: Academic Press, 1975, pp. 203–246
28. Kolattukudy, P.E., Agrawal, V.P.: Structure and composition of aliphatic constituents of potato tuber skin (suberin). Lipids **9**, 682–691 (1974)
29. Kolattukudy, P.E., Brown, L.: Fate of naturally occurring epoxy acids: A soluble epoxide hydrase, which catalyzes *cis*-hydration, from *Fusarium solani pisi*. Arch. Biochem. Biophys. **166**, 599–607 (1975)
30. Kolattukudy, P.E., Buckner, J.S.: Chain elongation of fatty acids by cell-free extracts of epidermis from pea leaves (*Pisum sativum*). Biochem. Biophys. Res. Commun. **46**, 801–807 (1972)
31. Kolattukudy, P.E., Croteau, R., Buckner, J.S.: The biochemistry of plant waxes. In: The Chemistry and Biochemistry of Natural Waxes. Kolattukudy, P.E. (ed.). Amsterdam: Elsevier North Holland Publ. 1976, p. 289–347
32. Kolattukudy, P.E., Croteau, R., Walton, T.J.: Biosynthesis of cutin: Enzymatic conversion of ω-hydroxyfatty acids to dicarboxylic acids by cell-free extracts of *Vicia faba* epidermis. Plant Physiol. **55**, 875–880 (1975)
33. Kolattukudy, P.E., Dean, B.B.: Structure, gas chromatographic measurement, and function of suberin synthesized by potato tuber tissue slices. Plant Physiol. **54**, 116–121 (1974)
34. Kolattukudy, P.E., Kronman, K., Poulose, A.J.: Determination of structure and composition of suberin from the roots of carrot, parsnip, rutabaga, turnip, red beet, and sweet potato by combined gas-liquid chromatography and mass spectrometry. Plant Physiol. **55**, 567–573 (1975)
35. Kolattukudy, P.E., Purdy, R.E.: Identification of cutin, a lipid biopolymer, as significant component of sewage sludge. Environ. Sci. Technol. **7**, 619–622 (1973)
36. Kolattukudy, P.E., Walton, T.J.: Structure and biosynthesis of the hydroxy fatty acids of cutin in *Vicia faba* leaves. Biochemistry **11**, 1897–1907 (1972)
37. Kolattukudy, P.E., Walton, T.J.: The biochemistry of plant cuticular lipids. Progr. Chem. Fats other Lipids **13**, 119–175 (1973)
38. Kolattukudy, P.E., Walton, T.J., Kushwaha, R.P.S.: Epoxy acids in the lipid polymer, cutin, and their role in the biosynthesis of cutin. Biochem. Biophys. Res. Commun. **42**, 739–744 (1971)
39. Kolattukudy, P.E., Walton, T.J., Kushwaha, R.P.S.: Biosynthesis of the C_{18} family of cutin acids: ω-hydroxyoleic acid, ω-hydroxy-9,10-epoxystearic acid, 9,10,18-trihydroxystearic acid, and their Δ^{12}-unsaturated analogs. Biochemistry **12**, 4488–4498 (1973)
40. Kroh, M.: An electron microscopic study of the behavior of *Cruciferae* pollen after pollination. In: Pollen Physiology and Fertilization. Linskens, H.F., (ed.). Amsterdam: North Holland Publ. Co., 1964, pp. 221–224
41. Kronman, K., Kolattukudy, P.E.: unpublished results
42. Lin, T.S., Kolattukudy, P.E.: Evidence for Novel Linkages in a Glycoprotein Involving β-Hydroxyphenylalanine and β-Hydroxytyrosine. Biochem. Biophys. Res. Commun. **72**, 243–250 (1976). Glucuronyl Glycine, a Novel N-terminus in a Glycoprotein. Biochem. Biophys. Res. Commun. **75**, 87–93 (1977)
43. Linskens, H.F., Haage, P.: Cutinase-Nachweis in phytopathogenen Pilzen. Phytopathol. Z. **48**, 306–311 (1963)
44. Linskens, H.F., Heinen, W.: Cutinase-Nachweis in Pollen. Z. Botan. **50**, 338–347 (1962)
45. Lu, A.Y.H., Ryan, D., Jerina, D.M., Daly, J.W., Levin, W.: Liver microsomal epoxide hydrase: solubilization, purification, and characterization. J. Biol. Chem. **250**, 8283–8288 (1975)
46. Macey, M.J.K., Stumpf, P.K.: Fat metabolism in higher plants XXXVI: long chain fatty acid synthesis in germinating peas. Plant Physiol. **43**, 1637–1647 (1968)
47. Martin, J.T., Juniper, B.E.: The Cuticles of Plants. New York: St. Martins Press, 1970, 295 pp.
48. Oesch, F.: Mammalian epoxide hydrases: inducible enzymes catalysing the inactivation of carci-

nogenic and cytotoxic matabolites derived from aromatic and olefinic compounds. Xenobiotica **3**, 305–340 (1973)
49. Oesch, F., Kaubisch, N., Jerina, D.M., Daly, J.W.: Hepatic epoxide hydrase. Structure-activity relationships for substrates and inhibitors. Biochemistry **10**, 4858–4866 (1971)
50. Priestly, J.H.: The cuticle in angiosperms. Botan. Rev. **9**, 593–616 (1943)
51. Purdy, R., Kolattukudy, P.E.: Depolymerization of a hydroxy fatty acid biopolymer, cutin, by an extracellular enzyme from *Fusarium solani f. pisi:* isolation and some properties of the enzyme. Arch. Biochem. Biophys. **159**, 61–69 (1973)
52. Purdy, R., Kolattukudy, P.E.: Hydrolysis of plant cuticle by plant pathogens. Purification, amino acid composition, and molecular weight of two isozymes of cutinase and a non-specific esterase from *Fusarium solani f. pisi.* Biochemistry **14**, 2824–2831 (1975)
53. Purdy, R., Kolattukudy, P.E.: Hydrolysis of plant cuticle by plant pathogens. Properties of cutinase I, cutinase II, and a nonspecific esterase isolated from *Fusarium solani pisi.* Biochemistry **14**, 2832–2840 (1975)
54. Rodriguez-Miguens, B., Ribas-Marques, I.: Investiguciones quimicas subre el corcho de *Solanum tuberosum* L. (Patata). Quimica **68**, 302–308 (1972)
55. Shaykh, M., Kolattukudy, P.E.: unpublished results
56. Shishiyama, J., Araki, F., Akai, S.: Studies on cutin-esterase II. Characteristics of cutin-esterase from *Botrytis cinerea* and its activity on tomato cutin. Plant Cell Physiol. (Tokyo) **11**, 937–945 (1970)
57. Soliday, C.L., Kolattukudy, P.E.: Isolation and characterization of a cutinase from *Fusarium roseum culmorum* and its immunological comparison with cutinases from *F. solani pisi.* Arch. Biochem. Biophys. **176**, 334–343 (1976)
58. Soliday, C.L., Kolattukudy, P.E.: Biosynthesis of Cutin: ω-Hydroxylation of Fatty acids by a microsomal preparation from germinating *Vicia faba.* Plant Physiol. **59**, 1977, in press
59. Soliday, C.L., Kolattukudy, P.E.: unpublished results
60. Walton, T.J., Kolattukudy, P.E.: Determination of the structures of cutin monomers by a novel depolymerization procedure and combined gas chromatography and mass spectrometry. Biochemistry **11**, 1885–1897 (1972)
61. Walton, T.J., Kolattukudy, P.E.: Enzymatic conversion of 16-hydroxypalmitic acid into 10,15-dihydroxypalmitic acid in *Vicia faba* epidermal extracts. Biochem. Biophys. Res. Commun. **46**, 16–21 (1972)

CHAPTER 16

The Intermolecular Structure of Some Plant Cutins

A.H.B. DEAS and P.J. HOLLOWAY

Cutin is a lipid polymer that is found in higher plants mainly as the chief structural component of the cuticular membrane. This membrane is an integral part of the epidermis and occurs therefore on the external surfaces of all leaves, fruits, stems and other aerial tissues where secondary thickening is not present. The presence of cutin within membranes can most easily be detected using the light microscope and transverse sections stained with the fat soluble dyes Sudan III or IV (see, e.g., Refs. [2, 13, 14]). This shows that the polymer is present mainly in the outermost region of the membrane but that it is also found in the "pegs" that penetrate between the anticlinal walls of the epidermal cells. Under the transmission electron microscope the situation is more complex, but sections through the membranes of most plants show an ordered structural arrangement summarized in diagrammatic form in Figure 1. (For micrographs, consult Refs. [3, 6, 9, 15].) In modern terminology [15] the lamellate region is referred to as the "primary cuticle", which is laid down first in development, and this is structurally discrete from the reticulate region or "secondary cuticle", which develops between the primary cuticle and primary epidermal cell wall only when periclinal extension growth of the latter is complete. No further development of the primary cuticle takes place once the secondary cuticle has started to form. Although there is as yet no specific electron microscopic stain for cutin, it is generally agreed that it occurs in close association with wax in the primary region of the membrane and in close association with carbohydrate in the secondary region. In some plants cutin may also encrust the epidermal cell wall itself, i.e., the wall becomes cutinised. The location of cutin within the membrane and in particular its relationship to other cuticular components is an important consideration for any structural work with cutin, because unlike many other biopolymers, it cannot be obtained entirely free from other components because of its insoluble nature. The starting point for all chemical investigations is therefore the membrane itself.

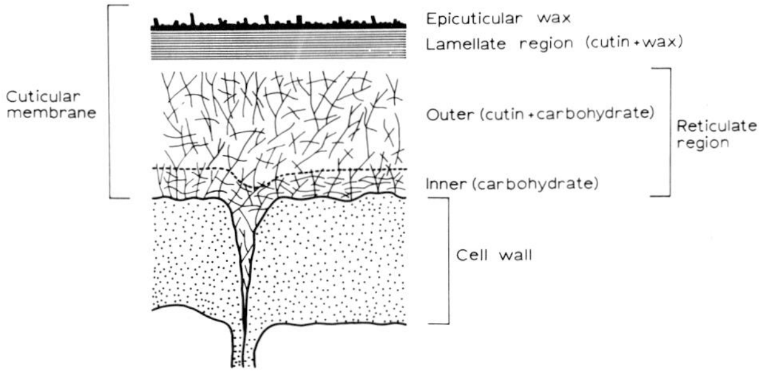

Fig. 1. Diagrammatic representation of a transverse section through a typical mature cuticular membrane as seen using the transmission electron microscope. (From Ref. [10])

It must be emphasised here that plants vary considerably in the thickness and degree of development of their cuticular membranes and also in the amount of cutin they contain. Cuticle thickness measured at the centre of the outer periclinal wall varies from about 1 up to 13 µm, e.g., Oleander leaf 11 µm, tomato fruit 4.5 µm, apple leaf 2 µm [13]. The cutin content as determined by chemical means is equally variable and ranges from as much as 1.5 mg/cm^2 in some apple fruit cultivars to as little as 10 $\mu g/cm^2$ in spinach and lettuce leaves. Significant differences may also occur on different regions of the same plant, such as between upper and lower surfaces of the same leaf and between leaves and fruits of the same plant.

Because of their insoluble nature, structural investigations of plant cutins have relied mainly on the analysis of monomers released after alkaline hydrolysis (aqueous or alcoholic), alkanolysis (NaOMe–MeOH, MeOH–HCl, BF_3–MeOH), or $LiAlH_4$ reduction, such depolymerization signifying that the polymers are essentially polyesters. However, the existence of some ester links with other membrane components, such as carbohydrate, cannot be entirely discounted. Further support for the polyester structure is obtained from infrared spectra of isolated membranes which exhibit strong carbonyl absorption (1730–1735 cm^{-1}). The characteristic monomers of cutin polyesters are a series of C_{16} and C_{18} hydroxyfatty acids, the chemistry and biosynthesis of which have been studied extensively (reviews by Refs. [11 and 12]).

The exact structure of cutin is unknown, but it is of considerable interest firstly because of the important biological role played by the cuticle, i.e., its value to the plant, and secondly, its possible influence on the penetration of agricultural chemicals applied to plant surfaces, i.e., its value to man in terms of efficient crop protection. Purely physical methods such as X-ray diffraction [16] and the polarising microscope (e.g., Ref. [5]) indicate that the structure of cutin is not regular or crystalline, although such results must be treated with caution because they have always been obtained with cuticular material and not with pure cutin. The lamellate region of membranes, however, seen in the electron microscope could be a manifestation of the intermolecular structure of the cutin present, but this has yet to be proved.

Most current views about cutin structure are simply based on the results obtained from monomer analyses and then assuming a random association in the intact polymer. The models that have been proposed [4, 7] are pure speculation. Other workers have also suggested that the degree of development of cuticular membranes is correlated with the abundance of OH groups in their constituent cutin monomers, which in turn governs the extent of cutin polymerization [1, 11]. This is certainly not true, because cutins of plants with thin membranes can comprise monomers with either a high or a low OH content. A similar situation is also found for well-developed or thicker membranes. Monomer analyses, however, do show clearly that cutin cannot be of the same intermolecular structure in all plants and that very marked differences must occur between some plant species. This can be easily shown by considering the structures of some typical cutin monomers (Table 1). In any polyester, intermolecular structure is ultimately governed by the number and position of esterifiable groups, and the monomeric composition of many plant cutins

Table 1. Typical cutin monomers and their ability to form polyesters

Cutin Monomers Capable of Cross-linking a Polyester Polymer	
[a] $HO{-}(CH_2)_6{-}\overset{OH}{\overset{\vert}{C}H}{-}(CH_2)_8{-}CO_2H$	10,16-Dihydroxyhexadecanoic (and other positional isomers)
$HO_2C{-}(CH_2)_5{-}\overset{OH}{\overset{\vert}{C}H}{-}(CH_2)_8{-}CO_2H$	7-Hydroxyhexadecane-1,16-dioic (and other positional isomers)
[a] $OHC{-}(CH_2)_6{-}\overset{OH}{\overset{\vert}{C}H}{-}(CH_2)_7{-}CO_2H$	9-Hydroxy-16-oxohexadecanoic (and other positional isomers)
[a] $HO{-}(CH_2)_8{-}\overset{OH}{\overset{\vert}{C}H}{-}\overset{OH}{\overset{\vert}{C}H}{-}(CH_2)_7{-}CO_2H$	9,10,18-Trihydroxyoctadecanoic (and its Δ 12 analogue)
$HO{-}(CH_2)_5{-}\overset{OH}{\overset{\vert}{C}H}{-}\overset{OH}{\overset{\vert}{C}H}{-}CH_2{-}\overset{OH}{\overset{\vert}{C}H}{-}\overset{OH}{\overset{\vert}{C}H}{-}(CH_2)_7{-}CO_2H$	9,10,12,13,18-Pentahydroxyoctadecanoic
Cutin Monomers Capable of Forming Only a Linear Polyester	
Monobasic	
α, ω-Dibasic	
[a] ω-Hydroxymonobasic viz. C_{16}, C_{18}, C_{18} $\Delta 9$, C_{18} $\Delta 9,12$	
[a] $HO{-}(CH_2)_6{-}\overset{O}{\overset{\Vert}{C}}{-}(CH_2)_8{-}CO_2H$	16-Hydroxy-10-oxohexadecanoic (and other positional isomers)
[a] $HO{-}(CH_2)_8{-}\overset{\diagup O \diagdown}{CH{-}CH}{-}(CH_2)_7{-}CO_2H$	9,10-Epoxy-18-hydroxyoctadecanoic (and its Δ 12 analogue)

[a] Major components of certain plant cutins

does show an excess of OH over COOH groups. Any cross-linking will be influenced mainly by the availability of secondary OH groups, and it is obvious that epoxy-, oxo- and ω-hydroxy monomers (Table 1) alone cannot participate in this process. Thus cutins that contain large proportions of such monomers must be predominantly linear, i.e., estolides. Most plant cutins, however, are usually mixtures of monomers containing secondary OH groups and others that do not have groups capable of cross-linking.

No direct attempts have yet been made to study the intermolecular structure of cutin, but we have recently attempted to study this problem in a few plants. Our basic approach was to "label" any free OH groups in the cutin matrix prior to depolymerization using chemical methods, the qualitative and quantitative composition of the depolymerization products then giving us information on the nature and extent of interesterification of the various monomers in the original polymer. Initial experiments were carried out with synthetic hydroxyfatty acids and esters, and subsequently with isolated membranes of the C_{16} type known to have a high cutin content—the simplest type of cutin to work with initially. The choice of predepolymerization reactions was governed by the stability of the products formed in relation to the depolymerization method used; we routinely use NaOMe-MeOH (methanolysis) [8]. Reactions involving the use of strongly alkaline reagents could not be utilised because they would also cause cutin breakdown. Three reactions were selected and tested:

1. $R{-}OH \xrightarrow{SOCl_2-C_5H_5N} R{-}Cl \xrightarrow{NaOMe-MeOH} R{-}OMe$

2. $R{-}OH \xrightarrow{BF_3\ \text{etherate}-CH_2N_2} R{-}OMe$

3. $R{-}CH_2OH \xrightarrow{\text{oxidation}} R{-}CO_2H$

$$\begin{matrix}R_1\\ \end{matrix}\!\!>\!CHOH \xrightarrow{\text{oxidation}} \begin{matrix}R_1\\ \end{matrix}\!\!>\!C{=}O$$

$$(R_1)(R_2)CHOH \xrightarrow{\text{oxidation}} (R_1)(R_2)C{=}O$$

i) $CrO_3{-}H_2SO_4{-}Me_2CO$
ii) $CrO_3{-}C_5H_5N$ complex

With monomeric compounds all reactions gave satisfactory conversions to the desired products, but with the $SOCl_2$ method forcing conditions were necessary to give the methyl ether, and Cl groups derived from secondary OH groups underwent dehydrohalogenation on treatment with NaOMe to yield the corresponding olefin. The $SOCl_2$ method was therefore rejected at this stage.

The predepolymerization reactions with cuticular membranes were more difficult, and a summary of the procedures we used is given in Scheme 1.

Discs (1–2 cm^2) epidermal tissue, dewaxed $CHCl_3$
Cuticular membranes detached by treatment with
ammonium oxalate-oxalic acid solution at 40° C

↓

Isolated membranes exhaustively extracted with MeOH
Dried at 40° C {ground to fine powder Wiley Mill}

↓

Predepolymerization reactions

↓

Depolymerization under anhydrous conditions with 0.01 M NaOMe-MeOH

↓

Analysis of depolymerization products TLC, GLC
Structural determination GC–MS (TMSi ethers)

Scheme 1. Summary of procedures used in the structural determination of plant cutins

Table 2. Oxidation–methanolysis products obtained from all the possible esterified combinations of 10,16-dihydroxyhexadecanoic acid

Polymeric form	Oxidation–methanolysis product
$HO{-}(CH_2)_6{-}CH(OH){-}(CH_2)_8{-}CO_2{-}CUT$	$HO_2C{-}(CH_2)_5{-}C({=}O){-}(CH_2)_8{-}CO_2Me$
$CUT{-}O{-}(CH_2)_6{-}CH(OH){-}(CH_2)_8{-}CO_2{-}CUT$	$HO{-}(CH_2)_6{-}C({=}O){-}(CH_2)_8{-}CO_2Me$
$CUT{-}O{-}(CH_2)_6{-}CH(O{-}CUT){-}(CH_2)_8{-}CO_2{-}CUT$	$HO{-}(CH_2)_6{-}CH(OH){-}(CH_2)_8{-}CO_2Me$
$HO{-}(CH_2)_6{-}CH(O{-}CUT){-}(CH_2)_8{-}CO_2{-}CUT$	$HO_2C{-}(CH_2)_5{-}CH(OH){-}(CH_2)_8{-}CO_2Me$

The main problem was achieving a complete reaction of all free OH groups, incomplete reaction being subsequently indicated by a monomer composition showing a much higher number of apparently esterified OH groups than the COOH groups available. These difficulties are probably a consequence of the heterogeneous nature of the membrane starting materials (Fig. 1). With the BF_3 etherate–CH_2N_2 and CrO_3-H_2SO_4 methods we could not always achieve a maximal level of reaction even after treatments of up to 7 days, but with the CrO_3–pyridine complex consistent reactions could be obtained after 18 h of treatment, i.e., no further oxidation of the cutin occurred after this time. This reagent was subsequently used in all our investigations of cutin structure.

The oxidation–methanolysis products, that is, the "labelling" produced, for all esterified combinations of 10,16-dihydroxyhexadecanoic acid are shown in Table 2 to illustrate the basis of the method. Its success or failure depends very much on carrying out the depolymerization (methanolysis) under anhydrous conditions because if any hydrolysis of methyl esters initially formed by transesterification occurs then false structural information will be obtained. In addition, because COOH groups are products of the predepolymerization reaction, the absence of any significant amounts of free acids in the starting materials also has to be verified. This was confirmed in all cases by analysis of the monomers obtained from $LiBH_4$ reduction, a reagent which will reduce ester but not COOH groups.

The results we obtained for the intermolecular structure of four plant cutins are given in Table 3. They show that in cutins from tomato fruits (*Lycopersicon esculentum*), rose-hips (*Rose canina*), and black currant leaves (*Ribes nigrum*) esterification occurs chiefly through the primary OH groups of the monomers and that a significant proportion (up to 40%) are also cross-linked through secondary OH groups. The excess OH groups predicted from monomer analyses are chiefly secondary. The results for lemon (*Citrus limon*) fruit cutin confirm its predominantly linear structure. We have not attempted to construct a new

Table 3. Polymeric form of dihydroxyhexadecanoic acid and related C_{16} acids in four plant cutins

Polymeric form	% Total monomers			
	Tomato fruit	Rose-hip fruit	Black currant leaf	Lemon fruit
CUT–O–$(CH_2)_{15}$–CO_2–CUT	5	7	18	14
CUT–O_2C–$(CH_2)_5$–CH(OH)–$(CH_2)_8$–CO_2–CUT	1	6	–	2
CUT–O_2C–$(CH_2)_5$–CH(O–CUT)–$(CH_2)_8$–CO_2–CUT	4	5	5	1
HO–$(CH_2)_6$–CH(OH)–$(CH_2)_8$–CO_2–CUT	2	5	3	2
CUT–O–$(CH_2)_6$–CH(OH)–$(CH_2)_8$–CO_2–CUT	48	50	38	25
CUT–O–$(CH_2)_6$–CH(O–CUT)–$(CH_2)_8$–CO_2–CUT	36	24	30	5
HO–$(CH_2)_6$–CH(O–CUT)–$(CH_2)_8$–CO_2–CUT	3	3	4	1
CUT–O–$(CH_2)_6$–C(=O)–$(CH_2)_8$–CO_2–CUT	2	1	2	51

cutin model from these results, but in all four cutins the esterification patterns of the dihydroxyhexadecanoic monomers are somewhat similar, suggesting some form of common structural arrangement in the cutin matrix. In this connection it was also interesting to find that the intermolecular structure of the C_{16} monomers of tomato fruit cutin changed little during growth of the fruits from 2 to 5 cm in diameter, a period of rapid expansion and cutin synthesis, indicating that the formation of cutin probably does not simply involve the random interesterification of monomers. The new cutin appears to have the same structure as that initially formed.

References

1. Baker, E.A.: The morphology and composition of isolated plant cuticles: New Phytologist **69**, 1053–1058 (1970)

2. Baker, E.A., Batt, R.F., Roberts, M.F., Martin, J.T.: Studies on plant cuticle: progress report: Rep. agric. hort. Res. Stn. Univ. Bristol 1961, pp. 114–123 (1962)
3. Chafe, S.C., Wardrop, A.B.: Fine structural observations on the epidermis. II. The cuticle: Planta **109**, 39–48 (1973)
4. Crisp, C.E.: The biopolymer cutin: Ph.D. thesis, University of California (1965)
5. Frey-Wyssling, A.: *Submicroscopic Morphology of Protoplasm and its Derivatives,* Amsterdam: Elsevier, 1948, p. 183
6. Hallam, N.D., Juniper, B.E.: The anatomy of the leaf surface: in *Ecology of Leaf Surface Micro-organisms.* Eds. Preece, T.F., Dickinson, C.H., London: Academic Press, 1971, pp. 3–37
7. Heinen, W., van den Brand, I.: Enzymatische Aspekte zur Biosynthese des Blatt-Cutins bei *Gasteria verricuosa*-Blättern nach Verletzung: Z. Naturforsch. **18b**, 67–79 (1963)
8. Holloway, P.J., Deas, A.H.B.: Epoxyoctadecanoic acids in plant cutins and suberins: Phytochemistry **12**, 1721–1735 (1973)
9. Jarvis, L.R., Wardrop, A.B.: The development of the cuticle in *Phormium tenax*: Planta **119**, 101–112 (1974)
10. Jeffree, C.E., Baker, E.A., Holloway, P.J.: Origins of the fine structure of plant epicuticular waxes: in *Microbiology of Aerial Plant Surfaces.* Eds. Dickinson, C.H., Preece, T.F., London: Academic Press, 1976, pp. 119–158
11. Kolattukudy, P.E., Walton, T.J.: The biochemistry of plant cuticular lipids: in *Progress in the Chemistry of Fats and Other Lipids.* Ed. Holman, R.T., Oxford: Pergamon Press, Vol. 13, part 3, 1972, pp. 121–175
12. Martin, J.T., Juniper, B.E.: *The Cuticles of Plants,* London: Edward Arnold, 1970
13. Norris, R.F.: Penetration of 2,4-D in relation to cuticle thickness: Am. J. Botany **61**, 74–79 (1974)
14. Norris, R.F., Bukovac, M.J.: Structure of the pear leaf cuticle with special reference to cuticular penetration: Am. J. Botany **55**, 975–983 (1968)
15. Sargent, C.: The occurrence of a secondary cuticle in *Libertia elegans* (Iridaceae): Ann. Bot. **40**, 355–359 (1976)
16. Wilson, L.A., Sterling, C.: Studies on the cuticle of tomato fruit I. Fine structure of the cuticle: Z. Pflanzenphysiol. **77**, 359–371 (1976)

Subject Index

T.C. Moore
Research Experiences in Plant Physiology

A Laboratory Manual

With 23 figures. XI, 462 pages. 1974
ISBN 3-540-06424-9

Distribution rights for India:
Allied Publishers, New Delhi

Research Experiences in Plant Physiology is a valuable textbook for introductory courses in plant physiology. It consists of 25 carefully selected, comprehensive exercises, each presented as a separate unit complete with introductory text, materials and methods, references, recommendations for scheduling, and report forms.
Each exercise has been class-tested; students and instructors can be sure all experiments will yield the expected results. The manual is perforated so that exercises can be handed in.
This manual is designed to confront students with actual research-type investigations with heavy reliance on the team approach. Most of the exercises were, in fact, developed directly from published research papers. While the book covers a broad spectrum of topics, there is a definite emphasis on growth and development.
Research Experiences in Plant Physiology is the fruit of many years of conscientious teaching in the laboratory. The author, an authority on hormonal regulation of growth and development of angiosperms, is the recipient of several awards for outstanding teaching of both graduates and undergraduates.

D. Hess
Plant Physiology

Molecular, Biochemical, and Physiological Fundamentals of Metabolism and Development

‹Springer Study Edition›

248 figures. XV, 333 pages. 1975.
ISBN 3-540-06643-8

Distribution rights for India:
UBS Publishers Distributors Ltd.
New Delhi

An elementary introduction to the metabolic and developmental physiology of higher plants from the point of view of molecular biology. The outstanding feature of the text is that equal weight is given to metabolism and development.

Contents: Control of Character Formation by Nucleic Acids. – Photosynthesis. – Carbohydrates. – Biological Oxidation. – Fats. – Terpenoids. – Phenols. – Amino Acids. – Alkaloids. – Porphyrins. – Growth During Division. – Differential Gene Activity as Principle of Differentiation. – Regulation. – Polarity and Unequal Cell Division as Fundamentals of Differentiation. – Longitudinal Growth. – The Formation of Seeds and Fruit. – Germination. – The Vascular System. – Flower Formation.

Encyclopedia of Plant Physiology New Series

Editors: A. Pirson; M.H. Zimmermann
Distribution rights for India:
UBS Publishers Distributors Ltd. New Delhi

Volume 1: Transport in Plants I

Phloem Transport

Editors: M.H. Zimmermann, Harvard University, Petersham, Mass., USA; J.A. Milburn, University of Glasgow, Glasgow, Great Britain

With contributions by M.J.P. Canny, J. Dainty, A.F.G. Dixon, W. Eschrich, D.S. Fensom, D.R. Geiger, W. Heyser, W. Höll, J.A. Milburn, T.R.F. Nonweiler, M.V. Parthasarathy, J.S. Pate, A.J. Peel, S.A. Sovonick, D.C. Spanner, P.M.L. Tammes, M.T. Tyree, J. van Die, H. Ziegler, M.H. Zimmermann

93 figures. XIX, 535 pages. 1975
ISBN 3-540-07314-0

Volume 2 (in Parts): Transport in Plants II

Editors: U. Lüttge, TH Darmstadt, Germany; M.G. Pitman, University of Sydney, Sydney, Australia

Part A: Cells

With a foreword by R.N. Robertson

With contributions by W.J. Cram, J. Dainty, G.P. Findlay, T.K. Hodges, A.B. Hope, D.H. Jennings, U. Lüttge, C.B. Osmond, M.G. Pitman, R.J. Poole, J.A. Raven, F.A. Smith, N.A. Walker

97 figures, 64 tables. XVI, 419 pages. 1976
ISBN 3-540-07452-X

Part B: Tissues and Organs

With contributions by W.P. Anderson, E. Epstein, A.E. Hill, B.S. Hill, T.C. Hsiao, W.D. Jeschke, A. Läuchli, U. Lüttge, J.S. Pate, M.G. Pitman, E. Schnepf, R.M. Spanswick, R.F.M. van Steveninck, J.F. Sutcliffe

129 figures, 45 tables. XII, 475 pages. 1976
ISBN 3-540-07453-8

Volume 3: Transport in Plants III

Intracellular Interactions and Transport Processes

Editors: C.R. Stocking, University of California, Davis, Calif., USA; U. Heber, University of Düsseldorf, Düsseldorf, Germany

With contributions by J. Brachet, R.H. Falk, H. Fock, R.J. Graesser, M.D. Hatch, H.W. Heldt, P. Matile, R.E. McCarty, H.H. Mollenhauer, D.J. Morré, S. Murakami, C.B. Osmond, C. Schnarrenberger, C.R. Stocking, H. Strotmann, D.A. Walker, A. Wiemken, R.H. Wilson, D. Woermann

123 figures. XXII, 517 pages. 1976
ISBN 3-540-07818-5

Volume 4:

Physiological Plant Pathology

Editors: R. Heitefuss, University of Göttingen, Göttingen, Germany; P.H. Williams, University of Wisconsin, Madison, Wis., USA

With contributions by J.R. Aist, P.J. Allen, H.G. Basham, D.F. Bateman, J.M. Daly, H.M. Dekhuijzen, V.H. Dropkin, J.M. Duniway, A.H. Ellingboe, F. Frič, W.H. Fuchs, D.G.Gilchrist, R.N. Goodman, R. Heitefuss, E.C. Hislop, D.S. Ingram, T. Kosuge, J.A. Kuć, B.B. Lippincott, J.A. Lippincott, K. Maramorosch, J.E. Mitchell, G.F. Pegg, D. Pitt, K. Rudolph, R.P. Scheffer, E. Schlösser, F. Schönbeck, K.J. Scott, R.C. Staples, I. Uritani, H.E. Wheeler, G. Wolf, Z. Yaniv, C.E. Yarwood, G.A. de Zoeten

92 figures. XX, 890 pages. 1976
ISBN 3-540-07557-7

Volume 5: Photosynthesis I

Photosynthetic Electron Transport and Photophosphorylation

Editors: A. Trebst, Bochum, Germany; M. Avron, Rehovot, Israel

135 figures. Approx. 820 pages. 1977
ISBN 3-540-07962-9

Springer-Verlag Berlin Heidelberg New York